BUT IS IT SCIENCE?

Frontiers of Philosophy

Peter H. Hare, Series Editor

Advisory Board

Norman E. Bowie (University of Delaware)

Antony Flew (University of Reading/Bowling Green State University)

Jesse Kalin (Vassar College)

E. D. Klemke (Iowa State University)

Alice Ambrose Lazerowitz (Smith College)

Joseph Margolis (Temple University)

Robert G. Meyers (State University of New York at Albany)

Gerald Myers (Graduate Center, City University of New York)

Sandra B. Rosenthal (Loyola University, New Orleans)

T. L. Short (Kenyon College)

Richard A. Watson (Washington University)

BUT IS IT SCIENCE?

THE PHILOSOPHICAL
QUESTION IN THE
CREATION/EVOLUTION
CONTROVERSY

EDITED BY
MICHAEL RUSE

Prometheus Books
59 John Glenn Drive
Amherst, NewYork 14228-2197

Published 1996 by Prometheus Books

Inquiries should be addressed to
Prometheus Books
59 John Glenn Drive
Amherst, New York 14228–2197
VOICE: 716–691–0133, ext. 207
FAX: 716–564–2711
WWW.PROMETHEUSBOOKS.COM

04 03 02 01 00   7 6 5 4 3

Library of Congress Cataloging-in-Publication Data

But is it science? : the philosophical question in the creation/evolution controversy /
    edited by Michael Ruse.
        p. cm.
    Includes bibliographical references. (p.   )
    ISBN 1–57392–087–8 (alk. paper)
    1. Evolution (Biology)—Philosophy. 2. Creationism—Philosophy. I. Ruse,
Michael.

QH360.5.B87 1996
575'.001—dc20
                                                                    96–4218
                                                                    CIP

Printed in the United States of America on acid-free paper

Preface

The battle between science and religion is not yet over. In recent years, as part of a general turn towards conservative doctrines, supporters of extreme, literal interpretations of the Bible have been striving mightily to have their world picture—"creationism"—declared a legitimate alternative to conventional scientific claims, particularly to those centering on the "evolutionary" origins of the earth and the denizens thereof. Most visible have been the literalists' successes in securing passages of laws in two states (Arkansas and Louisiana) mandating the teaching of creationist beliefs in schools, alongside the claims of orthodox scientists.

As it happens, these particular political triumphs proved short-lived, for (as you will learn from the pages of this book) the Arkansas law was soon thrown down, and recently (by a 7-2 vote) the U.S. Supreme Court has done the same for the Louisiana law. (The ruling came down on June 19, 1987.) But, it would be naive to think the creation/evolution clash is now finished. By their own admission, creationists are trying different tactics to achieve their ends. These moves include the pressuring of textbook publishers, school boards, individual teachers, and the like. Nor are such efforts proving fruitless. In my own province of Ontario, for instance, it was recently reported that although high-school students are to be taught the size and temperature of the sun, questions to do with its age are beyond the curriculum. Textbooks are also to avoid terms like "evolution" (*Toronto Globe and Mail*, July 25, 1987).

I myself have been much involved in the fight against creationism and I hold strong views on the subject. I hope this book will turn you—if you need turning—toward my way of thinking. However, let me rush to assure you that these pages do not try to do what several authors have already

5

done extremely well. This is not a straightforward assessment of the merits (or demerits) of evolutionism and of creationism. Rather, the aim here is to concentrate on one particular theme (or set of themes) that has kept surfacing through the creation/evolution controversy—a theme that I, as a philosopher, find both fascinating and important. In particular, I want to raise the question of status. What precisely is the status of the various knowledge claims that are being forwarded, supported, and criticized in the debate? Are they, at least some, part of science? part of religion? part of something else?

Different people have given different answers to these questions, and as the answers have convinced, or failed to convince, so important decisions have been made. Yet, there still are unresolved issues—intriguingly, many such issues between people whom you might have thought would be more in accord. Therefore, since the creation/evolution clash is not finished, it seems worthwhile to go back over the key philosophical question: But is it science? Is evolution science? Is creationism science? But if they are not science, then what are they?

Most straightforwardly, one would write a monograph attacking the issue, and this was my first inclination—especially given my sneaking suspicion that collections of already-published writings are the lazy person's way of producing a book. But, for once, I think a collection of readings and articles is permittable and, indeed, has special merits of its own. Philosophical issues are best resolved by letting all parties have a free and open debate. Since, as mentioned, I have also participated in the debate, holding strong views, it seems only fair and proper that those who hold equally strong views to the contrary should be allowed to make up their cases, in their own words. Moreover, I would like to think that the things that I and others have already written, in the heat of exchange, have a certain immediacy that would be lost in the attempt at a disinterested overview.

As a matter of historical fact, the focal point of the queries about the science/religion pretensions of evolution/creation claims was the ruling of Judge William Overton against the constitutionality of that already-mentioned Arkansas law that mandated the teaching of creationism. Picking up on my advice as an expert witness, the judge firmly ruled that, by reasonable criteria of what constitutes a science, creationism (or, as its supporters like to call it, "creation science") does not constitute a genuine science. By implication, the judge allowed that evolution is such a science. However, comforting though such a ruling was to us partisans for the cause of evolutionism, the judge's opinion not only upset (one presumes) the creationists, but it also raised the ire of a number of well-known philosophers. They strongly felt that the judge had been given bad philosophical advice, and that any short-term victory by evolutionists was bound to crumble beneath so inadequate a foundation.

This disagreement is the spur to this collection. Obviously, however, it would be unduly limiting (not to say grotesquely self-indulgent) were any collection to focus exclusively primarily on me and my critics. There are important general issues here, about the nature of science and about how anyone can hope to provide an adequate account of origins, particularly organic origins. I have, therefore, tried to structure this collection with the broader issues in mind. Moreover, I have deliberately tried to let supporters of the most varied opinions have their say. Consequently, the collection is less a unified whole, endorsing a common theme, and more a workbook, to let you go through the various positions and draw your own conclusions. As I have said, as a participant I hope you will end agreeing with me; but, as editor I have tried very consciously to stay apart from the fray.

The structure of the collection is simple. There are four main sections, each with its own introduction. First, we have a historical section, where the nineteenth-century background to the coming of evolutionism, and to today's debates, is presented. It is shown that there have always been questions about whether an understanding of origins can be truly scientific. Next are given discussions of today's thinking about evolutionary biology, and particularly about the status of Charles Darwin's mechanism of natural selection. Third, we move on to creationism and to attacks on its claims to be genuine science. Finally, the philosophers take over exclusively, as we argue about what constitutes a science, and whether either evolutionism or creationism can satisfy proposed criteria. A glossary of technical terms and suggestions for further reading conclude the collection.

Somewhat diffidently, I have started the collection with a prologue telling of my own personal foray into the creation/evolution controversy. At best it will tell you why I care so passionately about these matters; at worst it will save a host of explanatory footnotes later in the book. Less diffidently, I have relied on my own writings more heavily in this volume than would be proper in a normal collection. Much of what I have to say (especially in the early sections) is not particularly controversial. Where the battle does begin, it seemed important that I not hide behind the names of others since I am a major subject of criticism.

Contents

PART FOUR: THE PHILOSOPHICAL AFTERMATH

Prologue

A Philosopher's Day in Court

Michael Ruse

On Sunday, December 6, 1981, I found myself on a flight south, from Toronto, Ontario, to Little Rock, Arkansas. That week I was to appear as an expert witness in a U.S. Federal Court case. What was the case? Why would I, a philosophy professor from a small town in Canada, be summoned? What happened? Where are we today? These are some of the questions I hope to answer.

ACT 590

My story is about evolutionary biology, and as always when dealing with that topic the best place to begin is in 1859. It was that year which saw the publication of Charles Robert Darwin's major work, *On the Origin of Species by Means of Natural Selection, or the Preservation of Favoured Races in the Struggle for Life.*[1] In the course of some 450 pages, Darwin wrote vigorously in support of two major theses. First, he argued to the actual occurrence of *evolution,* that is, to the claim that all organisms (including ourselves) are descended by a slow natural process, gradually modifying from "one or a few" original forms. Second, Darwin supplied a *mechanism* for the evolutionary process: natural selection.

Darwin started with the potential Malthusian population explosion which exists everywhere in the living world. Given the obvious constraints due to limited supplies of food and space, we get a universal struggle for existence;

From *Science and Creationism,* edited by Ashley Montague, pp. 311–42. Copyright © 1984 by Ashley Montague. Reprinted by permission of Oxford University Press.

13

more particularly, we get a struggle for reproduction. Drawing on the analogy of artificial selection as practiced by animal and plant breeders, Darwin then went on to argue that the struggle fuels a form of "natural" selection. Given enough time, this force leads to full-blown evolution.

> How will the struggle for existence . . . act in regard to variation? . . . Let it be borne in mind in what an endless number of strange peculiarities our domestic productions, and, in a lesser degree, those under nature, vary; and how strong the hereditary tendency is. . . . Can it, then, be thought improbable, seeing that variations useful to man have undoubtedly occurred, that other variations useful in some way to each being in the great and comj lex battle of life, should sometimes occur in the course of thousands of generations? If such do occur, can we doubt (remembering that many more individuals are born than can possibly survive) that individuals having any advantage, however slight, over others, would have the best chance of surviving and of procreating their kind? On the other hand, we may feel sure that any variation in the least degree injurious would be rigidly destroyed. This preservation of favorable variations and the rejection of injurious variations, I call Natural Selection. (Darwin, 1859, pp. 80–81)[2]

In Victorian Britain, Darwin's ideas had a somewhat mixed reception. Many people recoiled from everything that he wanted to claim, refusing to have any truck whatsoever with filthy evolutionism. However, among the intelligentsia—scientists obviously, but extending right across the spectrum even to liberal clergymen—evolution per se was acceptable and accepted. Long before Darwin's *Origin* was published, the English had come to realize that a literal reading of Genesis—six days of creation, short time-span for Earth (about 6000 years, as calculated from the genealogies of the Bible), universal flood—was simply not tenable. The empirical facts speak against such a reading. Hence, the idea of evolution, binding and explaining so many different aspects of the organic world, was welcomed with enthusiasm.[3]

Where people had trouble was with Darwin's mechanism of selection. They simply could not see how "blind" law could lead to the intricate adaptations which we see around us in the world—the hand, the eye, the beautiful colors of the butterfly, and, above all, those qualities of intelligence and morality which raise us humans up above the apes. Here, critics felt, one simply has to add "something more." Thus we find Herbert Spencer relying on a full-blown Lamarckian inheritance of acquired characteristics; Darwin's "bulldog" Thomas Henry Huxley turning to large variations, "saltations"; and religious people of all stripes insisting that every now and then God gives His handiwork a little shove.

> We must suppose the idea of *Jumps* . . as if for instance a wolf should at some epoch of lupine history take to occasionally littering a dog or a fox among her cubs. Through such a process we introduce *mind, plan, design,* and to the . . . obvious exclusion of the haphazard view of the subject and the casual concourse of atoms.[4]

This general sense of unease about Darwin's mechanism of selection is something which has persisted down through the years even to the present. Virtually no active scientist today wants God to intervene personally in the course of evolution (except, perhaps, when it comes to immortal souls), but still all sorts of different rival mechanisms are proposed. One young researcher has recently argued for a form of Lamarckism, and there is a very articulate group of paleontologists pushing a form of neo-Saltationism. Stephen Jay Gould and others argue that every now and then evolution takes a leap forward, followed by periods of relative unchange. (This is the theory of "punctuated equilibria.")[5]

But for all of the troubles of selection, in Darwin's homeland at least, evolution had a relatively smooth ride. With scientists, liberal clergymen, and other educated laypeople all early converts to evolution, no formally organized opposition was really able to make much progress. In fact, Darwin was very lucky, for it was shortly after publication of the *Origin* that universal school education became available in Britain. And those very people who were involved in setting up such education were often the *same* people most concerned to spread evolutionism! For instance, T. H. Huxley, the chief spokesman for descent with modification, was a founding member of the incredibly influential London School Board. One can well imagine what the little East Enders got in their classes!

> In particular he [Huxley] advocated the teaching of "the first elements of physical science"; "by which I do not mean teaching astronomy and the use of the globes, and the rest of the abominable trash—but a little instruction of the child in what is the nature of common things about him; what their properties are, and in what relation this actual body of man stands to the universe outside of it." "There is no form of knowledge or instruction in which children take greater interest."[6]

In the U.S., particularly in the South, matters were otherwise. In times of stress and unhappiness, people frequently look to simplistic doctrines for support and comfort. Obviously, after the Civil War, in the South, one did indeed have such times of stress and unhappiness. Naturally, people turned to the most obvious place of consolation: the Holy Bible. In particular, Genesis gave people a sense of where they are and where they belong. Leviticus analogously gave people a guide for moral conduct. Thus, before long there was the flourishing of a peculiarly strong, indigenous brand of biblical literalism.

One consequence was that zealots started to monitor carefully the teaching and contents of school science classes. Woe and behold anyone who dared to step beyond the strictest bounds of the Old Testament. Those sufficiently foolish as to suggest that Adam may not have sprung up from mud soon found themselves in want of a job.

It was after the First World War that Fundamentalists (as biblical literalists were now called) scored some of their most striking successes. Several states of the Union, including Tennessee, passed laws prohibiting the teaching of evolutionism. As is well known, this led to a famous trial, when a young

Dayton, Tennessee schoolteacher, John Thomas Scopes, let himself be prosecuted for teaching evolution. Matters soon took on a carnival air, attracting the attention of the whole nation. Three-time presidential candidate William Jennings Bryan led the prosecution, and noted freethinker and devastating advocate Clarence Darrow appeared for the defense. Refused permission to introduce experts on evolution, Darrow had the brilliant idea of cross-examining Bryan on the literal truth of the Bible. Before long, Darrow had Bryan tied into knots as Bryan tried to defend such an unexpected doctrine.

Thus, although Scopes was found guilty—he had, after all, confessed to teaching evolution—evolutionists rightly felt they had won a moral victory. Moreover, thanks to the savage pen of *Baltimore Sun* reporter H. L. Mencken—who referred to the good people of Dayton as anthropoid rabble—many other states wisely and quietly shelved their proposed antievolutionary "monkey laws."[7]

Thus matters rested for long years, although it was not until the 1960's that the U.S. Supreme Court finally overturned the Tennessee law, ruling the teaching of evolution constitutional. I have a friend who grew up in Tennessee in the 1950's and 1960's. Her father worked at Oak Ridge National Laboratory. She tells me that in her high school—which may well not have been typical—they read Darwin. But in literature classes, not in science! At the local library, the copy of the *Origin* was missing from the Great Books Series. It was kept under the desk, along with all the other dirty books. Naturally enough, it was the most-read book of the whole series.

Fundamentalism lost some of its virulence after the 1920's, but in the 1960's it started to grow again. It was the Russians, of all people, who had a major part in this. In 1957 they put up Sputnik. This simply terrified America, which saw itself behind Russia, both in science and in technology. Typically, therefore, lots of money was thrown at the problem, and this (wisely) included money directed towards the improvement of American science education. As a consequence, high-powered committees were struck and fine new up-to-date textbooks produced.[8]

Naturally enough, the biology textbooks took evolution for granted. Unfortunately, when children started to bring these new books home, trouble developed all over again. Evolutionary biology was too much for biblical literalists, who felt that they simply had to do something about the situation.

But times had moved on since the days of the original Scopes Trial. By the late 1960's, evolution could no longer be the direct issue, given the Supreme Court's ruling on the constitutionality of its teaching. Furthermore, in the past half century, the Court has ruled with increasing force that the First Amendment's separation of church and state means precisely that. In particular, one may not teach religion as religion in schools. One certainly may not teach it in biology classes.

This puts modern-day Creationists in a bind. They would like to exclude evolution, but they cannot. They would like to include Genesis, but they cannot. As a compromise, therefore, they try to slide Genesis into classrooms, sideways. They argue that—Surprise! Surprise!—all of the claims of Genesis can be supported by the best principles and premises of empirical science. In other words, *as scientists,* people can argue for instant creation of the universe, separate ancestry for man and apes, short time-span for the Earth (between approximately 6,000 and 20,000 years), and a universal flood over everything, at some later date.[9]

Hence, we have the growth of "Scientific Creationism" or "Creation-science." An institute has been set up; a college; a museum; and an organization of Creation-scientists. Full membership in the latter demands that one have a graduate degree in the sciences; however, most of the 500 full members have nonbiological degrees in such areas as mining. Additionally, many Creationist books, magazines, and films have been produced.[10] Most successfully, leading Creation-scientists—notably Henry M. Morris and Duane T. Gish—travel the campus circuit, debating with local evolutionists on "Creation *versus* Evolution." The orthodox scientists tend not to be that skilled in public debate, and sometimes have lost their tempers at the Creationists' lies and underhand tricks. Hence, the Creationists usually garner major propaganda value from these circuses.

The cry today is for "equal time" or "balanced treatment" in the schools between what the Creationists refer to as "Evolution-science" and their own "Creation-science." It is a cry which has a powerful appeal, for it seems that it is those who would deny the Creationists' demands who are the bigots. Why should Creationists not have the chance to make their case? Indeed, surely the best principles of education demand that children be exposed to all kinds of ideas, and not just to those of one group, however powerful. The denial of the rights of minorities to make their case smacks of fascism.

Many people outside of the Creation-science movement have responded favorably to this plea for equal time for Creationist ideas in the schoolroom. One such respondent is no less a person than the incumbent President of the United States. When he was on the campaign trail, he spoke as follows:

> Well, it is a theory, it is a scientific theory only, and it has in recent years been challenged in the world of science and is not yet believed in the scientific community to be as infallible as it once was believed. But if it was going to be taught in the schools, then I think that also the biblical theory of creation, which is not a theory but the biblical story of creation, should also be taught.

Finally, in 1981, the breakthrough hoped for by the Creationist movement became a reality. The state legislature of Arkansas considered a bill requiring of its teachers that if they talk at all of origins in the classroom, then they must talk of Creation-science as well as evolution. The pertinent parts of Act 590 (as it was called) were as follows.

> (1) "Creation-science" means the scientific evidences for creation and inferences from those scientific evidences. Creation-science includes the scientific evidences and related inferences that indicate: 1) Sudden creation of the Universe, energy

and life from nothing, 2) The insufficiency of mutation and natural selection in bringing about development of all living kinds from a single organism, 3) Changes only within fixed limits of originally created kinds of plants and animals, 4) Separate ancestry for man and apes, 5) Explanation of the Earth's geology by catastrophism, including the occurrence of a world-wide flood, and 6) A relatively recent inception of the Earth and living kinds.

(b) "Evolution-science" means the scientific evidences for evolution and inferences from those scientific evidences. Evolution-science includes the scientific evidences and related inferences that indicate: 1) Emergence by naturalistic processes of the Universe from disordered matter and emergence of life from non-life, 2) The sufficiency of mutation and natural selection in bringing about development of present living kinds from simple earlier kinds, 3) Emergency [sic] by mutation and natural selection of present living kinds from simple earlier kinds, 4) Emergence of man from a common ancestor with apes, 5) Explanation of the Earth's geology and the evolutionary sequence by uniformitarianism, and 6) An inception several billion years ago of the Earth and somewhat later of life.

(c) "Public schools" mean public secondary and elementary schools.[11]

Act 590 did not originate in Arkansas. Indeed, it was a "model bill," drawn up by out-of-state Creationists. Their hope was that sympathetic politicians in several southern states would respond favorably to this model. In Arkansas, their hopes came to fruition, for a Fundamentalist senator steered it through both houses with no opposition whatsoever. Thus, the bill came before Governor Frank J. White. On March 19, 1981, apparently without taxing himself so far as to read it, White signed the bill, and so Act 590 became law.[12]

PREPARATION

"Hi! My name's David Klasfeld. I'm with the New York law firm of Skadden, Arps. I've been told that you might be able to help me." And so, I got personally involved with the fight against Act 590.

The American Civil Liberties Union is an organization dedicated to the preservation and support of the Constitution. As soon as Act 590 was passed, it sprang into active opposition, first at the state level and then at the national office in New York City. As noted, the First Amendment to the Constitution separates church and state. Fairly obviously, the ACLU saw Act 590 as a clear violation of this separation. Eventually, although separation of church and state was always the main thrust of the case, the ACLU added two other reasons why it thought Act 590 unconstitutional: that Act 590 infringes on the teachers' "academic freedom" to teach his/her subject properly, and that the Act is unconstitutionally vague. (The ACLU were not the plaintiffs as such. They were acting for Arkansas individuals, who claimed their civil liberties were being infringed. Many of the actual plaintiffs were clergymen.)

In fact, as the ACLU geared up for action, within the state of Arkansas itself there was growing opposition to Act 590. Of all groups, one that was most upset was the Junior Chamber of Commerce! Arkansas, like many sunny states in the South and Southwest, is busily trying to attract high-technology industry, such as computer firms. But the last thing that a bright young computer engineer wants is transference to a state where his kids have to learn Creationism as a matter of course. Hence, Arkansas businessmen began to sense that Act 590 might prove a disaster for the well-being of the state. They inquired whether the Act could be quietly shelved, but it was too late. Nothing could be done. Therefore, as the ACLU began to prepare its side of the case for the families, the office of the Attorney General of Arkansas began to prepare the defense. The case was to come to trial at the beginning of December before a Federal judge, William J. Overton, sitting on his own without a jury.

The ACLU does not have a large staff of its own. In a major case, such as this promised to be, it looks for help. This time, help came from an apparently unlikely source. The New York law firm of Skadden, Arps, Slate, Meagher, and Flom is a huge organization, with 250 lawyers and some 400 support staff. It specializes, very successfully, in aiding the biggest of American corporations in their aims of swallowing up all competitors, or in aiding the just-less-than-biggest companies in avoiding being swallowed up by predators. Just as the Arkansas case was coming to trial, it was helping Marathon Oil in its successful escape from the jaws of Mobil Oil.

For reasons which were never made absolutely clear, Skadden-Arps agreed to let about ten of its young lawyers go to work for the ACLU, *pro bonum* (that is, for nothing). I cannot believe that this was an act of purely disinterested altruism. Probably, the truth is that even the most successful of firms need to look to their image. Skadden-Arps has to go out with everyone else, recruiting the cream of the top law schools. Involvement with the ACLU case in Arkansas would certainly help to soften the picture of an enterprise concerned solely with money and power.

Whatever the reasons, the ACLU got together its men, one woman partner and a number of female legal assistants. And very bright people they were too. I have never encountered so sharp and enthusiastic and hardworking a group of young folk in my whole life. They could pick out the flaw in any argument instantly. They could start work after supper, build a case, take all of the components to pieces, put them back together again properly, and have everything typed, in multiple copies, on one's desk by nine o'clock the next morning. The ACLU looked for the best, and Skadden-Arps gave it that.

Since the ACLU was working with the plaintiffs—the ones objecting to the law—it had to make the positive case. It had to show why the law was unconstitutional. It was decided to make as broad a case as possible, because it was obvious that "Scopes II" would attract a great deal of media attention— newspapers, radio, and television. The more that Creation-science could be shown to be the travesty that it is, the more public opinion could be brought against it. Hence, the less success Creationists would be likely to have in the future.

To this end, the ACLU decided to divide its case into three parts. The first part of the case would be devoted to *religion*. Expert witnesses would be sought to prove that Creation-science is no more than Genesis by another name. Second, the case would move to *science*. It would be shown that Act 590 is thoroughly confused about evolution and that Creation-science has no right whatsoever to be called "science." Third, the case would conclude with *education*, and various witnesses would be called to show the difficulties of translating Act 590 into actual classroom practice.

Where did I come in? I am neither a theologian nor a scientist. I am not an expert on education. I am not an American.[13] And, as the state took some pains to point out, I have never taken a biology class in my life. This gap in my education, incidentally, was not due to a conscious act on my part. I grew up and was educated in England. It is, or at least was in the 1950's, virtually impossible for anyone to specialize in mathematics and physics—as I did—and yet take courses in biology. Regretfully, biology, like Spanish and geography, was for those who could not really handle the "hard" subjects.[14]

I am a historian and philosopher of science, and my claim to fame— more modestly, that which made the ACLU interested in me—was that I have written extensively on evolutionary theory. Moreover, when David Klasfeld contacted me in September 1981, I had just completed a manuscript which dealt in detail with evolutionary ideas.[15] Most pertinently, I had completed an in-depth, highly critical study of Creation-science. I was therefore already primed to fight for the ACLU.

But why a historian and philosopher at all? The ACLU decided that it wanted a historian to give the judge some of the general background to the story of man's quest for ultimate origins. Most particularly, they wanted to show that, contrary to Creationist complaints, evolutionists have not simply walked in and excluded all opposition. Creationism had a good run for its money—about 2000 years. As noted earlier, Creationism began to collapse of its own accord *before Darwin*. Hence, the ACLU wanted someone who could talk about men like the Reverend Adam Sedgwick, Professor of Geology at Cambridge and a deeply committed Christian. In 1831, the year in which Darwin graduated from Cambridge, Sedgwick publicly declared that one can no longer take Genesis as a literally true account of Earth's history. Most particularly, Sedgwick openly conceded that there is simply no evidence of a worldwide flood, and that he was working with a time frame far greater than that of the traditional 6000 years.[16]

An historian, therefore, could show the Court that juxtaposing Creationism against evolution sets up a false dichotomy. Creationism was simply judged not to work as a science anyway. And a philosopher could then go on to hammer home the nonscientific nature of Creationism. It is the job of the philosopher to look critically at science from afar, asking such questions as: What is the nature of a scientific theory? Are all sciences similar in logic and structure? What is the relationship between theory and evidence, and why do scientists sometimes change their minds?

Obviously, therefore, a philosopher can ask many pertinent questions about Creation-science. Having given a general account of science, one can then move right into an analysis of Creation-science, showing where it fails as science. Of course, this in itself does not show Creation-science to be a religion. There are many things which are neither science nor religion. Philosophy, for instance. But demolition of Creation-science as science would be an important first step. Furthermore, seeing where Creation-science falls down as science can surely give important clues as to its true religious nature.

For reasons such as these, the ACLU sought out someone with historical and philosophical training. As I have explained, I myself attracted attention, because I had already worked on the very issues that the ACLU wanted to put before the judge. I readily agreed to lend a hand, although I did not then realize just how hard I was going to have to work.

I should say now, unequivocally, that I found it a terrifically exciting experience to work for the ACLU. Also, now that the fight is over in Arkansas, I look back with a great sense of personal satisfaction. Following Socrates, I believe that the unexamined life is not worth living. For me, philosophy is a consuming passion, and not just a job. However, one does not usually get the immediate, tangible sense of the worth of one's work as was afforded by the Arkansas trial. It was good to be able to take the theoretical training of years and apply it to an actual practical problem, fighting something which I believe to be a real intellectual and moral evil.

In the course of preparing for the trial, I made two trips to New York. The first was to meet the lawyers and to get acquainted with the general facts of the case. For two days my brains were picked constantly about the nature of science, until I was fairly reeling (a condition that was alleviated when I took time out to see a World Series game, including Billy Martin in action). Returning to Guelph, I continued to stay busy. In the next month I got three requests for position/discussion papers. These were to be brief essays touching on pertinent points, written clearly and directly so that the lawyers concerned in the case could grasp essential factors about the history and philosophy of evolution and Creation-science. In them I state that I believe the key distinguishing factor about science to be its appeal to and reliance on *law*: blind, natural regularity. Everything else follows from an unpacking of this notion: explanation, prediction, testing, confirmation, falsifiability, tentativeness. Judged by these various criteria, evolutionary theory is a genuine scientific theory, whereas Creation-science is just not science.

Although, as noted, evolution was not on trial, it was clearly at the back of everyone's mind. The lawyers, therefore, wanted to know as much as possible about it, and this was why I dealt so fully with it in my discussion papers. I feel no need at all to apologize for my discussion papers, but I would point out that, necessarily, they were written at great speed: two days for the first and a day each for the second and third. Therefore, they could not be researched, but had to be written "off the top of my head." This, of course, was what the lawyers wanted: elementary lectures, not contributions to the literature. I would like to pride myself that the points I made, particularly in the first paper, went straight through into Judge Overton's decision. (See especially, section IV[C].)[17]

Before I returned to New York, I went off to a previously arranged conference in East Germany. On the way, stopping over in England, I spoke to an elderly zoologist, L. Harrison Matthews, who wrote the introduction to Darwin's *Origin* in the Everyman Edition.[18] In phrases which have been seized on by Creationists, Matthews argues that belief in Darwinism is like a religious commitment.[19] This was going to be used by the State of Arkansas, who would argue that belief in Creation-science is logically identical to belief in evolution. Hence, since one can teach the latter, one should be allowed to teach the former. (A more rigorous conclusion would be that since both are religion, neither should be taught. But no matter.)

Would Matthews recant? He was happy to do so, and wrote me a strong letter about the misuse that he felt Creationists had made of his introduction. Reading between the lines, I got the strong impression that what motivated Matthews in his introduction was not the logic of evolutionary theory at all. He wanted to poke the late Sir Gavin de Beer in the eye. De Beer was a fanatical Darwinian, and Matthews was dressing him down for the undue strength of his feelings!

On to East Germany, where the scientists from the other side of the Iron Curtain talked nonstop about Creationism. Nothing I could say would prevent them from tying Creationism in directly with Western capitalism. Creationism is the paradigmatic example of what goes wrong if you are not Marxist-Leninist! Of course, the men arguing this way really were not that motivated by ideological purity. By tying in Creationism with capitalism, at the same time noting that they themselves were not Creationists, they hoped to curry favor with their own superiors. "Look at us," they were saying. "We and we alone do proper science. Therefore encourage us." Given the predicament of scientists in the Soviet bloc, it is hard to blame them.

Upon returning to New York, I met for the first time Jack Novik, the lawyer from ACLU who was to lead me in direct examination. Tall, dark, heavily mustached, Novik is everything one expects of a bright, aggressive, young New York lawyer. He has a fantastic ability to cut through to the central point in a complex discussion and a total inability to relax. He thinks, eats, and breathes his cases. When you are in a jam, you could not have a better partner.

I was made to bone up on some twenty or so Creationist texts. Fortunately, there is not much variation on the same theme. Then, for one whole exhausting day I had my deposition taken by the assistant attorney general of Arkansas, David Williams. He was undoubtedly the toughest of the defense team, and by the time he had finished with me I was as limp as a rag. For the uninformed, having your deposition taken is pure reciprocal altruism in action. The lawyers from the other side are allowed to examine your witnesses in order to find out what these witnesses will say. You let this happen, because then you can have a crack at their witnesses. A witness obviously cannot lie, but he can be as unhelpful as possible, simply by not volunteering any information other than that directly demanded. Your own lawyer sits next to you (a court reporter is also present, taking everything down), and he can monitor for fairness. Like all professors I talk too much, and Jack Novik kept me quiet simply by making me more terrified of his wrath than of Williams's.

The deposition ranged widely. What were my religious beliefs? (Agnostic with flashes of deism.) Did I care about my children's religious beliefs? (I do, to the extent of sending my children to private Anglican schools.) Had I ever heard of Sir Karl Popper? (Yes.) Did I think him the greatest philosopher of science that had ever lived? (No.) Was I aware that some people so considered him? (Yes.) Did I think that one could have morality in a world of evolution? (Yes.) How do I regard morality? (I intuit moral values as objective realities. [Don't ask me what that means. Fortunately, Williams didn't ask me either, and so we moved on.])

There was one point at which I ran into rather heavy weather. In the first book I wrote, in defending the synthetic theory of evolution (i.e., the Darwin-Mendel synthesis), I asserted somewhat passionately that it "was established beyond all reasonable doubt."[20] Did this not show my own dogmatism? Did it not show that, contrary to my own claims about the tentative nature of science, I simply could not conceive of modern evolutionary theory being thrown over?

I sweated for a minute or two and then realized that the phrase could be turned to advantage. Scientists obviously don't keep questioning their theories every day of the week. When something is reasonably well established, they accept it. Similarly, in the realm of law one tries to reach a verdict which will be accepted without constant question: "it is beyond reasonable doubt." But because someone has been found guilty, it does not follow that the case can never be reopened. If important new evidence comes up, then there are ways of looking again at verdicts, and perhaps of overturning them. Similarly in science. Even if something is established "beyond all reasonable doubt," there is always the possibility that new evidence will make one reopen the case.

I think Williams realized what a tight spot he had me in. But when we came to trial, Novik and I brought out this analogy in direct examination, using it for our own purpose. We heard no more on the subject.

Finally my deposition came to an end, and with it nearly all of the pretrial work. There was only one more thing to be done: preparing what lawyers refer to as one's "Q's and A's." This is the script for the direct examination (thus "Questions and Answers"). Novik produced a 50-page script, and I was told to learn it. Since it was based on my own position papers, it was hardly new. But I confess that I did shiver rather at the thought of blanking out at some crucial moment. Suppose I forgot what a law is! I really pored over that script. It was made very clear to me by the ACLU that its witnesses were expected to perform without notes.

THE TRIAL

I arrived in Little Rock at four o'clock in the afternoon of December 6 and was taken straight to the hotel: the Sam Peck, a comfortable but plain building in downtown Little Rock, directly opposite the huge greystone Federal Building that contains the law courts. Since the religion experts were to appear first, it was agreed that I should arrive with them in order to see the kinds of points they would make. We wanted no gaffes, with me saying the very opposite to the previous witness.

The religion witnesses were Bruce Vawter, a Catholic priest and Old Testament expert; George Marsden, an historian who has specialized in the study of development of Fundamentalism; Dorothy Nelkin, a sociologist of science, who has studied the ways in which Creationists operate and how they have kept evolutionary ideas at a minimum in school texts; and Langdon Gilkey, a rather trendy and superarticulate theologian from Chicago Divinity School. As a side business, Gilkey works for the IRS, flying off to places like California, seeing if weird new sects qualify as religions and thus merit tax benefits.

There were no lawyers to be seen. Later we found that they were all huddled in last-minute preparations, still squabbling about who was to do what. Frankly, the Arkansas trial meant a lot to those young people, and a fair number of selfish genes were working hard to ensure that their survival machines had maximum time in court. Adding complexity to complexity, we not only had the New York lawyers involved, but now we had two Little Rock law firms too.

Fortunately or foolishly, a hospitality room had been set up in the hotel (i.e., unlimited free liquor). Religion witnesses and lonely philosopher tucked in, and by the time that the lawyers came back to earth an hour or so later, no one was feeling much pain. To be honest, I think all of the witnesses were feeling a little scared. How would we do the next day? Would we be ripped apart on cross-examination, as in a Perry Mason show? Would the judge be hostile or difficult, as in the original Scopes trial?

The lawyers were furious. They had hoped to work us hard all evening. As it was, we got through supper in a comfortable haze, and any discussion was at a pretty low level. Later in the evening, Novik, Klasfeld, and I tried running through my testimony. The rehearsal was an absolute disaster. I forgot points and—far worse—simply would not get to the crux of what I was supposed to be saying. I would wander on and on like the worst kind of teacher. Nevertheless, looking back, I don't feel all that sorry. All of the witnesses had been chosen because they would be able to perform well in open court, as well as for obvious expertise. I think people rather forgot how we too needed to relax and be reassured. Even those (especially those?) who appear most confident and self-assured in public have preperformance fears. I know that thanks to several large gins I myself settled in much more rapidly than I would have done otherwise.

On Monday we started off. The courtroom was a large oak-paneled hall on the fifth floor of the Federal Building. The judge, impressive in his simple black gown, sat high at a desk at one end. Before him was the court reporter, speaking nonstop into her machine. Then came the witness box, isolated before the two large tables used by the opposing lawyers. The plaintiffs (the ACLU's side) had at least three times as many people as the defense (the State of Arkansas). To one side were some twenty or so reporters, including several artists busily sketching faces for the evening newscasts. On the other side sat the citizens of Arkansas who had brought the original complaint against Act 590.

Then, in the main body of the hall one saw the various witnesses, together with the several experts who were helping the two sides. We had the support and advice of the distinguished evolutionists Niles Eldredge and Joel Cracraft. The attorney general was being aided by leading Creation scientist, Duane Gish. Finally, behind the direct participants in the trial was the audience of about two hundred: schoolchildren, bused in for lessons in "civics"; Fundamentalist ministers, with that overgroomed look which is their trademark; and any number of others.

Following brief opening statements in which both sides laid out their main claims (the ACLU was against Act 590, and Arkansas was for it!), we got down to business.[21] Our first witness was the Methodist bishop of Arkansas, who said he was all in favor of religion but not in schools. This was a theme to be repeated again and again. No one is more upset by the Creationist movement than orthodox American churchmen, Christians and Jews. They hold the First Amendment separation of church and state very dear. So many of their ancestors came to freedom in America, driven from Europe because of religious persecution.

Hence, orthodox churchmen do not want religion—any religion—taught in schools. In such a mingling of church and state, we see the road leading to such religiously torn countries as Northern Ireland. Moreover, the churchmen loathe Creationism. They do not view it as the only true form of religion. Rather they see it as a perverted blasphemy. God did not give us our reason just to have us hide our heads in the arid, comforting sands of Genesis. The Bible is not a work of science, and to pretend otherwise is to lose its true meaning. The Holy Writ is the story of God, man, and the relationship between the two. What does God expect of us? What promise does He hold out for us in the future? The Bible is a work of spiritual and moral significance.

After the bishop came the expert religion witnesses. They explained in very short order that the description of Creation-science in Act 590 is no more and no less than the story of Creation in Genesis. Indeed, they explained this in rather too short order, for whereas we had expected these witnesses to take two full days, it suddenly seemed that they might be through before the first day was over. I was packed off back to the hotel to put on a tie. Then, Jack Novik, David Klasfeld, and I went to a quiet room to run through my testimony. At long last things started to come right. Questions and answers flowed smoothly, one following on the next—a combination of theater and philosophical symposium. At the end, David Klasfield said quite simply and truthfully: "Mike! That was the first time I heard you do it better than I could have done it."

Tuesday morning came—rather clearly for me. I woke at three A.M. and was unable to go back to sleep. Finally, after *The Lone Ranger* and *Sunrise Semester,* it was seven o'clock and time for breakfast with Novik. I realized suddenly that, for all his outer confidence, he was as tense as I was. This was his big moment too. Nine o'clock finally arrived, we rose for the judge, I took the stand, and the show was on.

With cross-examination, I was up in front of everyone for three and a half hours in the morning and a little more in the afternoon. Direct examination went like a dream as Jack and I looked at each other, words coming straight to our lips, ideas bubbling out, and one point after another getting answered and hammered right home.

Q: What is science?

A: Science is an attempt to understand the physical world, primarily through law, that is, through unbroken natural regularity.

Q: Would you elaborate on that please?

A: Yes indeed. Understanding in science means explanation and prediction. Through this comes test, confirmation, and the potential to falsify. This means that a crucial mark of science is that it is tentative.

Q: Let's stop now and go through these various attributes of science, one by one, giving examples to explain them to the court.

And so we did explain and give examples, and the judge wrote everything down. He listened intently and would nod when he had grasped and noted a point. Then we would go on to the next topic.

We gave the judge a short history of evolutionary theory and of the growth of the idea of science separate from religion. We explained the essential attributes of science. Then we turned our attention to Act 590. Those crucial passages quoted earlier were examined word by word, line by line. Consider 4 (a) I "[Creation-science includes ideas implying] sudden creation of the universe, energy, and life from nothing." Why is there no description of how this creation will occur? In the corresponding phrase about evolution, 4 (b) I, it is explicitly stated that origins are "naturalistic," that is, governed by normal law. Clearly, Creation-science implies that origins are nonnaturalistic, that is, miraculous. Origins come through some sort of supernatural forces. This force is not a scientific notion. It is, however, a religious notion. (Notice how here we were able to make the point that Creation-science is not merely not science, but is religion.)

Moving on through the various clauses, we noted the talk of originally created "kinds" in the description of Creation-science. "Kind" is not a regular taxonomic term at all. But it does occur in the Genesis description of Creation. Analogously, it is very odd that a worldwide flood is singled out for special attention. If this is not a direct reference to Noah's Flood, why not talk of other natural disasters also? Why not talk of the Chicago fire or the San Francisco earthquake? And most particularly, what on earth could one mean by "a relatively recent inception of the earth and living kinds"? As I noted, and as virtually every other ACLU witness noted, this really is a meaningless statement. Is "relatively recent" a million years, the 6000 years traditionally calculated from the genealogies of the Bible, or (as Christopher Robin would have it) a week ago Friday?

At the same time as Novik and I went over the clauses describing Creation-science, we also told the Court a little about the supposed rival, "evolution-science." First we noted that "evolution-science" is both a name and a concept unknown outside Act 590 and the Creationist literature. Biologists just don't roll everything together in one overall hybrid. The ultimate origins of life and the subsequent evolution of life were separated by Darwin (he never mentions the former in the *Origin*). They have been kept separate ever since. I am sure that there are many more questions about ultimate origins than about evolution. Even if one had no idea whatsoever about how life originated, it could still be reasonable to believe in evolution.[22]

Second, we noted that Act 590 gets evolutionary theory wrong. No one—certainly not Darwin—has ever thought that natural selection is a sufficient mechanism for evolution. Today, everyone allows some randomness in evolution—for instance, that which comes in genetic drift. Third, the very juxtaposition of evolution with Creation-science—what Creationists call the "two-model approach"—is itself fallacious. One cannot prove Creationism by disproving evolution. There are alternative positions, as the State was to find to its embarrassment. For instance, recently the physicists Fred Hoyle and N. C. Wickramasinghe have proposed an earth-picture supposing life to come here, through the ages, from outer space.[23] This is not conventional evolution, but neither is it Creation-science. Apart from anything else, Hoyle and Wickramasinghe believe the earth to be very old. The attorney general was to call Wickramasinghe as a witness. When he had finished, he had done more damage to the State's case than to the ACLU case. The judge was left mystified as to why Wickramasinghe was called.[24]

Novik and I were almost through. But, as in the marriage at Cana, the best was left until last. We turned finally to the voluminous Creation-science literature, and for some twenty-five wonderful minutes showed its total failure as science. With Gish in the audience, I delighted in reading the following passage from *Evolution? The Fossils Say No!* The ACLU team could not have made the point more nicely if it had written the passage itself.[25]

CREATION. By creation we mean the bringing into being of the basic kinds of plants and animals by the process of sudden, or fiat, creation described in the first two chapters of Genesis. Here we find the creation by God of the plants and animals, each commanded to reproduce after its own kind using processes which were essentially instantaneous.

We do not know how God created, what processes He used, for God used processes which are not now operating anywhere in the natural universe. This is why we refer to divine creation as special creation. We cannot discover by scientific investigations anything about the creative processes used by God.

Incidentally, many people have suggested that Steve Clark, the Arkansas attorney general, threw the State's case by not using as witnesses the major Creation-scientists, such as Gish and Morris. The simple fact of the matter is that he couldn't. Can you imagine what an ACLU lawyer would have done with Gish on the stand? How lingeringly and how lovingly they would have dealt with that statement!

We next looked at questions of explanation and prediction. We showed that in Creation-science these never occur in a genuine scientific way. I talked of the facts of homology—nonfunctional isomorphisms between organisms of different species. Evolution explains them as a result of common descent. Creation-science has no answer at all. Lamely, its supporters suggest that homology is irrelevant, because all classification is arbitrary anyway. Just imagine. We are supposed to accept that the classification of organisms into birds, whales, dogs, men, etc., is arbitrary.[26]

We emphasized the nontentative nature of Creation-science. Again and again, the Creationists assert dogmatically that one must accept their position. It is not open to doubt and debate, as all genuine science must be. I read from the manifesto that all Creation-scientists must sign when they join the Creation Research Society. They have to affirm their belief in the literal truth of every word of the Bible. Whatever this may be, this is not the way in which science is done. Can you imagine every evolutionist having to sign a statement accepting the literal truth of Darwin's *Origin?*[27]

Finally, we came to the question of honesty. Novik and I wanted to prove not merely that Creation-science is not science, but that is a dishonest and thoroughly corrupt enterprise, violating every standard of intellectual integrity. Again and again, I was able to show that statements made by eminent evolutionists are lifted by Creation-scientists and quoted out of context. Evolutionists are made to say the very opposite to what they intended.

A classic case of such distortion occurs in a recent Creationist book, *Creation: The Facts of Life* by Gary E. Parker. Repeatedly, Parker refers to "noted Harvard geneticist [Richard] Lewontin's views" that things like the hand and the eye are the best evidence of God's design. Can this really be so? Has the distinguished author of *The Genetic Basis of Evolutionary Change*[28] really thrown over Darwin for Moses? When you actually look at the source of Lewontin's views, you find that you must wait awhile for Lewontin's full conversion.[29] What he says, in fact, is that *before Darwin* people believed that such phenomena as the hand and the eye were evidence of God's direct design. Now we accept evolution through natural selection.

The theory about the history of life that is now generally accepted, the Darwinian theory of evolution by natural selection, is meant to explain two different aspects of the appearance of the living world: diversity and fitness. . . . . It was the marvelous fit of organisms to the environment, much more than the great diversity of forms, that was the chief evidence of a Supreme Designer. Darwin realized that if a naturalistic theory of evolution was to be successful, it would have to explain the apparent perfection of organisms and not simply their variation. . . . The modern view of adaptation is that the external world sets certain "problems" that organisms need to "solve," and that evolution by means of natural selection is the mechanism for creating these solutions.

But the reader learns nothing of this. Parker makes Lewontin say the very opposite. And on this high note we ended my direct testimony. Cheerfully, I referred to the Creation-scientists as "sleazy." And that term certainly gave a keen edge to cross-examination, as Williams and I crossed wits for the rest of the morning and into the afternoon. He wanted to make me seem stupid. I dearly wanted to return the compliment. Actually, however, I never got a sense of personal bitterness. There were a lot of jokes cracked through the trial, and I confess I rather grew to like the members of the defense team. Certainly, Steve Clark was the epitome of the gracious southern gentleman, who always had a friendly word when you bumped into him.[30]

And so, Williams and I went over and over the territory again. "What about Popper?" "Well, what about him?" "Hasn't he said such and such?" "Yes he has, but since then he's changed his mind, and he didn't know what he was talking about in the first place." "But that's just opinion, isn't it?" "Yes, but I think it's a good opinion. I'd be happy to go into matters in some detail, if you'd like, Mr. Williams." "No, I don't think we need bother about that just at the moment, thank you, Dr. Ruse."[31]

My religious beliefs gave us all several minutes' entertainment. What did I mean by "agnostic"? What did I mean by "deist"? Were these claims consistent with what I'd said in my deposition? Wasn't my equivocation just one more proof of the godless, atheistic nature of evolutionism? Finally, half in embarrassment, half defiantly, I blurted out: "I'm sorry, Mr. Williams! Surely you can see that I'm not an expert witness on my own religious beliefs!" The burst of laughter that greeted my confession squelched any further discussion, and so we moved on.

One thing which occupied us for quite some time was the manuscript of my book, *Darwinism Defended*. (The defense had been given copies of everything that I have written. The Xeroxing bills in the Arkansas trial were quite colossal.) In that book, not only do I examine Creation-science in some detail, but I explain why I do not think it should be taught in schools. I argue that free speech does not imply that crazy ideas belong in the biology classroom. Nevertheless, quoting John Stuart Mill, I defend the right of anyone to believe anything they like.

Q: Dr. Ruse. In your new book you quote John Stuart Mill. Would you please read to the court the words that you quote.

A: "If all mankind minus one were of one opinion, and only one person were of the contrary opinion, mankind would be no more justified in silencing that one person, than he, if he had the power, would be justified in silencing mankind."[32]

Q: Don't you think you're being inconsistent in saying that? At the same time you want to ban Creation-science?

A: I don't want to ban Creation-science. Anyone who wants to can believe it and say it out loud.

Q: But not in the schools?

A: But not in the schools. Creation-science is religion and has no place in biology classrooms.

Finally, I was finished. We had covered just about everything under the sun, with the possible exception of L. Harrison Matthews's claims about the religious nature of Darwinism. When Williams saw the scathing letter that Matthews wrote to me about Creationism, he decided not to introduce Matthews into the testimony.

I was tired but satisfied. It was now the turns of others. And magnificent turns they were, too. Francisco Ayala gave a brilliant exposition of population genetics and of modern work on evolutionary mechanisms. Similarly, Stephen Jay Gould put us all right on the fossil record, showing the dishonest stupidity of those who think the record speaks against evolution. Ayala and Gould were nice complements, for the first is an ardent Darwinian and the second has led the attack on conventional Darwinism. By putting the two together, the ACLU neatly defused a major Creationist misrepresentation, namely that differences between evolutionists over mechanisms imply that evolution itself is in doubt. Both men stood strongly for one of the greatest of all ideas. To hear Ayala talking lovingly of his fruit flies and Gould of his fossils was to realize so vividly that it is those who deny evolution who are anti-God, not those who affirm it.

A major plank in the Creationist attack on science is based on thermodynamics. Creationists argue that life could not have originated and then evolve, because order can never come from disorder. Hence, the ACLU called Harold Morowitz of Yale, who talked about the second law of thermodynamics, showing why it does not disprove a possible natural origin of life or evolution. (Simply, the second law applies only to closed systems and this earth is an open system.)

Rounding out the science witnesses was G. Brent Dalrymple of the U.S. Geological Survey. He gave a quite brilliant disquisition on methods of dating the earth. One would not think that such a topic could be all that intrinsically interesting, but Dalrymple gave this assumption the total lie. He held us absolutely spellbound as he talked of various dating techniques and how geologists compensate for weaknesses in one direction by strengths from another. My sense was that Dalrymple was so good and so firm that he rather broke the back of the State's case. He had checked all of the Creationist arguments, and showed in devastating detail the trail of misquotations, computational errors, out-of-date references, and sheer blind stupidity which allows the Creationists to assign the earth an age of 6000 years. After Dalrymple, the State seemed far less ready to tangle with witnesses.

The science witnesses had done their job. Then came the most moving testimony of all, as ordinary schoolteachers from Arkansas explained how they simply could not teach the travesty of Creation-science. I shall never forget the man who cried out under cross-examination: "Look, sir! I'm not a martyr or anything! But I just can't teach that stuff. I'm not a scientist. I'm a science teacher. I'm like a traffic cop, directing ideas down from scientists to schoolchildren. My pupils respect me. All teachers are like parents in a way. How can I go into my classroom, spreading ideas that I know to be wrong? My students will despise me, and I'll not be able to live with myself."

What a man! How one would have loved to have had him as a teacher. How one would love one's children to have him as a teacher.

The ACLU case drew to a close. Then it was time for the defense to make its case. Unfortunately, I had to leave. Family, students, and exams called me back to Guelph. Would that I could have been there as the ACLU lawyers took on the State's witnesses. I would love to have heard the expert who allowed that he believed in flying saucers because he had read about them in the *Reader's Digest*. His explanation was that they are emissaries from Satan. In case this sounds too ridiculous to be true, I should say that Morris himself is on record as believing that the canals of Mars are an aftereffect of the fight between Satan and the Archangel Michael. He thinks that evolution is undoubtedly a pit dug for the unwary by the Evil One.

Generally, none of the State's witnesses were able to make any progress in showing that Creation-science is not religion but genuine science. One after another, they admitted that they believe what they believe for religious reasons.[33] One witness was so intimidated by what he saw in Court before he himself was to appear that he left on the next plane. He did not wait to be demolished. The only defense witness of any scientific caliber was Wickramasinghe, and I have explained already why the State probably wished they had never called him in the first place.

All the Creationist arguments crumbled. "Were the Creationists excluded by the professional scientific community?" "Yes, they were." "Had any Creationist submitted any articles to an established scientific journal in the past twenty years?" "No, they had not." "Did the Creationists agree that scientists should have open minds?" "Yes, they did." "Was there any evidence whatsoever that would make the witnesses alter their own opinions?" "No, there was not." "Had the witnesses done original research on their claims, rather than simply combining the evolutionists' literature?" "No, they had not." And so on, and so on, and so on.

Creationists have gotten a lot of mileage in recent years claiming that they have a valid case which should be heard. A case which merits "balanced treatment" in the classroom. When they had their day in Court—literally—they had nothing to offer.

DECISION AND RETROSPECT

Judge Overton handed down his decision on January 5, 1982.[34] In no uncertain terms, he ruled that Creation-science is religion and therefore is constitutionally barred from the classroom. Act 590 may not be enforced. Quoting Justice Felix Frankfurter, Overton concluded that "good fences make good neighbors."[35] The State of Arkansas decided not to appeal, and so a great victory has been won.

But as the months slip by, thoughts of the Arkansas trial stay with me. What is there to be said in retrospect, and also prospectively in looking forward?

First, as I have said, personally I found the Arkansas trial a very rewarding experience. To stand in defense of the nobility of science, along with Gilkey, Dalrymple, and the Arkansas teachers, was a once-in-a-lifetime privilege. Several professional colleagues have since criticized me for participating, arguing that one cannot really make philosophical points clearly in a courtroom: one has to ignore all sorts of subtleties and distinctions. But without in any way conceding that I had to compromise, let me say simply that when the discipline of Socrates, John Locke, John Stuart Mill, Bertrand Russell, and Jean-Paul Sartre can no longer get up and defend right from wrong, then indeed it has collapsed ignobly and become irrelevant.

Moreover, I took particular satisfaction in doing something for America. At a personal level, many of my best friends are Americans, I was a student for some years at a U.S. college, and American publishers and journals have taken up my ideas and put them before the public. At a general level, for all the faults (and they are there), only a fool or a knave would deny the worth of American democracy—for the whole world, as well as for America itself. Going to Arkansas was partial payment of a large debt.

Second, let me pay full compliment to the fine people of Arkansas: judges, lawyers, church people, teachers. They got up and fought. Act 590 was a wretchedly stupid law to have passed in the first place, but once the Act became law, good people came out and opposed it. One of the plaintiffs, a clergyman, said to me: "Everyone thinks we're red-necks, but we're not. We don't want this kind of thing in our schools any more than anyone else does. We're just ordinary people." There was nothing at all ordinary about the people I met.

Third, let me warn against complacency. We have won a great victory. We have not yet won the war. The State of Louisiana has also passed a Creation-science law, so the ACLU must do battle there. No doubt other states will pass similar laws. And informally, Creation-scientists have announced their intentions to fight at the levels of individual teachers and school boards. There will be a long, hard, unglamorous slog before evolution gets fair, unfettered treatment in the classrooms of the U.S. And the same, I am afraid, holds true of my own country, Canada, where Creationist ideas are already making some headway in the schools.

Those of us who know better must roll up our sleeves and fight for the truth and for our children's education. Remember, although we may be little higher than the apes, we are also little lower than the angels. What better way could there be of showing this than by opposing Creation-science? No brute ever played Beethoven. No brute ever inquired into his past, opening up the magnificent picture that Darwin bequeathed to us.

EPILOGUE

Let me end on a lighter note. By the evening of the third day of the trial, it was clear to all that the ACLU was building an absolutely devastating case. The attorney general's lawyers were simply being brushed aside by the might of our science forces. Indeed, I have mentioned how, after Dalrymple's testimony, the defense had virtually given up fighting. In actual order of presentation, Gould was the final science witness. I have never seen such a disappointed man as he when, after only half an hour's cross-examination, Williams turned to the judge and said, "No further questions, your honor." Steve had been looking forward to a cozy afternoon putting us all right on the gaps in the fossil record, and he was finished before he had begun: rather than talking of gaps in the record, he was condemned forever to be one.

With the scientists cowing the defense, for us the tension broke. That night, nearly everyone on the plaintiff's team—lawyers, assistants, witnesses, assorted friends on the fringe—went out to eat in a restaurant—to talk, to drink, to play. Toward the end of the evening, someone started singing, and before long we were all joined in chorus. Inevitably we launched into hymns. My experience is that liberals almost always have a good church background, and that under the influence this comes to the fore. An angelic member of the Skadden-Arps contingent led us in that beautiful hymn, "Amazing Grace." We all held forth: paleontologist, philosopher, leading counsel for the ACLU. We came to the line which talks of worshiping God for 10,000 years. That sort of time span was a little too close for comfort. We broke off and looked at each other in embarrassment. Then we broke into uncontrollable laughter.

It was a good moment.

NOTES

1. London: John Murray.
2. Darwin, *Origin*, pp. 80–81.
3. The classic account of the reception of the *Origin*, showing how rapidly people did accept evolution, is A. Ellegard, *Darwin and the General Reader* (Goteborg: Goteborgs Universitets Arsskrift, 1958).

4. This is from an unpublished letter, written and sent by the philosopher/scientist John F. W. Herschel to the geologist Charles Lyell, April 14, 1863. I discuss the various proposed supplements and alternatives to Darwinian selection in my book *The Darwinian Revolution: Science Red in Tooth and Claw* (Chicago: University of Chicago Press, 1979).

5. For full details of modern evolutionary controversies, see my *Darwinism Defended: A Guide to the Evolution Controversies* (Reading, Mass.: Addison-Wesley, 1982).

6. L. Huxley, *The Life and Letters of Thomas H. Huxley* (New York: Appleton, 1900), I, 366.

7. The Scopes Trial has been written about extensively. See, for instance, L. S. de Camp, *The Great Monkey Trial* (New York: Doubleday, 1968).

8. D. Nelkin, *Science Textbook Controversies and the Politics of Equal Time* (Cambridge, Mass.: M.I.T. Press, 1977). Nelkin points out that although Scopes may have been a moral victory for evolution, at another level evolution suffered badly. Textbook publishers simply took evolution out of school texts. Even if it was not always illegal to teach evolution, as a matter of fact it was often not taught.

9. The "classic" work, which seems acknowledged as the founding text of the new Creationist movement, is John C. Whitcomb, Jr. and Henry M. Morris, *The Genesis Flood* (Nutley, N.J.: Presbyterian and Reformed Publishing Co., 1961).

10. The standard work, produced by a large team of Creation-scientists, is H. M. Morris (ed.), *Scientific Creationism* (San Diego: Creation-Life Publishers, 1974). A very popular work (over 130,000 copies sold) is Duane T. Gish, *Evolution?—The Fossils Say No!* (San Diego: Creation-Life Publishers, 1972). A recent book is Gary E. Parker, *Creation: The Facts of Life* (San Diego: Creation-Life Publishers, 1980).

11. Ark. Stat. Ann. §80-1663, *et.seq.*(1981 Supp.).

12. These were facts that were to emerge at the trial of Act 590. They are sketched in the memorandum opinion drawn up by the presiding judge, William R. Overton.

13. Canadian newspeople had utmost difficulty in thinking that perhaps someone from north of the border could make a useful contribution in Arkansas. Actually, however, my nationality never became an issue. It was of course irrelevant, but I suspect the main reason is that the State wanted to use a Canadian witness also. Incredibly, the witness turned out to be the daughter of a colleague in one of my departments.

14. Had I been really bright, I should have specialized in classics.

15. *Darwinism Defended.* Other pertinent books by me include *The Philosophy of Biology* (London: Hutchinson, 1973); *The Darwinian Revolution;* and *Is Science Sexist? And Other Problems in the Biomedical Sciences* (Dordrecht: Reidel, 1981).

16. Presidential address to the Geological Society. *Pro. Geol. Soc. Lond.,* I, 281-316.

17. Several witnesses suggested definitions of science. A descriptive definition was said to be that science is what is "accepted by the scientific community" and is "what scientists do." The obvious implication of this description is that, in a free society, knowledge does not require the imprimatur of legislation in order to become science. More precisely, the essential characteristics of science are: (1) it is guided by natural law; (2) it has to be explanatory by reference to natural law; (3) it is testable against the empirical world; (4) its conclusions are tentative, i.e., are not necessarily the final word; and (5) it is falsifiable. Creation-science as described in section 4(A) fails to meet these essential characteristics.

18. London: Dent, 1971.

19. "Belief in evolution is thus parallel to belief in special creation—both are concepts which believers know to be true but neither, up to the present, has been capable of truth" (p. x). This is quoted in Morris, *Scientific Creationism,* p. 6, n.l.

20. *The Philosophy of Biology,* Chapter 6. I was talking of the theory which came together in the 1930's, combining Darwinian selection with Mendelian genetics. I was defending it against a lot of vitalistic arguments, and within the context of the discussion, I think the conclusion still holds.

21. Brief accounts of the trial were written for *Science* by Roger Lewin ("Creationism on

the defensive in Arkansas," [1982] 215, 33–34; "Where is the science in Creation science?" *Science* [1982] 215, 142–146).

22. This is certainly not to claim that no one has any sensible ideas about natural causes for ultimate origins. For more on this subject, see my *Darwinism Defended*.

23. *Evolution from Space* (London: Dent, 1981).

24. Technically speaking, Creationists and the State of Arkansas confuse "contraries" with "contradictories." With contradictories, if one side is right then the other side is wrong, and vice versa. (This pen is red/This pen is not red.) With contraries, both can be wrong but both cannot be right. (This pen is blue/This pen is yellow.) The "two-model approach" assumes that evolution and Creation are contradictories, whereas they are really contraries.

25. As noted, a number of Creationists do now admit that Creation science is not genuine science. They argue that evolution is not genuine science either, and so since the latter may be taught, the former should be taught also. But as also noted, the correct inference is surely that *nothing* should be taught about origins. Since I do not accept all of the premises, I do not accept the conclusion either.

26. See Morris, *Scientific Creationism*, pp. 71–72.

27. The full statement is as follows: "1. The Bible is the written Word of God, and because we believe it to be inspired thruout [sic], all of its assertions are historically and scientifically true in all of the original autographs. To the student of nature, this means that the account of origins in Genesis is a factual presentation of simple historical truths. (2) All basic types of living things, including man, were made by direct creative acts of God during Creation Week as described in Genesis. Whatever biological changes have occurred since Creation have accomplished only changes within the original created kinds. (3) The great Flood described in Genesis, commonly referred to as the Noachian Deluge, was an historical event, world-wide in its extent and effect. (4) Finally, we are an organization of Christian men of science, who accept Jesus Christ as our Lord and Savior. The account of the special creation of Adam and Eve as one man and one woman, and their subsequent Fall into sin, is the basis for our belief in the necessity of a Savior for all mankind. Therefore, salvation can come only thru [sic] accepting Jesus Christ as our Savior." (Px 115).

28. New York: Columbia University Press, 1974.

29. The misquoted article, entitled "Adaptation," is from an excellent issue of *Scientific American* (September 1978, pp. 213–30), devoted entirely to evolutionary thought.

30. I keep harping on the fact, but one thing which sticks in my mind is the extreme youth of all of the lawyers. Novik, Klasfeld, and Clark were all about 35. Several lawyers, including Williams, were under 30. Overton was an old man; like me, he is in his early 40's!

31. The point about Sir Karl Popper is that at one point he thought Darwinian evolutionary biology fails the test of being a genuine science because it is not falsifiable. (This is Popper's "criterion of demarcation" between science and nonscience. Popper argued that Darwinism is a "metaphysical research program." Recently, Popper has allowed that he was wrong and that Darwinism is indeed genuine science. See K. R. Popper, "Darwinism as a metaphysical research programme" in P. A. Schilpp, ed., *The Philosophy of Karl Popper* (LaSalle, Ill.: Open Court, 1974); letter to the editor, *New Scientist* (1980), 87, 611. See also my *Is Science Sexist?*

32. J. S. Mill, *On Liberty* (New York: Norton, 1975), p. 18. First published 1859.

33. See Lewin, "Where is the science in Creation science?"

34. For a brief summary, see R. Lewin, "Judge's ruling hits hard at Creationism," *Science* (1982) 215, 381–84. For the full text, see W. R. Overton, Memorandum on Rev. Bill McLean et al. LR C 81 322 (January 5, 1982), reprinted below.

35. *McCollum v. Board of Education*, 333 U.S. 203, 232 (1948).

Part One

The Nineteenth-Century Background

Introduction

Part One aims to spell out, in a little more detail, some of the historical facts alluded to, very briefly, in the opening paragraphs of the "Prologue." In particular, I want to show you that right from the coming of Darwin's theory—before, even—there was concern and debate around the status of the various claims made about origins. What precisely would be the nature of a scientific approach to the beginnings of organisms? Could one ever hope to satisfy such an approach? Does science itself reveal its own limitations showing that no scientific answers about origins can ever be given?

The section can best be thought of as falling into three parts. First, we have ideas leading up to Darwin's own achievements. The creation stories from Genesis and the argument from design, as given in Archdeacon Paley's *Natural Theology,* set the scene. I might note, parenthetically, that today's scholars date the second creation story before the first. The second appears at the time of King David (he resigned 1000–961 B.C.), and with its twin emphases on God's having created man, yet also ceded man power over the earth, represents an attempt to define the proper role of a leader to a population unused to (and somewhat dubious about) having a king. The first story appears during the Babylonian exile (post-587 B.C.), picks up on other Near Eastern creation stories, and tries to reconvince the Jews of both the power of their God and of their own special place in His heart.

Such subtleties of scholarship were unknown to thinkers at the beginning of the nineteenth century; but, in "The Relationship between Science and Religion in Britain, 1830–1870," following the primary readings, I explain how Darwin came into an intellectual community that, although much concerned with the question of organic origins because of its religious implications, was becoming more and more aware that naive, literalistic readings of the Bible would not do. Such readings put far too much strain on the facts as then known.

This is not to say that the pre-Darwinian community felt at liberty to theorize and conclude precisely as their fancies took them. Constraints had to be met. First, it was demanded that one stay true to the principles of science. I speak in my article of two factions, "liberals" and "conservatives," although in a way they might more readily be characterized as "empiricists" and "rationalists," for the former were ardent to argue *from* experience (to beliefs about God and the like) whereas the latter were ardent to argue *to* experience (from beliefs about God and the like). But, either way, both factions agreed that the best kinds of explanations are causal, exhibiting so-called "true causes," *verae causae*. Second, in dealing with origins, in line with the natural theology of which Paley was the classic expositor, one had to respect the teleological nature of organisms. One had to see organisms not as randomly thrown together, but as integrated functioning entities, marked above all else by having features that are adapted to life's needs.

For all of their differences, almost everyone in Britain agreed that the earliest attempts at evolutionary theorizing failed these two tests. This applied particularly to the most notorious speculator of them all, the French biologist Jean Baptiste de Lamarck. At the beginning of the nineteenth century, he had argued that organisms "transform" in a chain or ladder of development, from primitive blobs up to humans. This transformation was supposedly aided, in part, by a mechanism that assumed the inheritance of acquired characters. However, in the opinion of just about all, these "Lamarckian" speculations were a methodological failure from beginning to end. There was neither a *vera causa* nor an adequate understanding of design.

In the second part of the section, we move to Darwin's contribution, particularly that given in his *Origin of Species*. In "Charles Darwin and the *Origin of Species,*" backed by pertinent passages from Darwin himself, I argue that it was Darwin's genius to respond to both of the demands that his elders and teachers put on solutions to the problem of origins. Darwin did not so much break with the past as mold it to his own ends. Thus, on the one hand, he strove to put his theorizing on a *vera causa* basis. As I explain, the liberal/empiricist thinkers like John F. W. Herschel interpreted *verae causae* as causes of which one had analogical evidence (thus one argues from the already known to the hypothesized unknown). The conservative/rationalist thinkers like William Whewell interpreted *verae causae* as causes at the center of "consiliences of inductions," where disparate elements of our science are unified beneath one postulate (thus one argues from the hypothesized unknown to the already known). Darwin succeeded in his work (I believe) to such an extent that he satisfied both of the chief interpretations that others put on the *vera causa* notion.

On the other hand, through his mechanism of natural selection, Darwin hoped to speak not only to evolution, but also to the adaptedness of organisms. This latter point cannot be overstressed. The more I wrestle with the nature of evolutionary theory, past and present, the more I become convinced that what really distinguishes Darwin and his true supporters is the conviction, first, that adaptation is the key aspect of organic nature and, second, that natural selection is a mechanism adequate to an understanding of this adaptation. In one fundamental way, Darwin differed from Paley. After the *Origin,* there was no place for the Master Craftsman. In another equally fundamental way, there was no difference at all. Darwin was at one with Paley in seeing the world *as if* designed.

The third part of this section takes up two of the contemporary responses of the *Origin.* In the decade after Darwin published, there was a huge amount written, both in favor of and against evolutionism. Many of the criticisms were well taken. Darwin himself was much troubled by his inability to produce a coherent theory of heredity—of how the new variations (the "raw stuff" of evolution) might be produced and transmitted—and the critics were quick to magnify and exploit these troubles. Other criticisms worried Darwin less. Many critics harped constantly on the implications of evolution for our own species, but Darwin was always quite cold-bloodedly convinced that one must go the whole hog (or, rather, the whole primate). In 1871, in the *Descent of Man,* he focused explicit attention on humankind, but essentially his thinking was an expansion rather than an innovation.

What is interesting is the extent to which one finds thought and worries about methodology running through the responses of the ideas of the *Origin.* Is evolution the kind of approach that could, in principle, be scientific? Is natural selection an adequate cause? I have chosen two such responses where methodolgoical concerns have a front role. One response is critical and one response is favorable, although when you read them, you will see that there is slightly more to the story than simple agreement or disagreement.

Adam Sedgwick, Darwin's old friend and teacher, vehemently opposes evolution. Darwin has failed to be truly "inductive." One might, perhaps, be forgiven for wondering what Sedgwick means by this, but whatever it is, Darwin has not got it. Note, however, that Sedgwick is at one with Darwin on the significance of adaptation. It is just that he does not think that selection is adequate to the task. This, I suspect, is the real crux of Sedgwick's opposition. Thomas Henry Huxley, Darwin's great popular spokesperson, has a much more sophisticated grasp of methodology. He picks right up on the three key aspects of Darwin's argument. Note, however, that although Huxley clearly accepts evolution, he is much more reticent on natural selection, supposing even that evolution might go in jumps.

Huxley's agnosticism (his own word!) about selection becomes more understandable when you remember what I just said above. Huxley, unlike Darwin and Sedgwick, was never that moved by adaptation. He had never been under the spell of natural theology. Hence, he never accepted as a problem that which selection was intended to solve. Paradoxical though it may seem, I am led to conclude that although Huxley was a Darwinian emotionally and socially, intellectually there was always a chasm between the two men. We shall hear echoes from this divide when we turn to modern evolutionary thinking in the next section.

1

Genesis

CHAPTER 1

In the beginning God created the heaven and the earth.

2. And the earth was without form, and void: and darkness *was* upon the face of the deep. And the Spirit of God moved upon the face of the waters.

3. And God said, Let there be light: and there was light.

4. And God saw the light, that *it was* good: and God divided the light from the darkness.

5. And God called the light Day, and the darkness he called Night. And the evening and the morning were the first day.

6. AND God said, let there be a firmament in the midst of the waters, and let it divide the waters from the waters.

7. And God made the firmament, and divided the waters which *were* under the firmament from the waters which *were* above the firmament: and it was so.

8. And God called the firmament Heaven. And the evening and the morning were the second day.

9. AND God said, Let the waters under the heaven be gathered together unto one place, and let the dry *land* appear: and it was so.

10. And God called the dry *land* Earth; and the gathering together of the waters called he Seas: and God saw that *it was* good.

11. And God said, Let the earth bring forth grass, the herb yielding seed, *and* the fruit tree yielding fruit after his kind, whose seed *is* in itself, upon the earth: and it was so.

12. And the earth brought forth grass, *and* herb yielding seed after his

kind, and the tree yielding fruit, whose seed *was* in itself, after his kind: and God saw that *it was* good.

13. And the evening and the morning were the third day.

14. AND God said, Let there be lights in the firmament of the heaven to divide the day from the night; and let them be for signs, and for seasons, and for days, and years:

15. And let them be for lights in the firmament of the heaven to give light upon the earth: and it was so.

16. And God made two great lights; the greater light to rule the day, and the lesser light to rule the night: *he made* the stars also.

17. And God set them in the firmament of the heaven to give light upon the earth,

18. And to rule over the day and over the night, and to divide the light from the darkness: and God saw that *it was* good.

19. And the evening and the morning were the fourth day.

20. And God said, Let the waters bring forth abundantly the moving creature that hath life, and fowl *that* may fly above the earth in the open firmament of heaven.

21. And God created great whales, and every living creature that moveth, which the waters brought forth abundantly after their kind, and every winged fowl after his kind: and God saw that *it was* good.

22. And God blessed them, saying, Be fruitful, and multiply, and fill the waters in the seas, and let fowl multiply in the earth.

23. And the evening and the morning were the fifth day.

24. AND God said, Let the earth bring forth the living creature after his kind, cattle, and creeping thing, and beast of the earth after his kind: and it was so.

25. And God made the beast of the earth after his kind, and cattle after their kind, and every thing that creepeth upon the earth after his kind: and God saw that *it was* good.

26. AND God said, Let us make man in our image, after our likeness: and let them have dominion over the fish of the sea, and over the fowl of the air, and over the cattle, and over all the earth, and over every creeping thing that creepeth upon the earth.

27. So God created man in his *own* image, in the image of God created he him; male and female created he them.

28. And God blessed them, and God said unto them, Be fruitful, and multiply, and replenish the earth, and subdue it: and have dominion over the fish of the sea, and over the fowl of the air, and over every living thing that moveth upon the earth.

29. AND God said, Behold, I have given you every herb bearing seed, which *is* upon the face of all the earth, and every tree, in the which *is* the fruit of a tree yielding seed; to you it shall be for meat.

30. And to every beast of the earth, and to every fowl of the air, and

to every thing that creepeth upon the earth, wherein *there* is life, *I have given* every green herb for meat: and it was so.

31. And God saw every thing that he had made, and, behold, *it was* very good. And the evening and the morning were the sixth day.

CHAPTER 2

Thus the heavens and the earth were finished, and all the host of them.

2. And on the seventh day God ended his work which he had made; and he rested on the seventh day from all his work which he had made.

3. And God blessed the seventh day, and sanctified it: because that in it he had rested from all his work which God created and made.

4. THESE *are* the generations of the heavens and of the earth when they were created, in the day that the LORD God made the earth and the heavens.

5. And every plant of the field before it was in the earth, and every herb of the field before it grew: for the LORD God had not caused it to rain upon the earth, and *there was* not a man to till the ground.

6. But there went up a mist from the earth, and watered the whole face of the ground.

7. And the LORD God formed man *of* the dust of the ground, and breathed into his nostrils the breath of life; and man became a living soul.

8. AND the LORD God planted a garden eastward in Eden; and there he put the man whom he had formed.

9. And out of the ground made the LORD God to grow every tree that is pleasant to the sight, and good for food; the tree of life also in the midst of the garden and the tree of knowledge of good and evil.

10. And a river went out of Eden to water the garden; and from thence it was parted, and became into four heads.

11. The name of the first *is* Pison; that *is* it which compasseth the whole land of Havilah, where *there is* gold;

12. And the gold of that land *is* good: there *is* bdellium and the onyx stone.

13. And the name of the second river *is* Gihon: the same *is* it that compasseth the whole land of Ethiopia.

14. And the name of the third river *is* Hiddekel: that *is* it which goeth toward the east of Assyria. And the fourth river *is* Euphrates.

15. And the LORD God took the man, and put him into the garden of Eden to dress it and to keep it.

16. And the LORD God commanded the man, saying, Of every tree of the garden thou mayest freely eat:

17. But of the tree of the knowledge of good and evil, thou shalt not eat of it: for in the day that thou eatest thereof thou shalt surely die.

18. AND the LORD God said, *It is* not good that the man should be alone; I will make him an help meet for him,

19. And out of the ground the LORD God formed every beast of the field, and every fowl of the air; and brought *them* unto Adam to see what he would call them; and whatsoever Adam called every living creature, that *was* the name thereof.

20. And Adam gave names to all cattle, and to the fowl of the air, and to every beast of the field; but for Adam there was not found an help meet for him.

21. And the LORD God caused a deep sleep to fall upon Adam, and he slept: and he took one of his ribs, and closed up the flesh instead thereof;

22. And the rib, which the LORD God had taken from man, made he a woman, and brought her unto the man.

23. And Adam said, This *is* now bone of my bones, and flesh of my flesh: she shall be called Woman, because she was taken out of Man.

24. Therefore shall a man leave his father and his mother, and shall cleave unto his wife; and they shall be one flesh.

25. And they were both naked, the man and his wife, and were not ashamed.

2

Natural Theology

William Paley

I. STATE OF THE ARGUMENT

In crossing a heath, suppose I pitched my foot against a *stone,* and were asked how the stone came to be there, I might possibly answer, that for anything I knew to the contrary it had lain there forever; nor would it, perhaps, be very easy to show the absurdity of this answer. But suppose I had found a *watch* upon the ground, and it should be inquired how the watch happened to be in that place, I should hardly think of the answer which I had before given, that for anything I knew the watch might have always been there. Yet why should not this answer serve for the watch as well as for the stone; why is it not as admissible in the second case as in the first? For this reason, and for no other, namely, that when we come to inspect the watch, we perceive—what we could not discover in the stone—that its several parts are framed and put together for a purpose, *e.g.* that they are so formed and adjusted as to produce motion, and that motion so regulated as to point out the hour of the day; that if the different parts had been differently shaped from what they are, or placed after any other manner or in any other order than that in which they are placed, either no motion at all would have been carried on in the machine, or none which would have answered the use that is now served by it. To reckon up a few of the plainest of these parts and of their offices, all tending to one result: We see a cylindrical box containing a coiled elastic spring, which, by its endeavor to relax itself, turns round the box. We next observe a flexible chain—artifically wrought for the

Extract from *Natural Theology,* first published in 1805.

sake of flexure—communicating the action of the spring from the box to the fusee. We then find a series of wheels, the teeth of which catch in and apply to each other, conducting the motion from the fusee to the balance and from the balance to the pointer, and at the same time, by the size and shape of those wheels, so regulating that motion as to terminate in causing an index, by an equable and measured progression, to pass over a given space in a given time. We take notice that the wheels are made of brass, in order to keep them from rust; the springs of steel, no other metal being so elastic; that over the face of the watch there is placed a glass, a material employed in no other part of the work, but in the room of which, if there had been any other than a transparent substance, the hour could not be seen without opening the case. This mechanism being observed—it requires indeed an examination of the instrument, and perhaps some previous knowledge of the subject, to perceive and understand it; but being once, as we have said, observed and understood, the inference we think is inevitable, that the watch must have had a maker—that there must have existed, at some time and at some place or other, an artificer or artificers who formed it for the purpose which we find it actually to answer, who comprehended its construction and designed its use. . . .

II. STATE OF THE ARGUMENT CONTINUED

Suppose, in the next place, that the person who found the watch should after some time discover, that in addition to all the properties which he had hitherto observed in it, it possessed the unexpected property of producing in the course of its movement another watch like itself—the thing is conceivable; that it contained within it a mechanism, a system of parts—a mold, for instance, or a complex adjustment of lathes, files, and other tools—evidently and separately calculated for this purpose; let us inquire what effect ought such a discovery to have upon his former conclusion.

I. The first effect would be to increase his admiration of the contrivance, and his conviction of the consummate skill of the contriver. Whether he regarded the object of the contrivance, the distinct apparatus, the intricate, yet in many parts intelligible mechanism by which it was carried on, he would perceive in this new observation nothing but an additional reason for doing what he had already done—for referring the construction of the watch to design and to supreme art. If that construction *without* this property, or which is the same thing, before this property had been noticed, proved intention and art to have been employed about it, still more strong would the proof appear when he came to the knowledge of this further property, the crown and perfection of all the rest.

II. He would reflect, that though the watch before him were *in some sense* the maker of the watch which was fabricated in the course of its movements, yet it was in a very different sense from that in which a carpenter, for instance, is the maker of a chair—the author of its contrivance, the cause of the relation of its parts to their use. With respect to these, the first watch was no cause at all to the second; in no such sense as this was it the a-uthor of the constitution and order, either of the parts which the new watch contained, or of the parts by the aid and instrumentality of which it was produced. We might possibly say, but with great latitude of expression, that a stream of water ground corn; but no latitude of expression would allow us to say, no stretch of conjecture could lead us to think, that the stream of water built the mill, though it were too ancient for us to know who the builder was. What the stream of water does in the affair is neither more nor less than this: by the application of an unintelligent impulse to a mechanism previously arranged, arranged independently of it and arranged by intelligence, an effect is produced, namely, the corn is ground. But the effect results from the arrangement. The force of the stream cannot be said to be the cause or the author of the effect, still less of the arrangement. Understanding and plan in the formation of the mill were not the less necessary for any share which the water has in grinding the corn; yet is this share the same as that which the watch would have contributed to the production of the new watch, upon the supposition assumed in the last section. Therefore,

III. Though it be now no longer probable that the individual watch which our observer had found was made immediately by the hand of an artificer, yet doth not this alteration in anywise affect the inference, that an artificer had been originally employed and concerned in the production. The argument from design remains as it was. Marks of design and contrivance are no more accounted for now than they were before. In the same thing, we may ask for the cause of different properties. We may ask for the cause of the color of a body, of its hardness, of its heat; and these causes may be all different. We are now asking for the cause of that subserviency to a use, that relation to an end, which we have remarked in the watch before us. No answer is given to this question, by telling us that a preceding watch produced it. There cannot be design without a designer; contrivance, without a contriver; order, without choice; arrangement, without anything capable of arranging; sub-serviency and relation to a purpose, without that which could intend a purpose; means suitable to an end, and executing their office in accomplishing that end, without the end ever having been contemplated, or the means accom-modated to it. Arrangement, disposition of parts, subserviency of means to an end, relation of instruments to a use, imply the presence of intelligence and mind. No one, therefore, can rationally believe that the insensible, inanimate watch, from which the watch before us issued, was the proper cause of the mechanism we so much admire in it—could be truly said to have constructed the instrument, disposed its parts, assigned their office, determined their order, action, and mutual dependency, combined their several motions into one result, and that also a result connected with the utilities of other beings. All these properties, therefore, are as much unaccounted for as they were before. . . .

The conclusion which the *first* examination of the watch, of its works, construction, and movement, suggested, was, that it must have had, for cause and author of that construction, an artificer who understood its mechanism and designed its use. This conclusion is invincible. A *second* examination presents us with a new discovery. The watch is found, in the course of its movement, to produce another watch similar to itself; and not only so, but we perceive in it a system or organization separately calculated for that purpose. What effect would this discovery have, or ought it to have, upon our former inference? What, as hath already been said, but to increase beyond measure our admiration of the skill which had been employed in the formation of such a machine? Or shall it, instead of this, all at once turn us round to an opposite conclusion, namely, that no art or skill whatever has been concerned in the business, although all other evidences of art and skill remain as they were, and this last and supreme piece of art be now added to the rest? Can this be maintained without absurdity? Yet this is atheism.

III. APPLICATION OF THE ARGUMENT

This is atheism; for every indication of contrivance, every manifestation of design which existed in the watch, exists in the works of nature, with the difference on the side of nature of being greater and more, and that in a degree which exceeds all computation. I mean, that the contrivances of nature surpass the contrivances of art, in the complexity, subtilty, and curiosity of the mechanism; and still more, if possible, do they go beyond them in number and variety; yet, in a multitude of cases, are not less evidently mechanical, not less evidently contrivances, not less evidently accommodated to their end or suited to their office, than are the most perfect productions of human ingenuity.

I know no better method of introducing so large a subject, than that of comparing a single thing with a single thing: an eye, for example, with a telescope. As far as the examination of the instrument goes, there is precisely the same proof that the eye was made for vision, as there is that the telescope was made for assisting it. They are made upon the same principles; both being adjusted to the laws by which the transmission and refraction of rays of light are regulated. I speak not of the origin of the laws themselves; but such laws being fixed, the construction in both cases is adapted to them. For instance, these laws require, in order to produce the same effect, that the rays of light, in passing from water into the eye, should be refracted by a more convex surface than when it passes out of air into the eye. Accordingly we find that the eye of a fish, in that part of it called the crystalline lens, is much rounder than the eye of terrestrial animals. What plainer manifestation of design can there be than this difference? What could a mathematical instrument maker have done more to show his knowledge of his principle, his application of that knowledge, his suiting of his means to his end—I will not say to display the compass or excellence of his skill and art, for in these all comparison is indecorous, but to testify counsel, choice, consideration, purpose?

3

The Relationship between Science and Religion in Britain, 1830–1870

Michael Ruse

It is almost a truism that when Charles Darwin's *Origin of Species* first appeared, in 1859, many people found its evolutionism to be unacceptable for religious reasons. They thought the theory of natural selection working by random variations conflicted with long-held and cherished beliefs about God and His relationship with man and the world. But although the general fact of the religious opposition to Darwinism is well-known, precise questions about the nature of the opposition—if indeed there was total opposition—have yet to be answered fully.[1] The present article seeks to go some way towards the asking and answering of such questions, although the discussion will keep to relatively sophisticated thinkers who took both science and religion seriously, and who were therefore concerned to achieve some harmony between the two. It will not deal with those who cared only for either science or religion.

To begin we shall consider the relationship between science and religion in Britain in the 1830s, for it was then that the positions which formed the background to the Darwinian debate were being articulated.[2] The discussion will center upon three basic questions. What, in the 1830s, was thought to be the proper relationship between science and revealed religion, particularly between science and the claims of the Bible? Second, what was thought to be the proper relationship between science and natural religion or theology, particularly the argument from design? Third, a question probably covered by the first two questions but conveniently kept separate is: to what extent

From *Church History* 44 (1975): 1–18. Copyright © 1975. Reprinted by permission of the publisher.

was it thought that a scientist could and indeed should admit miracles? (Many meanings have been put on the term "miracle" and this fact will not go unnoticed in the discussion.)

Having examined the positions of the 1830s, we shall then consider how in the next twenty years these positions matured and were sharpened, particularly in the face of challenges posed first by Robert Chambers' evolutionary speculations, and then by a controversy about extra-terrestrial life. Finally, we shall look at the exact threat from the *Origin* and at the religious responses to it. It will be my claim that much light can be thrown on religious reactions to Darwinism by our consideration of the relationship between science and religion in the thirty years prior to the *Origin*.

THE POSITIONS

My concern is with men who were in the forefront of science but who had also deep religious commitments.[3] These men were members of the established church. They were linked not just through friendship, but through membership in several scientific societies, probably the most vigorous of which was the Geological Society of London. They included John F. W. Herschel, the leading astronomer of the day (indeed the leading man of science) and the author of a deservedly popular book in the philosophy of science;[4] Charles Babbage, inventor of a calculating machine; Charles Lyell, well-known geologist; the Reverend Baden Powell, Savilian professor of geometry at Oxford; the Reverend Adam Sedgwick, Woodwordian professor of geology at Cambridge and canon at Norwich; and, probably the most interesting and influential figure of them all, the Reverend William Whewell, successively professor of mineralogy, and of moral philosophy at Cambridge, Master of Trinity after 1841 and author of works on the history and philosophy of science.[5]

From within this group we see in the 1830s two basic positions on the science-religion relationship being articulated. I shall call these the positions of the "liberals" and the "conservatives." Baden Powell and Babbage fall fairly clearly into the liberal camp, Sedgwick and Whewell into the conservative camp. The two major scientists, Lyell and Herschel, seem basically to have been liberals, but, as we shall see, in certain respects they had strong sympathies with the conservatives. No political connotations should be attached to my labels. For example Sedgwick, whom I call a conservative, was an ardent Whig. Let us turn first to the relationship between revealed religion and science.

The first major intimations of the liberal position came in Lyell's classic *Principles of Geology*,[6] in which he enunciated the so-called "uniformitarian" view of the earth, a view which sees the world considered geologically as having been in a kind of steady state of eruption and decay as far back as we can discern. What is important in this view is that there is no place for spasmodic upheavals, "catastrophes," involving greater turmoil and energy than at other times. And in arguing this way Lyell was going against geological orthodoxy. In the 1820s Britain's most authoritative geologist, the Reverend William Buckland, had strongly endorsed a catastrophist viewpoint, which Buckland thought harmonized nicely with revealed religion. Although he had to allow that geology shows the world far older than the Bible implies, he felt able to identify the last catastrophe with Noah's Flood.[7] By denying catastrophes and by arguing for slow, gradual changes Lyell was knocking away at the links Buckland had tried to forge between geology and the Bible, and Lyell did explicitly criticize attempts to force a direct connection between geology and the sacred narrative.[8] However, it seems true to say that primarily Lyell wanted to avoid theological controversy and to get on with the job of doing science. He left it to others to try to articulate the kind of position on science and revealed religion that he was presupposing.

It fell instead upon Baden Powell to state the liberal position in full form. As we shall see, it may have been a little too fully formed for Lyell himself. In a pamphlet entitled *Revelation and Science*[9] Baden Powell pushed to the limit the right of science to say almost exactly what it pleases, no matter what the Bible says. Taken literally, admitted Baden Powell, many claims in the Bible contradict scientific knowledge—the six day creation, the Flood, the motion of the sun and so on. Hence these claims cannot be literally true— they must involve "poetical imagery." More generally, he argued that it is a complete mistake to look to the Bible for scientific information. The Bible was concerned exclusively with man's moral and spiritual destiny, and Baden Powell quoted Herschel to the effect that "truth can never be opposed to truth"—why else would God have given us our reason if he had not intended us to use it?[10]

Turning now to those whom we have styled "conservatives," one finds that in many respects they agreed with the liberals on the science-revealed religion relationship. It was conceded that Buckland's attempt to find physical evidence of the Flood had failed, that the earth is very old, that, more generally, the aim of the Bible is not to teach science and that therefore the scientist should not feel obligated to take absolutely everything in the Bible literally.[11] The Bible was concerned with man and his destiny. So long as science did not impinge on that, science had free sway.[12]

Nevertheless, there were at least two differences between liberals and conservatives on the subject of revealed religion and science. For a start, it is probably true to say that the conservatives took more seriously than the liberals the Bible considered as man's earliest record of himself. Thus with respect to the Flood Whewell wrote, "it would be as absurd to disregard the most ancient historical record, in attempting to trace back the history of the earth, as it would be gratuitously to reject any other source of information."[13]

Second, and rather more importantly, it is probably true to say also that the conservatives would have felt less obligated than Baden Powell to explain away creation stories in terms of poetic imagery. Whewell for instance argued that in all historical sciences, like geology, as we trace back we come eventually to events, the like of which there is no modern counterpart. At this point he thought that, because scientists must judge the past in terms of the present, they go beyond science—"the thread of induction respecting the natural course of the world snaps in our fingers. . . ."[14] Apparently then, when it comes to beginnings, even as scientists men are quite free to refer to the Bible, because they have gone beyond science. Precisely where this leaves them with respect to a literal interpretation of the beginning of Genesis, Whewell left unsaid—he was far too wily to get himself caught in print on so controversial a subject—but he did seem open to a more literal approach than Baden Powell.

Despite these differences, however, it is clear that there was fundamental overlap between liberals and conservatives on the proper way for a scientist to regard the Bible. The Bible was not a work of science, and the scientists were not therefore to feel constrained by limited time-spans and so on. It was the interpretation of the Bible which had to be modified by the advances of science, not vice-versa. For both liberals and conservatives the Bible was primarily a work to do with man and his moral and spiritual destiny. However, as we conclude our initial look at the science-revealed religion relationship, it is important for us not to belittle this last fact—were any scientific theory to threaten man's place we ought to expect compromise and perhaps hostility. And this, of course, applies particularly to the more conservative thinkers who, as we have just seen, probably took the Bible a degree more literally than the liberals.

When we turn to natural religion and its relationship to science, we find the differences between liberals and conservatives beginning to widen. Because they wrote first, we shall begin with the conservatives.[15] We can distinguish two basic elements in the conservative position. First, it was believed that the organic world gives undeniable evidence of organization, in the sense of things being directed towards ends. Organisms have characteristics, adaptations, which serve certain functions; that is, they aid their possessors in living, reproduction and so on. This belief in organization towards ends was known as the doctrine of "final causes," and it was in itself a purely scientific doctrine. In particular, it was argued by the conservatives that any scientific theory about organisms must give full place to final causes—conversely, any theory which threatened final causes was inadequate to science. Second, it was believed by conservatives that final causes point indubitably to a wise Designer. A favorite question was taken from the anatomist Richard Owen's discussion of the adaptations of the kangaroo for feeding its young; these adaptations are, we are told, "irrefragable evidence of creative forethought."[16] In other words, only by presupposing the existence of a God can we get a full understanding of organic adaptation. It must however be emphasized that this understanding is one taking us beyond science. Neither the Bible (considered as more than just a record) nor a Designer could be brought into science. As scientists, men must be content with just the phenomenon of final cause. (Of course, as we shall see, this is not to say that as religious scientists the conservatives did not have strong extra-scientific reasons for their science being what it was.)

Before looking at the liberals' views on natural theology, several questions should be asked about the conservative position. Were the conservatives committed to seeing direct adaptation for ends in every facet of the physical world, and if not, what did this do to their natural theology? The answers are that the conservatives recognized that not everything served a direct end—the nipples of the male are classic examples of things which have no function—and consequently it was suggested that God might have had ends other than direct utility in view when He designed the universe. Whewell, for instance, suggested that God might have been striving for beauty, symmetry and similarity. In any case, argued Whewell, although final causes are sufficient to prove God's existence, they are not necessary. The very fact that things work according to law, even though men may see no ends, proves the existence of God. Hence science generally proves his existence.[17]

Liberals, particularly Herschel and Lyell, certainly did not outrightly reject views like these.[18] Everyone agreed that the recognition of organic adaptation was an essential aspect of biology and was moreover a clear mark of God's design. But we find, particularly in the thought of Charles Babbage and Baden Powell, a very much greater emphasis on the idea of God as Lawmaker—that the most overwhelming evidence of God's existence and wisdom was uninterrupted law. Babbage, for example, argued that if we have two phenomena giving clear evidence of design, but if we then find that although the one phenomenon has been directly created the other has been created through the medium of a machine—that is, through the working of law—we would obviously think that much more highly of a Creator who worked through the machine than one who worked directly.[19] Similarly, the incessant refrain of Baden Powell was that uninterrupted, all-sufficient law was the truest mark of creative intelligence.[20]

We see, therefore, that there was no outright contradiction between conservatives and liberals on the subjects of natural theology and its relationship to science. Both sides agreed that there are final causes and also that not everything exhibits final cause. Both sides agreed that final causes point to God, that God must have had several ends in view when designing, and that law in itself is evidence of God. Both sides also agreed that natural theology should be kept out of science, and it must be reiterated that the doctrine of final causes was in itself a purely scientific doctrine, whatever theological beliefs it was used to carry. But there was a major difference in emphasis. For the conservatives final causes were crucial to their natural theology; for the liberals (at least the extreme liberals) it was uninterrupted law which was crucial. Obviously this difference colored their respective views of the acceptability or unacceptability of certain scientific beliefs—beliefs which we shall encounter as we turn to our third basic question: to what exent was it believed a scientist can allow miracles?

It will be convenient to begin consideration of this question by explicating a few terms. The 1830s conception of "law" was, in certain respects, surprisingly modern. Herschel, the chief authority on these matters, saw laws as being universal, empirical statements "of what will happen in certain contingencies."[21] To use modern terminology, laws express the way things *must* be, they allow for counterfactuals—if *A* were to occur (even though it does not), then *B* must follow. Herschel distinguished upper level laws, "fundamental laws," from lower level derived laws, "empirical laws."[22] Newton's laws of motion and gravitation were the fundamental laws *par excellence,* Kepler's laws were prime examples of empirical laws. In their fundamental laws, scientists were obliged to make reference to causes, particular *verae causae*—causes "competent, under different modifications, to the production of a great multitude of effects, besides those which originally led to a knowledge of them."[23]

Let us turn next to the term "miracle," which has notoriously many meanings. Listing some of the main meanings in a rough decreasing order of strength we have: direct divine violation of established laws; divine acts which are in some sense outside of established laws; phenomena which are in some way, perhaps a divinely inspired way, "subjective"—that is, are person-dependent and do not involve normal laws of nature; and phenomena which may appear miraculous to us because unexpected or because of coincidence, but which are caused by and in accordance with normal laws of nature. Obviously, this last category might more properly be said to refer to "apparent miracle" rather than "genuine miracle."

No one, liberal or conservative, argued for miracles in the first sense. God was not about to violate Newton's laws; paradoxically, the conservatives were, if anything, even stronger than the liberals on this, for Whewell (unlike Herschel and Powell) argued that the denial of certain key laws of nature is "inconceivable."[24] However, everyone agreed that there were such things as miracles or at least possible candidates for miracles, and we find a split over which of the other senses of "miracle" truly apply.[25] In particular, although we find a suggestion (particularly by Herschel) that some miracles are "subjective,"[26] we find a split over the question of organic origins—are such origins to be interpreted as miracles involving direct divine intervention or are they to be interpreted as only apparently miraculous, but really subject to normal laws of nature?

To see why there should have been this split, it is necesssary first to see how the problem of organic origins arises. It stems from the facts that we see about us different species of organisms, that at the same time we see no direct evidence of new species being created, and that nevertheless we see also in the fossil record that new species did suddenly appear at different periods of the earth's history. The question then arises as to how these new species were created; in particular, did the beginnings of new species involve miraculous interventions by God?

It was the general consensus of the liberals that they did not, that organic origins were bound by laws, laws which still hold today. Thus Herschel, agreeing with the stand taken in the *Principles,* wrote to Lyell that "we are led, by all analogy, to suppose that he operates through a series of intermediate causes, and that in consequence the origination of fresh species, could it ever come under our cognizance, would be found to be a natural in contradiction to a miraculous process. . . ."[27] And if it were objected that we ought therefore to see new species being created today, Lyell gave elaborate calculations showing that the rate at which new species appear to have been introduced is so slow that the chances are very much against our ever having seen new species created.[28] Moreover, Babbage had a helpful analogical argument showing that even if something like a species' origin still seems miraculous, we can still reasonably put it in the final category of miracle given above— miracle working entirely within natural law. He showed how his calculating machine could be set to generate a series of natural numbers, from 1 up to 100,000,001, at which point instead of the expected 100,000,002, the machine would generate 100,010,002. Hence, although we may have much evidence of the way laws seem to work, we can have no guarantee that there may not be exceptions, apparently miraculous, but still law-governed.[29] Possibly organic origins are exceptions of this kind.

The hiatus in the liberal position, one which Whewell scornfully noted,[30] was that despite this general agreement in the rule of law, no one had the first idea as to which laws, particularly which causal laws, were involved in organic origins. At the beginning of the century, Lamarck had proposed an evolutionary theory based on needs and acquired characteristics, but it was Lyell who led the attack on the adequacy of this theory.[31] That there was law, all agreed—but which law, none knew.

The conservatives could not accept this line of reasoning at all. Although there may be laws in the mind of the creator, as far as we are concerned, "the power by which these varied forms were successively brought into being, resembles nothing of which we can see any vestige in the present world: it appears to belong, not to what we are accustomed to speak of as the laws of nature, but to that Supreme Will, which is their source and foundation."[32] And towards the end of the decade Whewell spoke of "supernatural influences" and of "acts . . . out of the common course of nature; acts which, therefore, we may properly call miraculous."[33]

We have seen that for the conservatives, Whewell particularly, there was no question of breaking existing laws of nature. But it is clear that more was being envisioned than just something subjective or natural-law-bound in the sense supposed by Babbage. Both from the language being used and by a process of elimination, the sense of miracle held by men like Sedgwick and Whewell was that of some kind of direct intervention by God, an intervention in some manner outside laws as known to us. They applied this sense to organic origins, and it is just this sense of miracle that many years later in reply to Darwin Sedgwick endorsed explicitly. He spoke of the "miracle" of creation, involving an hypothesis which "does not suspend or interrupt an established law of Nature," but which "appeals to a power above established laws, and yet acting in harmony and conformity with them."[34]

But how, as scientists, did Whewell and Sedgwick think they could refer organic origins to miracles, that is, to something beyond the reach of science?

Here it was argued that if species were created naturally in the past, we ought to see evidence of such natural creation in the present. This we do not; the evidence, if anything, points the other way. Hence, it is just not reasonable for us *as scientists* to suppose that species originated in a natural-law-bound manner. But why, one might ask, was a position like Herschel's unacceptable— that species originated in a natural-law-bound manner, if it just happens that we have not yet seen such an origination? Since the conservatives' argument so far has been the same as that used to give a meaningful sphere of authority to the Bible, one might think that the appeal to miracles was based merely on a reading of the Bible. But although, as we shall see, Sedgwick in particular was certainly not against using the Bible to flesh out his understanding of God's creative powers, no real attempt was made to harmonize the story of Genesis and God's creation of species. Rather the appeal to miracles was based on scientific grounds. More specifically, at this point it was the belief in final causes that was crucial. Whewell and Sedgwick felt that blind unguided law was just not the kind of thing which leads to intricate, organic adaptation. Unguided laws led to random disorder. But since there obviously was organization in the organic world, there must have been more than mere laws, that is, there must have been creative interventions. As Whewell put it, to adopt natural-law-bound origins for species is to sacrifice "that belief in the adaptation of the structure of every creature to its destined mode of being, which not only most persons would give up with repugnance, but which, as we have seen, has constantly and irresistibly impressed itself on the minds of the best naturalists, as the true view of the order of the world."[35] Hence, *as scientists* the conservatives felt they were forced to resort to miracles. Moreover, man in particular was singled out as just not the kind of thing which could occur through blind law.[36]

We have then two different positions on the organic origins problem defined and defended. As yet, no explicit mention of religion has been made in the premises of the arguments, and it seems fair to say that both liberals and conservatives felt that they were making their claims on non-religious grounds. (Strictly speaking, one ought to say that the conservatives made no appeal to religion until, on scientific grounds, they felt they had shown science alone inadequate.) The liberals appealed to a kind of meta-scientific belief in the universality of the rule of law, and the conservatives argued that final causes could not have come through unguided law. However, it is also true that these positions dovetailed nicely with their respective beliefs about theology, and since neither side was beyond pointing to this fact,[37] it seems fair to conclude that there were strong religious motives for the different positions taken on the organic origins question. Speaking roughly, the liberals' natural theology inclined them towards law-bound organic origins, whereas the conservatives' natural theology inclined them towards miracles, and their revealed theology made even more pressing a special origin for man.

The qualification of the last sentence was deliberate, because, as so often happens, the situation was not quite as clear-cut as the historian could wish.

We have noted that the liberals, particularly centrists like Herschel and Lyell, were not about to deny final causes altogether. We have noted also that everyone, liberal and conservative, felt that man calls for special treatment. One might expect, and indeed does find, that these concerns made themselves felt. Lyell, for example, worried continually in his *Principles* about the marks of adaptation in the organic world which he took to be evidences of a Designer. For this reason it seems that although Lyell wanted to bring organic origins beneath law, he wanted the laws to be of special kinds, namely, laws which were in some way divinely directed or guided (that is, more than just blind regularity).[38] Otherwise, Lyell seems to have felt that final causes would be jeopardized. Furthermore, when it came to the subject of man, Lyell, for all his uniformitarianism, wrote flatly that "a real departure from the antecedent course of physical events [can] be traced in the introduction of man."[39] At this point, apparently, a miracle of the kind favored by the conservatives was being envisioned.

It is probable, therefore, that the most accurate description of the science-religion relationship in the 1830s (on the intellectual level being considered) is that of a spectrum. No one wanted to base scientific claims on religious premises, but with respect to the organic origins question we find a range of opinions. At one extreme, some would feel their religion threatened if miracles were not invoked. Then there were those like Lyell, wanting to avoid miracles, but supposing special kinds of laws and thinking man an exception. And right at this other extreme would be someone like Baden Powell; for him God's rule of law was absolutely crucial. Man would therefore call for the absolute minimum of special interference, and the idea of special guided creative laws was quite unnecessary. For Baden Powell, all of God's laws were specially guided and thus there was no need to distinguish one set from another.

We have thus far seen the positions taken in the 1830s on the question of the relationship between religion and science. Let us now consider how these positions withstood the challenges of the next three decades.[40]

THE VESTIGES OF THE NATURAL HISTORY OF CREATION

The *cause célèbre* of the 1840s was the anonymously published *Vestiges of the Natural History of Creation*,[41] which we now know to have been written by the Scottish publisher Robert Chambers. In the *Vestiges* Chambers argued strenuously for an evolutionary view of the organic world, although it would hardly be true to say that he offered much by way of an evolutionary mechanism. He suggested that embryos can develop through various stages, in which they are progressively fish, reptiles and so on up to mammals, that normally an embryo will develop just as far as the stage the parent represents, but that every now and then for some unknown reason the embryo from a parent at one stage will take a jump up the progression, so we get an evolution from one form to another. Chambers could give no reasons for these jumps.

His main arguments were directed towards phenomena which he thought could be explained only on an evolutionary hypothesis. A prime example of such a phenomenon was the fossil record which he interpreted as being progressive in a way compatible only with an evolutionary hypothesis such as his. He also sought to demonstrate the general necessity of explaining through law. To this latter end, Chambers made copious references to the successes of the physical sciences, and he gave a detailed account of Babbage's claims about how even apparent exceptions to law might really be subject to law— Chambers' belief being that jumps from one species to another might be just such apparent exceptions.

Chambers wisely tried to avoid tangling with religion, although it is clear that (for all that his critics said) he had absolutely no intention of threatening either the existence of God or his evidences in the world. As far as revealed religion is concerned he admitted that some might find his views disturbing, but he suggested that he offered no more of a threat than geology and astronomy before him. Hence, as it had done before, revealed religion might be expected to move over and make room.[42] As far as natural religion was concerned, the reader was calmly referred to Paley and the *Bridgewater Treatises:* "The physical constitution of animals is, then, to be regarded as in the nicest congruity and adaptation to the external world."[43] And as we have just seen, Babbage was Chambers' authority on miracles.

A simplistic expectation would be that liberals would embrace Chambers and conservatives reject him, but this turns out to be only partially true. Nobody who knew much about science could accept all of Chambers, for some of his claims were ludicrously untenable. For instance, Chambers (unlike Darwin) was keen to show that the organic originally comes from the inorganic. To this end, claims were made about spontaneous generation, and analogies were freely drawn between frost patterns and living ferns, both of which moves drew justifiable scorn. But at least some of the liberals had other objections. Herschel, for instance, found Chambers' views philosophically unacceptable. Chambers likened his description of evolution to the Newtonian law of gravity.[44] In Herschel's eyes this was an unforgivable sin. Chambers, he believed with justification, had confused a description which yielded at best a phenomenal law with the paradigm of physical or causal law.[45]

Nonetheless, there is truth in our expectations. Baden Powell, for one, jumped aboard the evolutionary bandwagon and agreed with Chambers that the evidence from other sciences points to normal, natural causes and laws for organic origins.[46] With respect to revealed religion he pointed out that Genesis and geology are already at odds, and he suggested that an evolutionary theory hardly makes matters worse. With respect to natural religion he reiterated his stand that working by and through law accords more with design than does creative intervention. There was, of course, the problem of man, for as we have seen even Baden Powell thought there was something special about man. At this point Baden Powell compromised, suggesting that man *qua* physical and intellectual being was a product of evolution through law, and that only inasmuch as man was a spiritual being was a special divine intervention required. Did this not involve some kind of contradiction? Apparently not, for an assertion about man's spiritual nature "refers wholly to a *different order of things,* apart from, and transcending, any material ideas whatsoever: hence *it cannot be affected by any considerations or conclusions belonging to the laws of matter of nature.*"[47]

At the other end of the spectrum Sedgwick and Whewell rejected Chambers' speculations with vehemence.[48] They involved "rank, unbending and degrading materialism" leading to atheism.[49] Given our previous discussion we might expect two things. On the one hand, conservatives would find *Vestiges* religiously offensive—offensive probably to revealed religion and certainly to natural religion—and they would continue to plump for miracles. On the other hand, their arguments against *Vestiges* would be scientific. Both expectations are fulfilled. Chambers was not criticized for making the world too old or anything like that. But Sedgwick, for example, stated flatly that as a Christian he disliked *Vestiges* because it conflicted with the Bible. Chambers would derive man from a monkey whereas the Bible said that man (and woman) were made in the image of God.[50] Sedgwick added that *Vestiges' "*principles and language were invented and affirmed by those who did their best to cheat us out of our conceptions of a Creator, and denied the whole doctrine of Final Causes."[51] But to refute Chambers, Sedgwick turned from religion to science. He analyzed the fossil record in tedious detail to show that although there is a broadly progressive schema, it is not one which supports evolutionary hypotheses. To buttress his argument he pointed out that there are sharp breaks between species in the record and that new forms, for example, fishes, appear in the record if anything in their most developed state.[52] Similarly, Whewell pointed out that Chambers had done nothing to explain how final causes get into the organic world; if anything, his theory depending on apparently blind jumps pointed away from them.[53]

There was, however, one development or refinement in the conservatives' position. Hitherto, organic origins had been referred to miracles (in the sense explained earlier) and matters left at that. Miracles were still invoked, but it was now admitted that in some way law (natural, at least in the sense that we can see evidence of it) was relevant to organic origins. Whewell took note of the similarities between organisms of different species, for instance, between the skeletons of different mammals—things which had long proved troublesome to the doctrine of final cause—and he wrote that those who deny transmutation do not "doubt that this spectacle of analogies and resemblances implies the existence of laws in the mind of the Creator, according to which he has proceeded in the work of Creation."[54] Apparently, the point was that not only are these laws in the mind of God, but they are also laws we can see. Whewell, like everyone, had certainly been aware previously of the similarities, but earlier he had not put this construction on them. But he added that people like himself "cannot go onwards, with the writer of the *Vestiges,* from resemblance to sequence, and from sequence to causation.

They cannot venture to say that animals followed one another upon the earth in the order of their anatomical resemblances; and that their anatomical differences grew out of each other naturally by general laws."[55] In the terms introduced earlier, Whewell's position seems to have been that at the *phenomenal* level, some regularities can be discerned in organic origins, but that there are no natural *physical* or *causal* laws to refer them to. At that level, we must still appeal to direct divine intervention.

In taking a position like this, Whewell was influenced by the anatomist Richard Owen, who in the 1840s was articulating a neo-Platonist synthesis between the facts of organic adaptation on the one hand and the trans-specific similarities ("homologies") between organisms on the other.[56] Owen argued that God started with a kind of Platonic archetype or ground-plan, and somehow (although not necessarily in an exact order) had put ever more sophisticated versions of this into effect until finally He arrived at man. Obviously a position like this was attractive to a conservative like Whewell, for it had at its apex what we have seen was a (if not *the*) major concern of the would-be reconcilers, namely, man. Hence, it is probably true to say that inasmuch as there was any refinement in the views of someone like Whewell, the driving force should be seen not so much to extend the rule of law as to explain exceptions to the doctrine of final cause and to enhance further the special place of man.[57]

Although Owen himself seems not to have been totally opposed to the notion of some kind of law-bound causal mechanism for the creation of new species,[58] he stuck in his heels firmly when it came to man. In an unpublished letter advising Whewell not to reply to *Vestiges,* he wrote, "There can be no real danger in the book to him who believes of his first Parents that God created man in his own image, male and female; that He breathed into his nostrils the breath of life and he became a living soul."[59] One might perhaps question how literally this should be taken, but it does show strongly the extent to which revealed religion prejudiced people against evolutionary hypotheses. Christian scientists may not have worried too much about the age of the earth, but they did worry about man.[60]

THE PLURALITY OF WORLDS

In the 1850s it was Whewell who poured fresh fuel on the science-religion controversy, when he published anonymously a work arguing that this world of ours is the only one to carry living beings, particularly human-like beings.[61] This may seem a somewhat esoteric subject, but judging from the number of responses, it provided an engaging diversion in what would otherwise have been a lull between the *Vestiges* and the *Origin.*[62] And the implications which the plurality of worlds debate had for the evolutionary debate did not go unnoticed by those who worried about the science-religion relationship. Lyell, for instance, made repeated references to Whewell's views in some private notebooks he was keeping on the evolution question.[63]

Basically, Whewell's problems started, as we might expect, with man. At one point he had been favorably inclined towards the idea of inhabited other worlds.[64] But he began to fear that such a doctrine conflicts with man's special relationship with God as seen through the lens of Christianity. Specifically, Whewell thought Christ's coming to earth and his supreme sacrifice show that God has a special love for man; this love would be emasculated were we to find that God had just such a relationship with myriads of other beings all over the universe.[65] We see therefore that the conservative Whewell worried because a scientific claim conflicted with essential tenets of revealed religion. There is an obvious parallel here with the conservative Sedgwick's citation of the Bible against *Vestiges*. There is a parallel also in that Whewell thought the correct way to counter claims about inhabited other worlds is by providing scientific arguments pointing to earth's uniqueness. This he did by analyzing the kinds of conditions which appear to be necessary for life (as we know it), and by showing that these conditions seem not to obtain elsewhere in the universe. Thus, for example, the moon is barren and airless, a most unlikely home for anything resembling man.[66]

In the course of his attack on inhabited other worlds, Whewell managed to link his conclusion on this problem to his conclusion on the organic origins problem. In particular, he argued that an appeal to the law-bound nature of the universe will not support the natural occurrence of man-like beings all over the universe: "I know that the planet on which we live *is* a seat of life. I believe it has become so by acts out of the common order of the laws of nature; therefore I do not think it reasonable to assume, without any evidence, that other planets are also seats of life."[67] Thus, not only did Whewell show that he still stood strong on the question of miracles, but he showed quite explicitly that he had revealed religious motives for wanting to avoid evolutionary hypotheses. Once allow evolution and man's unique relationship with the Christian God is in danger.

So far, one might wonder why Whewell's ideas should have proven particularly controversial. The reason lay in the manner Whewell's conclusions cut across what many considered to be essential tenets of *natural* religion. If God designed wisely, then why on earth would He have troubled to create millions of uninhabited worlds? Without human-like beings fulfilling their destinies on them, they are pointless. Whewell knew he would be open to criticism of this kind and he tried to forestall it in two ways. First, as he had done in the 1830s, he showed that, other worlds apart, a natural religion which supposes that God always creates for the direct benefits of organisms faces grave problems. The symmetry of the snowflake, to choose a popular example, seems hardly to do anyone very much good. Nor does the pre-human existence of the earth with organisms, as shown indubitably by the geological record, seem to have done man much good. And in a passage which might have been drawn straight from the chapter on the struggle for existence in the *Origin of Species*, Whewell pointed out that the existence of most organisms seems hardly to have much direct point because they die before adulthood and reproduction.[68]

Then Whewell suggested other ends God might have had when designing. The earlier suggestions like symmetry and similarity made their reappearance. The claim that law in itself proves mind was given an extended treatment. And in a new move, Whewell suggested that this life of ours is one of intellectual trial as well as moral trial, that part of our task is to trace the laws God has put into the world, and that extra-terrestrial bodies serve the function of manifesting these laws that we might trace them.[69]

There seem to be two significant aspects to Whewell's argumentation. On the one hand, he was stressing the importance of man in the science-religion debate. His natural religion in particular was being more keenly focused on God's intentions for man; the whole universe, apparently, was provided for man's intellectual edification. On the other hand, despite his retrieval attempts, he was emphasizing the difficulties inherent in a natural religion which everywhere sees evidence of creative adaptive design. Whewell certainly did not want to deny that most aspects of organisms show indubitable signs of final cause, but he thrust forward those aspects which do not.

Reactions to Whewell's essay were fairly predictable. Baden Powell on the one side criticized Whewell for worrying at all about directly observable ends and accused him of "moral Ptolemaism" for wanting to rest his natural religion on anything but the evidence of uninterrupted law.[70] As might be expected, a key factor in Baden Powell's argument to the conclusion that there might well be life on other worlds was the suggestion that organic transmutation on this world makes likely organic transmutation elsewhere.[71] On the other side, Whewell was criticized for denying God's real intentions throughout the universe. One particularly vitriolic critic was Sir David Brewster, biographer of Newton, inventor of the kaleidoscope and long-time foe of Whewell. Brewster saw direct intention everywhere, and peopled everything with intelligent beings, even the sun! But although Brewster found Whewell's position religiously offensive, the revealing title of his reply (*More Worlds Than One: The Creed of the Philosopher and the Hope of the Christian*)[72] clearly shows how eager he was to demonstrate that it was as a scientist that he argued for inhabited other worlds. And so arguments were provided to show that, even given laws and causes as we know them, the moon and sun could indeed be inhabited, and so on. Science might indeed seem to contradict religion, but science had to be answered by science.

THE ORIGIN OF SPECIES

In his book[73] Darwin argued not only for the fact of organic evolution, but also for a particular mechanism of such evolution which was the natural selection of small heritable variations ("individual differences"). He did not, in the *Origin,* direct himself explicitly to questions of the conflict between science and religion; but, as he and everyone else knew, his theory did threaten certain religious tenets.

For a start, like uniformitarian geology (and indeed, like catastrophist geology) Darwin's theory required far more time than literalist readings of the Bible would allow. Second, although he did not give any extended treatment of man, he clearly intended to bring man in significant respects within the natural order. No one was under any illusions when, at the end of the *Origin,* Darwin wrote: "Light will be thrown on the origin of man and his history."[74] Third, he pushed out miracles—certainly miracles in the sense understood bv Sedgwick and Whewell. And fourth, he countered the view of final cause which saw organic organization in such a way that it could be explained only through direct divine intervention. Moreover, Darwin was not above making a few snide comments about explanations in terms of symmetry, noting that no physicist would ever resort to such reasons.[75] This is not to say that Darwin was trying to expel final cause from biology (understanding this in the purely biological sense of organisms being adapted or functionally organized towards ends); indeed, he accepted the belief fully, trying to explain organic adaptation through the law-bound process of natural selection. Nor is it to say that he was trying to deny the existence of God. In fact at this point Darwin seemed to accept the idea of God, but a God who put His intentions into effect through the medium of laws.[76] But, it is certainly the case that Darwin was taking an approach to final cause which could be (and indeed was) accepted by a scientist who wanted to have nothing at all to do with religion. Prior to Darwin, in the eyes of the average Englishman the facts of organic adaptation made slightly absurd claims that the world could have occurred through blind law. Even Hume had had to allow some force to the argument from design and J. S. Mill, for all his criticisms of orthodoxy, never relinquished the idea of design.[77] Darwin, however, opened the way for people like T. H. Huxley to jettison entirely the notion of a creative Designer,[78] even if most were unable and unwilling to do so.

In fairness both to Darwin's supporters and his critics, it should be emphasized that much of the reaction to Darwin's theory was purely scientific, and there were few strengths or weaknesses which went unnoticed.[79] Nonetheless, given previous discussion, it seems plausible to predict the following reactions to the theory. First, extreme liberals like Baden Powell would embrace the theory wholeheartedly, not because they would think it expels God from nature—far from it—but precisely because they would think it confirms God's presence in nature. (One must emphasize this point about Baden Powell: although he could embrace a scientific theory as enthusiastically as an agnostic like T. H. Huxley,[80] he was in his way as design-oriented as Whewell or Sedgwick.) Second, conservatives like Sedgwick and Whewell would reject the theory. They would not object to the great length of time the theory demands. They would object to the implications for man and would probably claim that the theory denies final causes (for all that Darwin said). However, although their hatred of the theory may have been religious, their explicit counter-arguments would be scientific. Third, middle-of-the-road liberals like Herschel and Lyell would probably not be wholly unsympathetic to Darwin's theory, but they too would have problems with man and final causes. Fourth, we may find that many younger men, although taught by the conservatives, would see certain attractions in Darwinism, and would thus be drawn to a more central stance themselves.

This final prediction is somewhat speculative, but is probably justified not merely because Darwinism gives plausible scientific answers to such problems as organic geographical distribution, but also because of the internal tensions we have seen appearing in the conservatives' position. In the 1830s they argued flatly for divine intervention. By the 1840s they were having to admit some lawful regularity (for organic origins), at least at the phenomenal level. In the 1830s despite exceptions there was a confidence in the conservatives' appeal to final causes. By the 1850s we find Whewell emphasizing the exceptions and Brewster forced into the belief that there are man-like beings on the sun. None of these points made a conservative position absolutely hopeless, but it was certainly much less compelling than it was thirty years previously.

All four predictions come true. First, Baden Powell did indeed accept Darwin's theory enthusiastically. He wrote of "Mr. Darwin's masterly volume on *The Origin of Species* by the law of 'natural selection' —which now substantiates on undeniable grounds the very principle so long denounced . . .—*the origination of new species by natural causes. . . .*"[81] The conservatives did indeed reject Darwin's theory, speaking of its materialism and its repudiation of final causes.[82] There were, however, no objections to Darwin's demands for vast time-spans and the like. Rather, the theory was rejected as a scientific theory because of its supposed inconsistency with such facts as the fossil record and so on. A classic statement of the conservative case appeared in the *Quarterly Review*, written by the Bishop of Oxford, Samuel Wilberforce, under the tutelage of Owen.[83] This response is particularly interesting because, by virtue of Wilberforce's celebrated clash with Huxley at the British Association, he is usually taken to be the archetypal religious opponent of Darwinism.[84] There was no place at all for crude biblical literalism in Wilberforce's response. His religious objections, he admitted openly, stemmed from Darwin's treatment of man and final causes,[85] and his arguments against Darwin's theory were purely scientific.

Third, men like Herschel and Lyell were attracted to a position like Darwin's, but felt that they had to do something about man and final causes. Although Herschel reputedly spoke of Darwin's theory as "the law of higgledy-piggledy,"[86] in fact, if one allowed some kind of teleologically directed source of new variation and a special place for man, he was "far from disposed to repudiate the view taken of this mysterious subject in Mr. Darwin's book."[87] Lyell agonized over the religious problems raised by Darwin's theory. On the one hand, he wanted desperately to explain organic origins through law and he recognized that this forced him into the camp of the transmutationists.[88] On the other, there were final causes, and there was man—moral, intellectual, spiritual man.

It is small comfort or consolation to me, who feels that Lamarck or Darwin have lessened the dignity of their ancestry, making them out to be with[t]. souls, to be told, "Never mind, you will be succeeded in unbroken lineal descent by angels who, like the Superior Beings spoken of by Pope, 'Will show a Newton as we show an ape.'"[89]

Eventually Lyell forced himself into a kind of acceptance of Darwin's theory, but he hedged this acceptance with all kinds of clauses about its compatibility with God's being ultimately responsible for final causes and with man's special place in the scheme of things.[90]

What is most interesting is the appeal of this mid-way stand both to those who like Lyell were Darwin-sympathizers, and to those, like St. George Mivart,[91] who were bitter Darwin-critics. Mivart, a convert to Catholicism, advocated a kind of saltatory evolution theory, one in which God could put his design into effect through a series of law-bound, but guided, jumps from one species to another.[92] Although Lyell would no doubt have given natural selection a greater role in the evolutionary process than would Mivart, as their contemporaries recognized, they were not that far apart.[93] When it came to man's soul, Mivart opted flatly for divine intervention and quoted the Bible in justification.[94]

Finally there is the question of the reactions to Darwinism by younger men—men educated at the great universities and who, because of the influence of men like Sedgwick and Whewell, might have been expected thirty years previously to have rejected Darwin's position entirely. We find a general shift to the left (in the language of my dichotomies); that whereas previously such men would have opted for miracles (meaning direct interventions), now they favored organic evolution through law (although this would be directed law with a special place for man). A good example was the Cambridge-educated Reverend Charles Kingsley, author, advocate, and historian who also dabbled in science. In the 1850s he insisted on miracles behind organic origins,[95] but by the end of the 1860s he firmly appealed to evolutionary law.[96] There were, I suspect, many who like Kingsley found that after the *Origin* the appeal to miracles (in the sense understood by the conservatives) was no longer a live option. One of the ways of reconciling science and religion was coming to an end.

CONCLUSION

That some found Darwinism religiously offensive has, I think, been amply demonstrated. What should also be clear is the fact that religious opposition to Darwinism was not uniform, that indeed some believers found that they could go part or practically all of the way with Darwin, and that these varied religious reactions are just what we would expect after studying the science-religion relationship in the thirty years prior to the *Origin*. We must therefore be careful in thinking of the *Origin* as a "watershed." In the purely scientific sense it clearly was, but from the viewpoint of the science-religion quarrel it was much less of one. Darwin's work certainly seems to have occasioned a general shift toward the view that evolutionism was compatible with science, and there is no doubt that by offering a naturalistic explanation of organic adaptation he made far more plausible the position of scientists like Huxley, who wanted to have no truck at all with religion. However, as we have seen, religious men, even religious Englishmen, had been dealing sympathetically with science long before the *Origin*, and in many respects the various attitudes taken towards the science-religion relationship were the same both before and after the *Origin*.

NOTES

1. The best analysis is in A. Ellegård, *Darwin and the General Reader* (Göteborg: Göteborgs Universitets Årsskrift, 1958). See also D. L. Hull, *Darwin and His Critics* (Cambridge, Mass.: Harvard University Press, 1973); H. E. Gruber and P. H. Barrett, *Darwin on Man* (New York: Dutton, 1974).

2. C. C. Gillespie, *Genesis and Geology* (Cambridge, Mass.: Harvard University Press, 1951); M. Millhauser, "The Scriptural Geologists. An Episode in the History of Opinion," *Osiris* 11 (1954): 65-86; M. J. S. Rudwick, *The Meaning of Fossils* (London: Macdonald, 1972).

3. W. F. Cannon, "Scientists and Broad Churchmen: An Early Victorian Intellectual Network," *Journal of British Studies* 4 (1964): 65-88, gives very full details of these men, their work and their relationships. See also E. W. Whately, *Personal and Family Glimpses of Remarkable People* (London: Hodder and Stoughton, 1889).

4. *A Preliminary Discourse on the Study of Natural Philosophy* (London: Longman, Rees, Orme, Brown, and Green, 1831).

5. *History of the Inductive Sciences*, 3 vols. (London: Parker, 1837); *Philosophy of the Inductive Sciences*, 2 vols. (London: Parker, 1840).

6. 3 vols., London: Murray, 1830-1833.

7. *Vindiciae Geologicae* (Oxford: Oxford University Press, 1820); *Reliquiae Diluvianae* (London: Murray, 1823).

8. Lyell, *Principles*, 1, chs. 2-4. But see also, M. Bartholomew, "Lyell and Evolution: An Account of Lyell's Response to the Prospect of an Evolutionary Ancestry for Man," *British Journal for the History of Science* 6 (1973): 261-303, esp. p. 267.

9. Oxford: Parker, 1833.

10. Baden Powell, *Revelation*, p. 35; quoting Herschel, *Discourse*, p. 9.

11. A Sedgwick, "Address to the Geological Society," *Proceedings of the Geological Society of London* 1 (1831): 281-316; Whewell, *Philosophy*, 2:137-157. Buckland himself took back his claims about evidence for the Flood (in a footnote!), *Geology and Mineralogy Considered with Reference to Natural Theology (Bridgewater Treatise 6)* (London: Pickering, 1836), 1:94n-95n.

12. W. Whewell, "Principles of Geology . . . by Charles Lyell . . . Vol. 1 . . . ," *British Critic* 9 (1831): 180-206, esp. p. 206.

13. Whewell, *History*, 3:602.

14. Whewell, *Philosophy*, 2:145.

15. A. Sedgwick, *Discourse on the Studies of the University* (Cambridge: University Press, 1833); W. Whewell, *Astronomy and General Physics (Bridgewater Treatise 3)* (London: Pickering, 1833). The *Bridgewater Treatises* (1833-1836) were eight commissioned works on natural theology.

16. Whewell, *Philosophy*, 2:83; quoting p. 348 of R. Owen, "On the Generation of the Marsupial Animals, with a Description of the Impregnated Uterus of the Kangaroo," *Philosophical Transactions* (1834), pp. 333-364.

17. Whewell, *Astronomy*, pp. 293-303.

18. See Bartholomew, "Lyell," esp. pp. 285-286; Herschel, *Discourse*, p. 4.

19. Charles Babbage, *Ninth Bridgewater Treatise: A Fragment*, 2d ed. (London: Murray, 1838), pp. 30-49.

20. Baden Powell, *The Connexion of Natural and Divine Truth* (London: Parker, 1838), pp. 113-204.

21. Herschel, *Discourse*, p. 98.

22. Whewell spoke of "physical" or "causal" laws, and "formal" or "phenomenal" laws.

23. Herschel, *Discourse*, p. 144.

24. Whewell, *Philosophy;* "On the Nature of the Truth of the Laws of Motion," *Transactions of the Cambridge Philosophical Society* 5 (1834): 149-172.

25. W. F. Cannon, "The Problem of Miracles in the 1830's," *Victorian Studies* 4 (1960): 5-32.

26. W. F. Cannon, "The Impact of Uniformitarianism, Two Letters from John Herschel to Charles Lyell, 1836-1837," *Proceedings of the American Philosophical Society* 105 (1961): 301-314, esp. p. 308.

27. Cannon, "Impact," p. 305. This passage was reprinted with endorsement by Babbage, *Treatise*, p. 226. See also Baden Powell, *Connexion*, p. 151.

28. Lyell, *Principles*, 2:183.

29. Babbage, *Treatise*, pp. 30-49.

30. Whewell, *History*, 3:588-589.

31. Lyell, *Principles*, 2:1-35.

32. W. Whewell, *"Principles of Geology . . . By Charles Lyell . . . Vol. II . . .,"* *Quarterly Review* 47(1832):103-132, esp. p. 125. See also Sedgwick, *"Address,"* p. 305; ". . . an adjusting power altogether different from what we commonly understand by the laws of nature. . ."

33. Whewell, *Philosophy*, 2:116; Whewell, *History*, 3:574.

34. "Objections to Mr. Darwin's Theory of the Origin of Species," *The Spectator* (1860), p. 285; reprinted in Hull, *Darwin*, p. 161.

35. Whewell, *History*, 3:574.

36. Sedgwick, "Address," p. 306.

37. For example, Babbage, *Treatise*, pp. 45-46; Whewell, *History*, 3:580-588.

38. Bartholomew, "Lyell," p. 286.

39. Lyell, *Principles*, 1:162.

40. Millhauser, "Spiritual Geologists," is very good on the less sophisticated positions on the science-religion relationship.

41. 1st ed., London: Churchill, 1844. See also M. Millhauser, *Just Before Darwin: Robert Chambers and "Vestiges"* (Middleton, Conn.: Wesleyan University Press, 1959); M. J. S. Hodge, "The Universal Gestation of Nature: Chambers' *Vestiges* and *Explanations,"* *Journal of the History of Biology* 5 (1972):127-151.

42. Chambers, *Vestiges*, pp. 387-390.

43. Ibid., p. 325. Whewell and Buckland, it will be remembered, had contributed to the *Bridgewater Treatises*.

44. Chambers, *Vestiges*, pp. 359-360.

45. J. F. W. Herschel, "Presidential Address to the British Association for the Advancement of Science, 1845," reprinted in J. F. W. Herschel, *Essays* (London: Longman, Brown, Green, Longmans, and Roberts, 1857), pp. 634-682, esp. p. 675.

46. *Essays on Inductive Philosophy* (London: Longman, Brown, Green, and Longmans, 1855), pp. 315-481. Although this essay (titled "On the Philosophy of Creation") did not appear until 1855, Baden Powell did respond favorably to Chambers' ideas in the 1840s. The exact date of Baden Powell's initial support for Chambersian ideas does not really matter—it happened before the *Origin*.

47. Baden Powell, *Essays*, p. 76 and p. 466.

48. A. Sedgwick, *"Vestiges . . . .,"* *Edinburgh Review* 82 (1845): 1-85; *Discourse*, 5th ed. (1850); W. Whewell, *Indications of the Creator* (London: Parker, 1845, 2d ed., 1846).

49. Sedgwick, *"Vestiges,"* p. 3.

50. Ibid., and p. 12.

51. Sedgwick, *Discourse* (5th ed.), p. xix; see also, for example, Sedgwick, *"Vestiges,"* pp. 11-12.

52. Sedgwick, *"Vestiges,"* p. 32 and p. 43.

53. Whewell, *Indications* (2d ed.), pp. 12-16. The same kind of argument was always invoked when the question of man arose. Although it was religion which made men desperately keen not to have to include man's origin in the natural course of events, scientific or pseudo-scientific reasons were advanced for the impossibility of such a natural origin. Man has a moral sense, reasoning power, and so on, all things which supposedly preclude origination through blind law.

54. Ibid., p. 13.

55. Ibid.

56. R. Owen, *On the Archetype and Homologies of the Vertebrate Skeleton* (London: Taylor, 1848); *On the Nature of Limbs* (London: Royal Institution, 1849). See also Rudwick, *Fossils*, pp. 207-214; and R. M. Macleod, "Evolutionism and Richard Owen, 1830-1868," *Isis* 56 (1965):259-280. Whewell and Sedgwick actually consulted Owen before responding to *Vestiges*. See R. Owen, *Life of Richard Owen*, 2 vols. (London: Murray, 1894), 1:252-255.

57. Baden Powell, conversely, liked Owen's position precisely because of the appeal to law, *Essays*, pp. 400-401.

58. Owen, *Limbs*, p. 86.

59. Whewell papers, Trinity College, Add Ms. a. 210[69] dated Feb. 14, 1844. This letter is pretty harsh on *Vestiges*, in contrast to the friendly letter Owen wrote to Chambers, in Owen, *Life*, 1:249-252.

60. Bartholomew, "Lyell," argues convincingly that Lyell was so concerned about man that he deliberately advocated a non-progressionist reading of the fossil-record, for he feared (despite Sedgwick) that a progressionist reading could well support evolutionism and hence threaten man's uniqueness.

61. [W. Whewell], *Plurality* (London: Parker, 1853, 3d ed., 1854). Everyone knew that Whewell was the author.

62. See I. Todhunter, *William Whewell*, 2 vols. (London: Macmillan, 1876), 1:184-210, which lists and discusses many of the responses.

63. L. Wilson, ed., *Sir Charles Lyell's Scientific Journals on the Species Question* (New Haven: Yale University Press, 1970), for example, pp. 99, 156. Darwin also read the *Plurality* and knew Whewell to be the author. (Information from an unpublished reading-list, Darwin Collection, University Library, Cambridge.)

64. Whewell, *Treatise*, p. 208. Few of the critics of *Plurality* missed the opportunity of suggesting that the anonymous author might consult this work with profit.

65. Todhunter, *Whewell*, 2:292, 294.

66. Whewell, *Plurality*, 3d ed., pp. 283-288.

67. Ibid., p. 44.

68. Ibid., pp. 342-343.

69. Ibid., pp. 376-377.

70. Powell, *Essays*, p. 239.

71. Ibid., p. 231.

72. London: Murray, 1854.

73. *The Origin of Species* (London: Murray, 1859).

74. Darwin, *Origin*, p. 488. This reference to man was deliberate. Darwin thought it dishonest entirely to conceal his views. See F. Darwin, ed., *Life and Letters of Charles Darwin*, 2 vols. (London: Murray, 1887), 1:94; 2:263-264.

75. Darwin, *Origin*, p. 453.

76. M. Mandelbaum, "Darwin's Religious Views," *Journal of the History of Ideas* 19 (1958): 363-378, is most valuable on this subject.

77. *Three Essays on Religion* (New York: Holt, 1874), p. 174.

78. Darwin himself, towards the end of his life, seems to have swung this way. See N. Barlow, ed., *Autobiography of Charles Darwin* (New York: Norton, 1969), esp. p. 87.

79. Hull, *Darwin;* P. Vorzimmer, *Charles Darwin: The Years of Controversy* (Philadelphia: Temple University Press, 1970).

80. "Agnostic" in this context may be slightly anachronistic, for the word was invented by Huxley (*Collected Essays* [London: Macmillan, 1901], 9:134).

81. This appears in his contribution, "On the Study of the Evidences of Christianity," to *Essays and Reviews* (London: Longman, Green, Longman, and Roberts, 1860), p. 139.

82. See Sedgwick's letter to Darwin in J. H. Clark and T. M. Hughes, *Life and Letters of the Reverend Adam Sedgwick,* 2 vols. (Cambridge: University Press, 1890), 2:356-359; his review of the *Origin* is reprinted in Hull, *Darwin,* pp. 159-166, and his final, unchanged 1872 views on man, in Clark and Hughes, *Life and Letters,* 2:468-469. Whewell criticized Darwin's theory to a correspondent, Todhunter, *Whewell,* 2:433-435; and he raised again the question of final causes, suggesting natural selection's inadequacies, in a new preface to the seventh edition of his *Bridgewater Treatise* (1863).

83. S. Wilberforce, *"On the Origin of Species . . .," Quarterly Review* 108 (1860):225-264.

84. L. Huxley, ed., *Life and Letters of Thomas H. Huxley* (New York: Appleton, 1900), 1:192-204.

85. Wilberforce, *"Origin,"* pp. 256-260.

86. Darwin, *Life,* 2:241.

87. J. F. W. Herschel, *Physical Geography* (Edinburgh: Black, 1861), p. 12n.

88. Wilson, *Lyell's Journals,* pp. 123, 246.

89. Ibid., p. 382.

90. Lyell, *Principles,* 10th ed. (1869), pp. 491-492.

91. Vorzimmer, *Darwin;* J. Gruber, *A Conscience in Conflict: The Life of St. George Jackson Mivart* (Philadelphia: Temple University Press, 1960).

92. *Genesis of Species* (London: Macmillan, 1870).

93. "It [*Genesis*] is a book that I think will please Sir Charles Lyell." Letter from A. R. Wallace to Miss A. Buckley, February 2, 1871, in J. Marchant, *Alfred Russel Wallace, Letters and Reminiscences* (London: Cassell, 1916), p. 288.

94. *Genesis,* 2d ed., pp. 325-326.

95. F. E. Kingsley, *Charles Kingsley* (London: King, 1877), p. 386, expresses support for Hugh Miller, a miracle-advocate, although p. 377 suggests that Kingsley may have had a slightly weaker notion of miracle than, say, Sedgwick.

96. C. K. Kingsley, "The Natural Theology of the Future," *Macmillan's Magazine* 23 (1871):369-378.

4

Charles Darwin and the "Origin of Species"

Michael Ruse

Charles Darwin, born in 1809, attended university first at Edinburgh and then (from 1828 to 1831) at Cambridge and spent the years from 1831 to 1836 circumnavigating the globe as naturalist on the H.M.S. *Beagle*. Very soon after returning to Britain, probably in the early spring of 1837, Darwin became an evolutionist, and in the fall of 1838 he hit on the mechanism of natural selection brought on by the struggle for existence. In 1842 he wrote a thirty-five-page sketch of his theory (hereafter referred to as the *Sketch)*, and in 1844 this was expanded to a 230-page essay (the *Essay)* (both are included in Darwin and Wallace 1958). None of this was made public, though he showed the *Essay* to [the botanist Joseph] Hooker. Darwin spent practically the whole of the next ten years concealing his evolutionism, as he labored to produce tomes on barnacle systematics. Only when this was done did he return full time to his evolutionary work, and in the mid-1850s he began a massive work on natural selection and evolution. He was interrupted by the arrival of [an essay written by Alfred Russel Wallace], and after this and short evolutionary extracts from Darwin's earlier writings had been published by the Linnaean Society (Darwin and Wallace 1958), Darwin rapidly wrote an "abstract" of his ideas. This abstract, *On the Origin of Species by Means of Natural Selection; or, The Preservation of Favoured Races in the Struggle for Life,* was published in November 1859. The private Darwin had joined the public Darwin.

From *The Darwinian Revolution: Science Red in Tooth and Claw* by Michael Ruse (Chicago: University of Chicago Press, 1979), pp. 160–98. Copyright © 1979 by the University of Chicago Press. Reprinted by permission of the publisher.

FIRST TO EVOLUTION

Before Charles Darwin was born, the name of Darwin was already famous, or notorious, for the idea of organic evolution. At the end of the eighteenth century, Charles's grandfather, Erasmus Darwin, had extolled in verse and prose the virtues and beauties of such a developmental world picture (Greene 1959; J. Harrison 1971). . . . . Charles Darwin himself, no doubt with truth, later wrote that "it is probable that the hearing rather early in life such [evolutionary] views maintained and praised may have favored my upholding them under a different form in my *Origin of Species*" (Darwin 1969, p. 49). But when he went to Cambridge, to become a clergyman, he did "not then in the least doubt the strict and literal truth of every word in the Bible" (1969, p. 57).

At the Cambridge of Whewell and Sedgwick, Darwin stood in no danger of forgetting about the organic origins question, and one can feel certain his elders made sure the grandson of "the celebrated Dr. Darwin" knew all the good reasons why his grandfather's wild hypotheses were just that. On the other hand, Darwin was plunged into a community that was feeling the tension between science and religion—a community in which leading members like Sedgwick and Whewell felt unable to take portions of the Old Testament absolutely literally. And toward the end of his undergraduate studies Darwin felt the stimulating effect of the empiricist philosophy of John Herschel, who took an even more liberal line on biblical truth. Perhaps partly because of Herschel's influence, on the *Beagle* voyage Darwin soon started to become a Lyellian geologist. Then in 1832 he received the second volume of Lyell's *Principles* (Darwin 1969, p. 77 n), with its detailed discussions of Lamarckian evolutionism, organic struggles for existence, and the ways organisms might be distributed around the world, its veiled hints about the naturalness of species creations, and its overwhelming evidence of organic extinction. . . .

Darwin was therefore looking at the organic world in the Lyellian way. In a field notebook of February 1835 we find him saying to himself: "With respect then to the *death* of species of Terrestrial mammalia in the S. Part of South America I am strongly inclined to reject the action of any sudden debacle.—Indeed the very numbers of the remains render it to me more probable that they are owing to a succession of deaths after the ordinary course of nature" (Darwin MSS, 42; quoted in Herbert 1974, p. 236 n). And the very next sentence shows Lyell is master here. "As Mr. Lyell supposes species may perish as well as individuals; to the arguments he adduces I hope the Caria of B. Blanca will be one more small instance of at least a relation to certain genera with certain districts of the earth. This co-relation to my mind renders the gradual birth and death of species more probable" (Herbert 1974, p. 236 n. By "gradual birth" in this context, Darwin is not hinting at evolutionism. He is referring to Lyell's constant creation of new species). In fact, Darwin was so far Lyellian in his thinking about the organic world that we find him making the ultra-Lyellian appeal to God's design in keeping the organic world in a steady state. If species are not being successively created to compensate for continual successive deaths of species, we must allow that on earth "the number of its inhabitants has varied exceedingly at different periods.— A supposition in contradiction to the fitness which the Author of Nature has now established" (Herbert 1974, p. 233 n).

But, as a Lyellian, in looking at the organic world Darwin was presented with the nagging question any Lyellian faced. If not by evolution, if not by miraculous special creations, then how on earth did one get a constant natural supply of new species? . . .

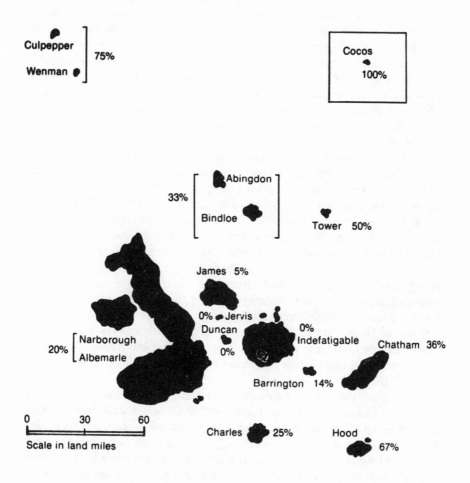

Figure 1. Percentage of endemic forms of finch on the various Galapagos Islands. Darwin, of course, did not have these percentages. From David Lack's *Darwin's Finches* (1947).

With his attention directed this way, Darwin apparently found three phenomena particularly striking and felt they were not adequately accounted for within the Lyellian world picture (Darwin 1969, p. 118). First, the Pampean formation contains fossils of armadillo species, now extinct, that resemble existing armadillos; second, very similar organisms replace each other going southward over the South American continent; and, third and most important, as Wallace was to recognize some twenty years later, organisms are distributed in a peculiar way on the Galápagos archipelago (see fig. 1). On different islands one gets different species of finches and tortoises, all very similar to each other and to South American forms. As a Lyellian, Darwin knew these species had to be naturally caused. From Lyell's *Principles* he could find no reason why they were distributed as they were. The problem was compounded because the Galápagos archipelago is of fairly recent geological origin and thus the organisms of the various islands seem more recent than the mainland forms.

In a sense, as Whewell pointed out in his *History* (1837, 3:589), not many options are open to the Lyellian faced with the problem of organic origins. Moreover, if he is honest with himself, one option recommends itself above all others. The Lyellian must believe in some kind of law-bound organic creation from nothing; or from something, organic or inorganic, widely different from that which is created; or he must believe in natural creation from similar forms. By any reasonable principle of parsimony the last choice is far preferable to the first two. But if this option is taken, though one may be loath to admit it, one is accepting a saltatory evolutionary theory. Although Lyell himself avoided posing this choice, Darwin was more forthright. When he returned to England he set about writing up his diary into a travel book, published as the *Journal of Researches*. This set him thinking again about the things he had seen, particularly in the Galápagos islands. And so in March or April of 1837 Darwin asked himself the question Lyell had refused to pose and, answering it in the obvious manner, moved across the divide and became an evolutionist. The Galápagos finches had to be explained as the natural law-bound product of one parent stock (Herbert 1974). Significantly, it was not until after the voyage, early in 1837, that the ornithologist John Gould convinced Darwin that the finches formed real species, not just varieties (Grinnell 1974, p. 262). . . .

It is plain that Darwin was thinking frantically and that nothing was very stable in his mind. In the summer of 1837, about three months after he became an evolutionist, he decided he could think more systematically if he kept notebooks devoted to the organic origins question. He kept these "species notebooks" for two years, right through the time when he discovered natural selection, and they are invaluable guides to tracing the minutiae of his thought. Let us examine these notebooks and see how, about eighteen months after his first rudimentary speculations, Darwin came upon the mechanism for which he is so famous.[1]

. . . AND THEN TO NATURAL SELECTION

From the first notebook it seems clear that Darwin's earliest speculations on the nature and causes of evolution did not last long. By midsummer 1837 the Lyellian worries about the possibility of gradually changing species had diminished. The Galápagos experience not only turned Darwin toward evolutionism, it influenced his thinking about the causes of evolution (Grinnell 1974). In particular, Darwin had in mind the model of a group of organisms, *isolated* from all others, evolving into a new species. "Let a pair be introduced and increase slowly, from many enemies, so as often to intermarry—who will dare say what result. According to this view animals on separate islands, ought to become different if kept long enough apart, with slightly differ[ent] circumstances.—Now Galápagos tortoises, mocking birds, Falkland fox, Chiloe fox.—English and Irish Hare—" (de Beer et al. 1960–67, B, p. 7). In such a model, as this passage shows Darwin recognized, the threat from external competition is eliminated; thus Lyellian fears for the adaptedness of changing species vanish. Darwin therefore felt free to posit gradual organic evolution based on small changes rather than sudden leaps. And he did switch to such minute changes, though we shall see remnants of saltatory changes in his thought for some time.

Having blocked off one Lyellian worry, Darwin continued to synthesize his position from Lyellian elements. Given our knowledge of Darwin as a geologist, this is what we would expect. First, Darwin thought that the constantly changing Lyellian world would create new isolated areas—the rising of the Galápagos archipelago, for example. But, second, he realized that isolation in itself would not lead to change. To affect organisms, one needs constant inorganic change in the isolated places. Once again Lyellian geology comes to the rescue. "We *know* world subject to cycle of change, temperature and all circumstances, which influence living beings" (B, pp. 2–3). Third, the Lyellian world view requires a whole new perspective on adaptation. It is no longer static, but is dynamic, with death the constant penalty for lack of success. "With respect to extinction we can easily see the variety of ostrich Petise may not be well adapted, and thus perish out, or on the other hand like Orpheus being favorable, many might be produced. This requires principle that the permanent varieties, produced by confined breeding and changing circumstances are continued and produce according to the adaptation of such circumstances, and therefore that death of species is a consequence (contrary to what would appear from America) of non-adaptation of circumstances" (B, pp. 37–39). Darwin drew a diagram (see fig. 2) showing just how extinction would lead to the distributions of species we find today, with big gaps between some genera but not between others.

Notice that even at this early point Darwin was thinking of irregularly branching, not linear, evolution. Indeed, he speculated that in the animal world we might get three main branches (for land, sea, and air), then have subbranches. "Organized beings represent a tree, *irregularly branched*" (B, p. 21), although "the tree of life should perhaps be called the coral of life, base of branches dead, so that passages cannot be seen" (B, p. 25; and see fig. 3). We know that a major implication of branching (considered as fundamental) is that man is no longer a measure of all things, and Darwin is aware of this. "It is absurd to talk of one animal being higher than another—we consider those where the [cerebral structure/intellectual faculties] most developed, as highest.—A bee doubtless would where the instincts were" (B, p. 74). More on this later. . . .

In the first species notebook, Darwin had tended to let variation look after itself. It occurred, and Darwin thought it was in some way a function of environmental conditions. But he had been concentrating on organisms evolving in isolation and thinking of the threat to an organism as coming primarily from outside its own species; though, as a passage quoted above shows, he was very conscious that adaptive failure spells extinction. Thus he had been able to shelve the question of adaptation, at least for newly evolving organisms. His first concern was to permit change—any kind of change. But about the beginning of 1838 Darwin began to worry in earnest. After all, isolation simply sets aside the problem of adaptation; it does not eliminate it. Even if they are protected from outside threats, isolated organisms like those on the Galápagos islands are building up adaptations, and as a good Lyellian—as well as the protégé of those rabid natural theologians Sedgwick and Whewell—Darwin knew this problem had to be solved. "With belief of transmutation and geographical grouping we are led to endeavor to discover *causes* of changes—the manner of adaptation (wish of parents??)" (de Beer et al. 1960–67, B, p. 227). Just how does one link new variation with the

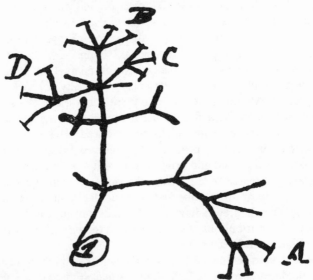

Figure 2. How species become separated through divergence and extinction. From Darwin's *Species Notebook* B.

Figure 3. The "coral of life." From Darwin's *Species Notebook* B.

fact of organic adaptation? "Can the wishing of the Parent produce any character in offspring? Does the mind produce any change in offspring? If so, adaptation of species by *generation* explained?" (B, p. 219). But this is just wild guessing, and so Darwin turned to the only possible source of information about new variations, the way they are passed on, and their relation to adaptation. He turned to the domestic world, the world of the animal and the plant breeder. . . .

During the summer of 1838 Darwin in no way actively promoted the analogy between the artificial and the natural worlds; but it is clear that his reading was not letting him forget that selection is the means of change in the domestic world or that analogies might be drawn from the natural world. Trying to understand heredity, variation, and adaptation, Darwin was reading deeply on breeding, and he encountered two pamphlets that described in some detail the process of artificial selection, its great effects, and the possibilities that such effects might be permanent—the very thing Lyell, and everyone else, was denying in support of the impossibility of evolution. Darwin responded enthusiastically to these pamphlets, by John Wilkinson (1820) and Sir John Sebright (1809). (Darwin's response is in C, p. 133; I discuss the pamphlets in detail in Ruse 1975*b*.) Although they still did not convince him that artificial selection is the clue to the natural mechanism of species change, it is plain that they were influencing him in that direction. . . .

We now come to September 1838. By then Darwin knew all about artificial selection and had even come across an idea of natural selection. But, . . . he was not thinking strongly of the analogy between them and probably did not consider natural selection a significant candidate as an evolutionary mechanism. He thought the key to adaptation lay in use and disuse, and he was ambivalent about selection and analogies from the domestic world. It would be a great thing if one could find a natural kind of selection, but how would one do this, and what guarantee would there be that it could cause indefinite evolution? But Darwin's study of the breeders was starting to push him in one very important aspect: it was showing him that organisms of the same species are not identical and that in the differences can lie the key to change. . . . Then, at the end of the month, Darwin read Malthus (1826) on population—the tendency to increase in number and the limited food and space all leading to an inevitable struggle for existence. Although he already thought of adaptation in a dynamic way, never before had Darwin realized what pressure there is on organisms or how critical and dynamic is the concept of adaptation: "Population is increase at geometrical ration in FAR SHORTER time than 25 years—yet until the one sentence of Malthus no one clearly perceived the great check amongst men" (de Beer et al. 1960–67, D, p. 135). Consequently, "One may say there is a force like a hundred thousand wedges trying [to] force every kind of adapted structure into the gaps in the economy of nature, or rather forming gaps by thrusting out weaker ones" (D, p. 135).

In some manner (to be discussed shortly), this reading of Malthus was enough. Darwin had all he needed. Malthus showed that the survival and reproductive differentials come from the desperate struggle for existence. To survive, an organism must have an adaptive edge not only over members of other species, but over members of its own species (Vorzimmer 1969). Darwin was able to connect this with what he knew about selection, which demands the very differential reproduction the struggle provides; and he saw that an organism's adaptive edge over other organisms could not be defined independent of the struggle. Adaptation was a function of the peculiarities of organisms that won out. Winning, and only winning, was what counted. Thus was born the idea of natural selection as an evolutionary mechanism. Some organisms win, because they have helpful characteristics the losers lack; and in the long run these characteristics add up to full-blown adaptations and significant evolutionary change. As Darwin later wrote, "Here, then, I had at least got a theory by which to work" (Darwin 1969, p. 120). And it is not long (27 November 1838) before we find him referring to his mechanism to explain change: "An habitual action must some way affect the brain in a manner which can be transmitted.—this is analogous to a blacksmith having children with strong arms.—The other principle of those children which *chance* produced with strong arms, outliving the weaker ones, may be applicable to the formation of instincts, independently of habits" (Gruber and Barrett 1974, N, p. 42).

But, despite what Darwin said, a mechanism is not a fully developed theory, and he still had much work to do before he could claim to have one. Let us turn now to the period from 1838 to 1844, the year the *Essay* was written.

DARWIN AND PHILOSOPHY

Whatever their differences, for both Herschel and Whewell Newtonian astronomy was the paradigm of a scientific theory. Throughout the species notebooks, up to and including the *Origin,* Darwin constantly followed the philosophers on this, claiming that any adequate solution to the organic origins question must satisfy the canons of Newtonian astronomy—but adding the corollary that Darwinian evolutionism will satisfy such canons whereas special creation will not! Thus, right in the middle of the first species notebook (summer 1837), Darwin wrote:

> Astronomers might formerly have said that God ordered each planet to move in its particular destiny. In same manner God orders each animal created with certain form in certain country, but how much more simple and sublime power let attraction act according to certain law, such are inevitable consequences— let animal be created, then by the fixed laws of generation, such will be their successors. Let the powers of transportal be such, and so will be the forms of one country to another.—Let geological changes go at such a rate, so will be the number and distribution of the species!! [B, pp. 101–2]

Of course at this time everyone was pushing Newton and the rule of law, so we ought not ascribe this general influence exclusively to the philosophers, though Darwin was impressed by them and their work. For instance, shortly after conceiving of natural selection as an evolutionary mechanism (August 1838), Darwin read with avid interest a review by Brewster of Comte's *Cours de philosophie positive.* What he would have got from this is that the aim of science is in the "positive" stage, that "the fundamental character of *Positive Philosophy* is to regard all phenomena as subjected to invariable natural laws," and that the best of all laws is the Newtonian law of gravitational attraction (Brewster 1837). This must surely have been a strong and (given the timing) crucial confirmation for Darwin of the importance of his "Newtonianism."[2] Nevertheless, the philosophers Herschel and Whewell went into far more detail than most about the precise implications of Newtonianism, and considering the extent to which Darwin followed their prescriptions—as he did in geology—direct influence seems plausible. Let us see how this is so.

There has long been a question about why Darwin reacted so enthusiastically to Malthus's work. That Darwin used Malthus's ideas in an essential way must be granted. But why was the actual reading of Malthus so important, since Darwin already knew all about the struggle for existence from the very detailed analysis in Lyell's *Principles* (Vorzimmer 1969)? Moreover, though he accepted Malthus's premises, Darwin wanted to stand Malthus's conclusion on its head: Whereas Darwin saw struggle as leading to change, Malthus essentially saw struggle as ruling out change!

Malthus, by concentrating on man, helped Darwin see struggle as intraspecific, not just as between groups or between a group and its environment.

But we must qualify this debt by pointing out that Malthus himself was more interested in man versus his environment, restricting his account of bloody intragroup struggle primarily to primitive man (Bowler 1976). Hence, Darwin probably got from Malthus what he needed to get. As I pointed out above, Darwin's key move to populational thinking—to recognizing critical variation within species—was probably much aided by his study of breeders' methods and results. We must therefore seek elsewhere for Darwin's major debt to Malthus. In light of Darwin's comment that his mechanism became clear as soon as he read Malthus (stressing the geometric ratio of population increase), and the way he presented his theory, undoubtedly the debt springs from Malthus's quasi-mathematical presentation of his subject. Right at the beginning of his *Essay* Malthus laid things out starkly: geometric population increases outstrip arithmetic increases in food supply. This leads to a crunch unless one invokes some sort of restraint (Malthus 1826, chaps. 1 and 2; see Ruse 1973). Helpfully, Malthus argued analogically from the nonhuman to the human world; so all Darwin had to do was argue back, dropping restraint and concluding that an organism struggles with everything, including its fellows. But the method of presentation was what counted, and the Herschel-Whewell influence puts this whole matter in context and makes Malthus's importance obvious. According to Herschel and Whewell's Newtonian philosophy, the best laws are quantitative, like the law of gravitational attraction. These laws do not occur in isolated splendor but are bound together in hypothetico-deductive systems. Darwin found Malthus's *Essay* new and striking not because of the truth of his premises—nearly everyone in the 1830s knew these and accepted them as indubitable—but because Malthus presented his ideas in lawlike, quantitative form, with a deductive approach. This was just what Darwin, soaked in the contemporary philosophy of science, was looking for. Not until Darwin read Malthus could he fit the struggle for existence into his mechanism of organic change—a mechanism that had to be scientific as he understood the word. Because of the way Malthus presented the struggle, Darwin realized it was sufficiently universal, inevitable, and powerful to cause indefinite organic change, unlike artificial selection (as he then thought).

As soon as Darwin had read Malthus, he started to think in terms of forces and pressures that were pushing organisms into the available and not-so-available gaps in the economy of nature, and this whole notion of force was central to the way the philosophers interpreted Newtonian physics. One must explain through causes, preferably *verae causae*. But the paradigm of a *vera causa* (for either Herschel or Whewell) was force. Hence, insofar as Malthus was able to show him something akin to a force at work on organisms, Darwin was becoming aware of something that would be a prime candidate for a scientific evolutionary mechanism. Because of the philosophical lens through which he viewed matters, he was highly receptive to the way Malthus presented the struggle for existence—because of his viewpoint, Darwin knew it had to be critical.

But this still takes us only to the mechanism. Toward the end of 1838,

Darwin had to start thinking about ways to develop his mechanism into a full theory of evolution. Once again we find the philosophical influence critical. Darwin was a highly professional young scientist. He felt it essential to put his theory, controversial as it would undoubtedly be, into a form that professional scientists would respect. People might reject his theory, but they would not reject it because of poor structure. What were the guidelines for a good theory and where would one find them? They were the ones specified by those arbiters of science Herschel and Whewell, and one would find them in the philosophers' conversation (at this time Darwin was mixing frequently with these men) and in their writings. In particular one must look at Herschel's *Discourse* and Whewell's recently published *History of the Inductive Sciences.*

At the end of 1838, this is just what Darwin was doing. He reread the *Discourse,* and went carefully through Whewell's *History* for the second time. Judging from his comments in the margin of the *History,* he was most interested in those things Whewell particularly admired in Newtonian astronomy and the arguments Whewell brought against organic evolutionism. Darwin wanted to make his theory as Newtonian as possible and to anticipate the worse attacks the critics could make. But how was he to do this? . . . Herschel and Whewell, though very close methodologically, came to differ over the concept of a *vera causa:* for Herschel, the empiricist, analogical argument from experience was what really counted; for Whewell, the rationalist, it was a consilience of inductions pointing upward. Darwin plainly strove to satisfy both these conceptions of *vera causa.*

Take first Herschel's analogical notion. In the context of discovery Darwin probably was most influenced, as he later said, by the artificial/natural selection analogy, though it did not play the overwhelming role he later implied it did. This is as may be. The important point is that once we have the notion of natural selection, we enter the context of justification; there was no need for Darwin to mention the analogy again. Indeed, there were good reasons he might refrain from mentioning it, as Wallace refrained twenty years later. For Lyell, Whewell, and everyone else, the *reductio* of evolutionism was that artificial selection does not lead to sustained, permanent change. But, though Darwin apparently shared this view just before discovering natural selection, no sooner had he read Herschel again than he suddenly started to emphasize the analogy. "It is a beautiful part of my theory, that domesticated races of organics are made by precisely same means as species" (E, p. 71; I take it that these "means" included, as they always did for Darwin, the inheritance of acquired characteristics). Moreover, Darwin soon thought of using the analogy as a way to present his theory publicly.

Varieties are made in two ways—local varieties when whole mass of species are subjected to same influence, & this would take place from changing country: but greyhound race-horse & poulter Pidgeon have not been thus produced, but by training, & crossing & keeping breed pure—& so in plants *effectually* the offspring are picked & not allowed to cross.—Has nature any process analogous— if so she can produce great ends—But how—even if placed on Isl^d. if &c &c—

make the difficulty apparent by cross-questioning—Here give my theory.—
excellently true theory. [E, p. 118]

This paradoxical about-face is paradoxical no longer when we realize
that Darwin was trying to present natural selection as a Herschelian *vera
causa*. We argue by analogy from actually experienced ways of changing
organisms to presumed ways. Moreover, Herschel's discussion of *verae causae*
in the *Discourse* shows added reason for Darwin's joy that so close an analogy
can be drawn between artificial and natural ways of producing change. Herschel
(1831, p. 149) argued that "if the analogy of two phenomena be very close
and striking, while, at the same time, the cause of one is very obvious, it
becomes scarcely possible to refuse to admit the action of an analogous cause
in the other, though not so obvious in itself." He then went on to explain
that the force we feel and cause when whirling a stone at the end of a string
points undeniably to the existence of an analogous force keeping the moon
in orbit as it whirls around the earth. This is just Darwin's situation. A causal
force, artificial selection, is directly perceived and caused by us. Therefore
we have the best evidence for the analogous causal force of natural selection.
(Although Darwin was going against Lyell here, Herschel and Lyell shared
the same *vera causa* concept, and so in another sense Darwin was being
thoroughly Lyellian: indeed, a Lyellian actualist. More generally, everything
in this section complements and reinforces the Lyellian influence; though in
the *Principles* Lyell was not hypothetico-deductive, and so we must go beyond
Lyell for philosophical/methodological influences on Darwin.)

By the beginning of 1839, then, Darwin was convinced that he should
exploit to the full the analogy between the domestic and natural worlds. From
his reading lists (a notebook in the Darwin collection labeled "Books to Be
Read"), we find that in the next few years he set to this task with gusto,
delving into the world of breeders and reading, for example, the classic works
on breeding by Youatt. Undoubtedly Darwin was convinced he should exploit
the domestic/natural analogy before he did much research. But, once convinced
of its importance, Darwin was in a peculiarly favored position, which helped
confirm his conviction that the opponents of evolutionism were mistaken when
they trotted out examples from the domestic world to disprove evolution.
Darwin was working at a time when scientific animal and plant breeding
was being developed and refined as never before, as agriculturalists tried to
keep up with the voracious demands of an exploding and increasingly urban
population. Coming as he did from Shropshire, the heart of agricultural
England, Darwin was able to draw extensively on local and family connections.
For instance, his uncle and (by early 1839) father-in-law Josiah Wedgwood,
with whom Darwin had a very close relationship, was one of England's leading
sheep breeders, as well as an officer of the Society for the Diffusion of Useful
Knowledge, which was then actively promoting the principles of scientific
agriculture. And the Darwin family itself had long been famous in Shrewsbury
as pigeon fanciers, a tradition Charles Darwin continued (Meteyard 1871).

Once convinced of the importance of the domestic/natural analogy, Darwin developed it to the hilt. In the *Sketch,* the *Essay,* and the *Origin,* Darwin began by discussing change in the domestic world, pointing to the importance of selection. Then, after noting variation in the wild and presenting the struggle for existence as a counterpart of man's desire for selection, Darwin argued by analogy to natural selection. Moreover, he repeatedly used the analogy to illustrate important points of this theory. . . .

So far we have considered Darwin's attention to Herschel's notion of a *vera causa.* Whewell's notion of a *vera causa* tied to a consilience was equally significant for Darwin,[3] though this will become plainer when we have laid out Darwin's completed theory of evolution. But, even before discovering natural selection as an evolutionary mechanism, Darwin showed that a consilient theory was his ideal. "Absolute knowledge that species die and other replace them.—Two hypotheses: fresh creations is mere assumption, it explains nothing further; points gained if any facts are connected" (de Beer et al. 1960-67, B, p. 104). Then as he labored toward the writing of the preliminary *Sketch* in 1842, Darwin worked hard to show how natural selection (and his other mechanisms) could be applied to explanations in behavior, paleontology, biogeography (geographical distribution), anatomy, systematics, embryology, and so on, all these areas being tied together chiefly by one mechanism. Just the thing a consilient theory should do, and just the thing that Darwin, sharing Whewell's theory ideals, thought a theory should do.

In 1842 and 1844, Darwin wrote out the preliminary versions of his theory, which in essence were very little changed in the *Origin.* We have a discussion of domestic selection. We get the struggle, then the analogical counterpart of domestic selection, natural selection. Then, along with discussions of the nature of sterility and the like, we get the mechanism of selection applied to all the problem areas mentioned above. In short, we have something Darwin himself could accept as a properly structured theory. Before we can continue chronologically, however, we must pause and go back. For Darwin's contemporaries, his teachers and his seniors in the scientific network, religious questions were a most important barrier to acceptance of an organic evolutionary theory. Why were they not a barrier for Darwin?

DARWIN AND RELIGION

At the center of revealed religion, based on faith and revelation, is the Bible, and we know that when he went up to Cambridge Darwin took the Bible literally. When he left Cambridge to join the *Beagle,* his faith was fairly orthodox, but during the voyage it started to crumble. Undoubtedly the major reason for this was Darwin's growing conviction that the Bible, particularly the Old Testament, was incompatible with science, particularly uniformitarian geology (Darwin 1969, p. 185). As Darwin became committed to science, he became more and more committed to the rule of law, which in turn excluded miracles. But for Darwin, Christianity without miracles was nothing (at least as a divinely inspired religion), and so his adherence to Christianity faded away (Darwin 1969, pp. 86-87). . . .

After the *Origin* was published, Darwin became something of an agnostic about the existence of God (Darwin 1969, p. 94; Mandelbaum 1958). Until that point it is probable that, although no Christian, he was neither atheist nor even agnostic. He was a deist of a kind—believing in an unmoved Creator who worked entirely through unbroken, unchanging law. He therefore accepted a natural religion based on reason and sense. This is the language Darwin used in the *Origin* (1859, p. 488), and it seems improbable that he was hypocritically using such a vocabulary for tactical reasons—he was going to ruffle the Christian feathers anyway. Certainly Darwin was a deist while he was discovering his theory, for he constantly used such language in his notebooks, when he was talking only to himself (Gruber and Barrett 1974, M, p. 154). . . .

In that aspect of natural religion centering on design, Paley was once again a starting point, for at Cambridge Darwin read the *Natural Theology* and was much impressed (Darwin 1969, p. 59). Darwin always accepted the major premise of the natural theologians of the 1830s, that the organic world must be understood in terms of its designlike appearance, its adaptedness, its relation to ends: we have seen this throughout Darwin's discovery of natural selection, a mechanism that itself focuses on the concept of adaptation. When discovering his theory Darwin took such adaptation as evidence of some designer (Gruber and Barrett 1974, M, p. 136), and at that time he was ready to allow that a purpose of earth's creation was man (de Beer et al. 1963, E, p. 49). In later years, however, Darwin seems to have felt that natural selection made the argument from design redundant to the point of untenability (Darwin 1969, p. 87), though until the end of his life he continued to have flashes that all this adaptation must spell design (F. Darwin 1887, 1:316 n).

Of course, even in the 1830s Darwin was not particularly conventional about natural religion, especially as it centered on adaptation. First, Darwin thought adaptation, designed or not, could be produced by normal, unbroken law. And since views like this were often linked to the picture of God as industrialist, it is interesting that, true to his rural connections, Darwin invoked its country cousin—God as farmer. In both the *Sketch* and the *Essay* Darwin explicitly likened natural selection to the work of a superbeing (Darwin and Wallace 1958, pp. 114-16). (Wallace went even further, arguing that selection is just like a self-regulating machine; Darwin and Wallace 1958, p. 278.) Second, Darwin showed his unconventionality by refusing to accept that organisms are essentially perfectly adapted to their environments. It was necessary to his organic world picture that organic adaptive failure be a regular phe-nomenon—not something that waits on catastrophes. Indeed, Darwin pushed to the limit the revolution about adaptation Lyell had started: From the absolute static property of the catastrophists, through the fallible property of Lyell, to his own thoroughly relativized dynamic phenomenon, where in any species some organisms must be adaptively inadequate at virtually any time (B, pp. 37-38, 90). The third point is minor but had an interesting consequence. Darwin veered from conventional thought on adaptation in that he could not consider anything in one species as provided solely for the benefit of some other species. He explicitly acknowledged that such an adaptation would refute his theory (Darwin 1859, p. 211).

Nevertheless, though he was putting a strain on adaptation, particularly as a bond between God and his creation (a fact he fully recognized; M, p. 70), this does not deny the great importance of adaptation for Darwin. He could not accept such adaptation as the product of divine intervention; but he knew from his scientific/religious background that any adequate biological theory must meet the problem head on, and he tried to do this with his mechanism of natural selection. Indeed, Darwin's sensitivity about adaptation had a somewhat paradoxical result. Darwin was educated and was doing his great creative work in the 1820s and 1830s, when the concept of design through adaptation was at its peak. By the time he published his work, . . . this strand of natural theology had been much buffeted. Hence, though his non-theological theory is often portrayed as taking the teleology out of biology, if anything Darwin was bringing it back in! Adaptation, with its orientation towards ends, was a more significant facet of the organic world for Darwin than it was for [T. H.] Huxley. Because the creative Darwin was a man of the 1830s rather than the 1850s, after the *Origin* biology was in a sense brought back to a teleology based on adaptation, from which it had started to stray.

THE LONG WAIT

We come now to the major puzzle in the Darwinian story. By the middle of 1844 Darwin had completed a version of his theory in the 230-page *Essay*. Although he made some changes when he wrote the *Origin,* they were comparatively minor. Why then did Darwin not publish his work at once, as soon as he had got something down on paper, rather than finish some geological work and then plunge into a massive, eight-year systematic study of barnacles? . . .

The true answer has to be sought in Darwin's professionalism, just as his success at finding and developing his theory must be sought there. Darwin was not amateur outsider like Chambers. He was part of the scientific network, a product of Cambridge and a close friend of Lyell, and he knew well the dread and the hatred most of the network had for evolutionism. If he had any doubts, the publication of *Vestiges* in the same year as he wrote his *Essay* confirmed them. Sedgwick raged against *Vestiges* for eighty-five pages in the *Edinburgh Review;* Whewell denounced it in *Indications;* and Herschel condemned it from the presidential chair of the British Association. Darwin knew his theory was much better than Chambers's—"better" in that it more adequately answered the problems as then understood—but it was evolutionary and materialistic nonetheless, and it was certainly not going to make its author very popular. When telling Hooker of his evolutionism, Darwin confessed that it was like admitting to a murder (F. Darwin 1887, 2:23). It was a murder—the purported murder of Christianity, and Darwin was not keen to be cast in this role. Hence the *Essay* went unpublished.

Was Darwin a coward? Not really. He could not have accomplished what he did without his background, firmly rooted in the scientific community. But because of this background, he could not do more. . . . Also, Darwin had no idea the delay would be so long. Stimulated by a strange barnacle he discovered while on the *Beagle,* Darwin decided to do a little work on barnacles (Darwin 1969, pp. 117-18). This project exploded into a full-length study, taking a great deal of time, and was further dragged out by the constant, severe illness that crippled him. Days, weeks, and months were lost when he was unable to work. From the pushy, vibrant young man of the 1830s, Darwin was reduced to an invalid. And so year after year the *Essay* on species lay untouched—with strict instructions that it be published in the event of his death (Darwin and Wallace 1958, pp. 35-36). Darwin had no desire to be ignored by posterity. . . .

The barnacle study finished, Darwin returned to the organic origins problem. At the urging of his friends, he started to write a colossal work on evolution (Stauffer 1975; Hodge 1977)—a kind of science by filibuster that would overwhelm his opponents by fact and footnote. This project was (luckily) interrupted in 1858 by the arrival of Wallace's paper. Dropping everything, in just over a year Darwin wrote the *Origin.* So we finally come to Darwin's major public pronouncement on organic evolution.

ON THE ORIGIN OF SPECIES

Darwin's first chapter, "Variation under Domestication," deals with animal and plant breeding. He argues that organisms with many different forms, like pigeons, have common ancestors and that the main reason for these diverse forms, besides use and disuse, is man's power of selection. The second chapter, "Variation under Nature" establishes the widespread variation in the wild. Darwin was concerned with very small variations, "individual differences," as opposed to larger changes, "single variations" (Darwin 1859, pp. 44-45). He showed more confidence in their ubiquitous existence in the *Origin* than in the *Essay,* perhaps in part because of his work on barnacles. In the *Essay* he explicitly allowed that natural evolution might occasionally occur through a saltation from one form to another (certainly not, however, across a species; Darwin and Wallace 1958, p. 150; Vorzimmer 1963). In the *Origin* all natural changes are smooth, fueled by individual differences.

The next two chapters are the crucial ones. First, we get the derivation of the "Struggle for Existence":

A struggle for existence inevitably follows from the high rate at which all organic beings tend to increase. Every being, which during its natural lifetime produces several eggs or seeds, must suffer destruction . . . otherwise, on the principle of geometrical increase, its numbers would quickly become so inordinately great that no country could support the product. Hence, as more individuals are produced than can possibly survive, there must in every case be a struggle for existence. . . . It is the doctrine of Malthus applied with manifold force to the whole animal and vegetable kingdoms. . . . Although some species may be now increasing, more or less rapidly, in numbers, all cannot do so, for the world would not hold them. [1859, pp. 63–64]

Then in the fourth chapter, "Natural Selection," Darwin goes on to derive his key mechanism:

How will the struggle for existence . . . act in regard to variation? . . . Let it be borne in mind in what an endless number of strange peculiarities our domestic productions, and, in a lesser degree, those under nature, vary; and how strong the hereditary tendency is . . . . Can it, then, be thought improbable, seeing that variations useful to man have undoubtedly occurred, that other variations useful in some way to each being in the great and complex battle of life, should sometimes occur in the course of thousands of generations? If such do occur, can we doubt (remembering that many more individuals are born than can possibly survive) that individuals having any advantage, however slight, over others, would have the best chance of surviving and of procreating their kind? On the other hand, we may feel sure that any variation in the least degree injurious would be rigidly destroyed. This preservation of favorable variations and the rejection of injurious variations, I call Natural Selection. [1859, pp. 80–81]

These arguments are not presented in absolutely formal fashion; but one can see the philosophers' influence (Ruse 1971, 1973, 1975a, c). The argumentation is certainly much closer to the hypothetico-deductive ideal than to anything, say, in Lyell. In the first argument he begins with the Malthusian premises that organic beings tend to increase at a geometric rate and that such increase outstrips supplies of food, which increase only arithmetically (not to mention space, which increases not at all). This leads to the struggle for existence. In the second argument he starts with this struggle as premise, adds in claims, based on analogy from the domestic world, that some of the variation in nature will help in the struggle and some will hinder, and thus neatly implies natural selection: organisms with useful heritable variations have a better chance of surviving and reproducing than organisms with injurious heritable variations.

Along with natural selection we get sexual selection, divided into selection through male combat and selection through female choice (Darwin 1859, pp. 87–90). Again Darwin justifies the claims by analogy to the domestic world, as he exploits to the full the *vera causa* justification of his evolutionary mechanisms. Then he illustrates and expands on his mechanisms, in particular natural selection. We are shown, for example, how scarcity of food might lead to competition among wolves, with victory going to the fastest. "I can under such circumstances see no reason to doubt that the swiftest and slimmest wolves would have the best chance of surviving, and so be preserved or selected . . . I can see no more reason to doubt this, than that man can improve the fleetness of his greyhounds by careful and methodical selection" (1859, p 90).

This chapter contains two important subsidiary discussions. First, Darwin shows that he has definitely relinquished geographical isolation as an essential element in evolutionary speciation. Relying entirely on the smallest of variations, Darwin reveals his fear that isolated populations, tending by their very nature to be small, might not be injected with enough new variation to cause significant change. He therefore gives up exclusive reliance on isolation in return for being able to argue that (splitting) speciation might occur between subgroups of large populations, where there definitely would be ample new variation. But even in large populations he seems to favor some kind of ecological isolation as helping speciation—"haunting different stations" or "breeding at slightly different seasons" (1859, p. 103).

Second is the "Principle of Divergence." Arguing analogically from pigeons, Darwin suggests that "the more diversified the descendants from any one species become in structure, constitution, and habits, by so much will they be better enabled to seize on many and widely diversified places in the polity of nature, and so be enabled to increase in numbers" (1859, p. 112). In other words, the reason there are so many kinds of species and so much splitting is that it confers a selective advantage. Darwin likened the principle of divergence to the "physiological division of labor" of the French biologist Henri Milne-Edwards, which claims that the more specialized the parts of the body (e.g., the stomach fitted for vegetable food alone), the more efficient the body is. In his *Autobiography* (1969, pp. 120–21) Darwin claimed that this principle of divergence came to him several years after the writing of the *Essay* (probably 1852), and while there is no reason to doubt that he did not consciously realize the problem (and the solution) till then, there nevertheless are strong intimations of both problem and solution in his earliest evolutionary writings (de Beer et al. 1960–67, E, pp. 95–97). This principle of divergence allowed Darwin to introduce his famous metaphor likening fossil and contemporary species to the dead and living branches of the Tree of Life (see fig. 4). Of course, the actual phenomenon of splitting and divergence (without cause or recognition of a need for cause) was not new for Darwin: it had been an integral part of his evolutionism from the beginning.

Next we get the chapter "Laws of Variation," a subject about which Darwin candidly admitted "our ignorance . . . is profound" (1859, p. 167). He seemed to think there were two basic kinds of variation. One type was more or less random (with respect to needs), and Darwin suspected these variations might be due to conditions impinging in some way on the reproductive system. Those of the other type do seem to involve a direct response to the environment: "I believe that the nearly wingless condition of several birds, which now inhabit

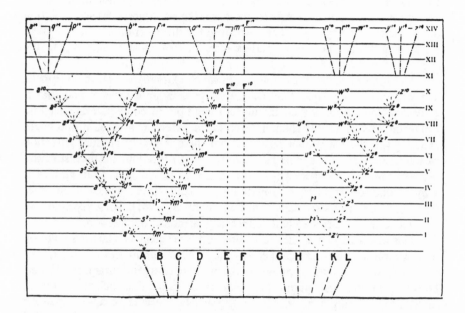

Figure 4. Darwin's diagram of descent (time moves upward). From the *Origin of Species*.

or have lately inhabited several oceanic islands, tenanted by no beast of prey, has been caused by disuse" (1859, p. 134). In later chapters we shall learn that Darwin had philosophical reasons for feeling dissatisfied with his treatment of heredity in the *Origin* and see how he strove to remedy matters.

The sixth chapter deals with "Difficulties on Theory." One problem Darwin tackles in this chapter is the absence of intermediate forms between species. Why is it that we do not "everywhere see innumerable transitional forms?" (1859, p. 171). Darwin posited that, excluding cases where speciation forms through isolation and where one would thus expect no intermediates, a group intermediate between two diverging varieties will tend to be smaller than the main groups because the intermediate groups will be in the rather narrow zones between the larger areas in which the main groups are adaptively diverging. These intermediate groups, because they are smaller, are more likely to be wiped out than the main groups, chiefly because they have less new variation on which selection can act and thus respond less readily to external challenge. Hence, generally we should not expect transitional forms.

This chapter also takes up the problem of organs that are highly perfected, like the eye. Although we cannot trace the eye's evolution, Darwin averred that the possibility of such evolution through natural selection is shown because in the Articulata we can set up a scale of living beings with eyes ranging from the simplest to the most complex (1859, p. 187). Hence, evolution could certainly have happened—the idea of transitory forms is not impossible—and, given the supporting evidence, there is no reason to believe such evolution did not occur. In any case, "The correction for the aberration of light is said, on high authority, not to be perfect even in that most perfect organ, the eye" (1859, p. 202).

A chapter titled "Instinct" follows, in which Darwin explains that instincts, like structures, vary; that they can be of great adaptive value; and that it is reasonable to think of them as subject to and produced by natural selection. Following this comes the chapter "Hybridism." Here he argues, as one might expect given the Darwinian position, that there is a very gradual gradation between perfect fertility and perfect sterility (p. 248), thus blurring the distinction between variety and species (p. 276). Darwin also suggests that sterility is a by-product of the laws of growth and heredity—it is "incidental on other differences, and not a specially endowed quality" (p. 261). These reproductive barriers between species are not deliberately fashioned by selection.

This discussion concludes just about every element of Darwin's solution to the problem of the origin of species—the problem of how and why organisms split into distinct groups as opposed to the problem of organic origins *simpliciter*. Darwin presented his solution piecemeal through the first half of the *Origin* rather than offering it all in one unit. Speciation occurs because organisms change in response to changing environments and because the more diversified and specialized they become the more efficiently they can exploit their environments. This can happen in isolation from other groups, but more often it involves geographically linked varieties of the same group responding to the different challenges of different zones. In these latter cases, intermediates disappear because their groups have less variation to draw on than the main groups. Sterility develops between groups as a result of other differences, not directly through selection.

We come next to the geological chapters: "On the Imperfection of the Geological Record" and "On the Geological Succession of Organic Beings." The first of these chapters is very Lyellian—explicitly so (p. 310). Darwin admits that, prima facie, the geological record raises major problems for an evolutionary theory like his own: he recognizes the objection that there was inadequate time for the slow process of natural selection and that the abrupt transition from one species to another (as revealed by the record) disproves evolution, as does the first appearance of life in all its sophistication. In reply to the first objection, Darwin counters that there was indeed enough time. One of the detailed examples on which Darwin stakes his case is the denudation of the Weald (in Sussex), which he calculates, estimating the time required for the sea to wear away a cliff, must have taken some three hundred million years (pp. 285-87; Burchfield 1974). This was the absolute time required for a phenomenon that (judging from the strata) had begun only in the latter part of the Secondary. One can therefore imagine, or perhaps cannot imagine, the total time available for evolution. In reply to the second objection, Darwin argues that the gaps in the fossil record can be explained by the inadequacy of the record—fossils were not deposited, we have failed to find the fossils, and so on. In any case, speciation occurs most frequently when the ground is being elevated, thus causing new stations (as on islands). But since this is not a time of deposition, no transitional fossils are being laid down (p. 292). Third, Why did life appear full-blown at the beginning of the Silurian? This problem, Darwin admits, worries him. It is not because the older the rock, the more metamorphosed it automatically becomes, and that all pre-Silurian rocks are so metamorphosed that they can show no fossils. The Silurian is too rich in fossils to allow that aging necessarily metamorphoses. Ever ready with a hypothesis, Darwin suggested that pre-Silurian organisms might have flourished on continents where oceans are now. And he guarded his hypothesis against contrary evidence like deep-sea borings by suggesting that the fossils of these organisms may now be metamorphosed by the great weight of the ocean above them! (pp. 306–10).

The second geological chapter was much more positive. For instance, it is necessary to Darwin's theory, unlike Lamarck's (and in certain respects Chambers's too), that there be no escalator-like evolution, where the loss of a particular species is compensated by the evolution of the same species at a later time. For Darwin, a species has only one chance, and he was pleased that the fossil record confirms this (p. 313). Similarly, he seized on the fact that throughout the world the fossil record shows parallel developments. New kinds of species, perfected by natural selection, would have an advantage over the old forms all over the world and thus would spread and leave similar fossil deposits worldwide (p. 327). Also, the fact that "the more ancient a form is, by so much the more it tends to connect by some of its characters groups now widely separated from each other" (p. 330) can easily be explained by descent with modification, as can the fact that "fossils from two consecutive formations are far more closely related to each other, than are the fossils from two remote formations" (p. 335). But what about the key question? What about progression? From background and previous hints we might expect that Darwin would take a position not unlike (the later) Owen, and this is what we find. Divergence was fundamental for Darwin, and since man is not ontologically different from the rest of creation, there could be no unilinear man-directed progression. We have divergent evolution leading to ever greater adaptive specialization. But within this framework Darwin was prepared to concede some vague subjective sense of progression: he too, as Lyell had always feared, linked progression and evolution. "The inhabitants of each successive period in the world's history have beaten their predecessors in the race for life, and are, in so far, higher in the scale of nature; and this may account for that vague yet ill-defined sentiment, felt by many paleontologists, that organization on the whole has progressed" (p. 345). . . .

The two chapters following the geological discussion, on "Geographical Distribution," contain some of Darwin's strongest cards, as our knowledge of his path to discovery might lead us to suspect. Here he gives some of the most striking facts of geographical distribution. For instance, it is well known that the Old and New Worlds have some very similar stations: "There is hardly a climate or condition in the Old World which cannot be paralleled in the New" (p. 346); yet the organisms of the two land masses are very different. Conversely, going down or up a continent like South America we find closely related organisms in different stations. "The plains near the straits of Magellan are inhabited by one species of Rhea (American ostrich), and northward the plains of La Plata by another species of the same genus" (p. 349). On top of this, where there are barriers we find different forms, though the absolute distances may be slight, whereas where there are no barriers we find similar forms, though they may be far apart.

Having set up his case, Darwin points out that all these phenomena could arise through Darwinian evolutionism and natural modes of transportation. And to support his general case that each new species was formed only once, Darwin provides detailed discussions of how organisms might or might not be transported around the globe, along with speculations on such things as the effects of the glacial period. As might be expected, the geographical distribution of the organisms of the Galápagos archipelago receives special attention, as he shows that Darwinian evolutionism, and only Darwinian evolutionism, can provide a satisfactory analysis. "We can clearly see why all the inhabitants of an archipelago, though specifically distinct on the several islets, should be closely related to each other, and likewise be related, but less closely, to those of the nearest continent or other source whence immigrants were probably derived" (p. 409).

Penultimately, we get a grab-bag chapter: "Mutual Affinities of Organic Beings: Morphology: Embryology: Rudimentary Organs." The natural system is explained as simply a function of common descent. Morphological problems dissolve in the same way. Take the classic homology between the hand of man, paw of the mole, leg of the horse, flipper of the porpoise, and wing of the bat. Naming Owen to support his position, Darwin argues that this problem is too much for the doctrine of final causes (p. 435; Owen had hardly adopted Darwin's solution!). But it can all be easily explained by a theory of descent with modification owing to natural selection. . . .

Turning to embryology, Darwin argues that the differences between embryos and adults are easy to explain in terms of the different selective pressures to which they are subject (pp. 439-50). He contends that new characteristics make themselves felt only in the adult form because embryos of different species do not feel different selective pressures as the adults do, and that this is why the embryos of species with widely different adults are often very similar. Selection forces adults apart, leaving embryos together. Darwin justifies this argument by detailed reference to related phenomena in the domestic world, and he points out that his explanation shows why the key to classification lies in embryology. Classification has its ultimate rationale in ancestry, with organisms classed according to the closeness of their lines of descent. But the embryos of modern forms tend to be like the embryos of ancestral forms; hence, two extant organisms with similar embryos probably have common ancestors, even though the adult forms are widely different (p. 449). . . .

Rounding out this chapter we find a brief discussion of "rudimentary, atrophied, or aborted organs." Again Darwin contends that only a theory of descent with modification can provide a satisfactory explanation, though it seems that he thought the main agency responsible for rudimentary organs was disuse rather than selection (p. 454). Finally comes a chapter titled "Recapitulation and Conclusion," noteworthy primarily for containing what must be the understatement of the nineteenth century: "Light will be thrown on the origin of man and his history" (p. 488). Virtually throughout the *Origin* Darwin had carefully avoided reference to that controversial being, man. But lest he be thought to have dishonorably concealed his views (1969, p. 130) right at the end he left little doubt that evolution through natural selection was intended to apply, without exception, to all organisms.

Note three final things about Darwin's work as a whole. First, Darwin intertwined the question of evolution itself with the mechanism for evolution. His cases for evolution and for natural selection are presented as a whole. To deal with them singly, one must separate them out oneself. Second, speculation about spontaneous generation is conspicuously absent. . . . Third, confirming earlier discussion, the overall nature of Darwin's theory was consilient. He once wrote of the *Origin* as "one long argument from the beginning to the end" (1969, p. 140), and in an important sense it is. Having derived his mechanism, Darwin applied it to many subdisciplines like behavior, paleontology, and biogeography. The theory as a whole has a fanlike appearance, with natural selection functioning as a Whewellian *vera causa* (see fig. 5).

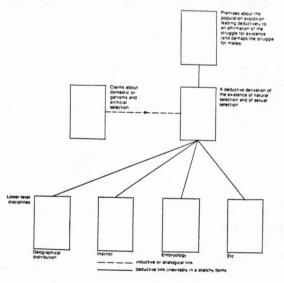

Figure 5. The three key structural elements of Darwin's theory: (1) the empiricist *vera causa*, from artificial selection; (2) the deductive core, leading to natural selection; (3) the rationalist *vera causa*, unifying the lower-level disciplines beneath natural selection.

NOTES

1. There are four notebooks directly related to the species question, marked by Darwin B, C, D, and E. These are transcribed and published in de Beer et al. 1960–67. There are also two notebooks, M and N, that deal essentially with man. These have been transcribed and published in Gruber and Barrett 1974. For convenience, I refer to Darwin's labels and pagination.

2. Without denying that reading Brewster's review of Comte must have given Darwin a strong psychological boost toward the rightness of Newtonianism, I cannot follow some recent commentators (e.g., Schweber 1977) in arguing that the review was *the* key philosophical influence on Darwin. Apart from anything else, Comte denied that science is a search for causes: Darwin followed Herschel and Whewell in considering this the ultimate aim of science. Theologically speaking, however, Comte's open atheism may have influenced Darwin.

3. The notion of a consilience, though not tied directly to a *vera causa*, also occurred in Herschel's *Discourse*. Whewell first wrote explicitly of a consilience, tying it to a *vera causa*, in the *Philosophy* (1840). I am not sure Darwin read this at this time, though he did respond enthusiastically to a detailed review by Herschel, explicitly discussing consilience and *verae causae:* "from Herschel's Review Quart. June 41 I see I MUST STUDY Whewell on Philosophy of Science" (Darwin, "Books to be Read"; Herschel's review was in the *Quarterly Review*). But, as he pointed out, Darwin knew Whewell's philosophy of science from the *History* and from personal contact. See also Thagard 1977, who shows that at some point Darwin may have read the *Philosophy*.

REFERENCES

Bowler, P. J. 1976. "Malthus, Darwin, and the Concept of Struggle," *Journal of the History of Ideas* 37: 631–50.

——. 1837. "Review of Comte's *Cours de philosophie positive,*" *Edinboro Review* 67: 271–308.

Burchfield, J. D. 1974. "Darwin and the Dilemma of Geological Time," *Isis* 65: 300–321.

Darwin, C. 1859. *On the Origin of Species by Means of Natural Selection.* London: Murray.

——. 1969. *Autobiography,* ed. N. Barlow. New York: Norton.

Darwin, C., and Wallace, A. R. 1958. *Evolution by Natural Selection.* Cambridge: Cambridge University Press.

Darwin, F. 1887. *The Life and Letters of Charles Darwin, Including an Autobiographical Chapter.* London: Murray.

de Beer, G., et al. 1960–67. "Darwin's Notebooks on Transmutation of Species," *Bulletin of the British Museum (Natural History)* history series, 2:27–200; 3:129–76.

Greene, J. C. 1959. *The Death of Adam.* Ames: Iowa State University Press.

Grinnell, G. 1974. "The Rise and Fall of Darwin's First Theory of Transmutation," *Journal of the History of Biology* 7: 259–73.

Gruber, H. E., and Barrett, P. H. 1974. *Darwin on Man.* New York: Dutton.

Harrison, J. 1971. "Erasmus Darwin's View of Evolution," *Journal of the History of Ideas* 32: 247–64.

Herbert, S. 1974. "The Place of Man in the Development of Darwin's Theory of Transmutation," Part I. to July 1837. *Journal of the History of Biology* 7: 217–58.

Herschel, J. F. W. 1831. *Preliminary Discourse on the Study of Natural Philosophy.*

London: Longmans, Rees, Orme, Brown, and Green.

———. 1841. "History . . . and Philosophy of the Inductive Sciences . . . by William Whewell . . ., *Quarterly Review* 135: 177–238.

Hodge, M. J. S. 1977. "The Structure and Strategy of Darwin's 'Long Argument,'" *British Journal of the History of Science* 10: 237–46.

Lack, David. 1947. *Darwin's Finches.* Cambridge: Cambridge University Press.

Lyell, C. 1830–33. *The Principles of Geology.* 1st ed. London: John Murray.

Malthus, T. R. 1826. *An Essay on the Principle of Population.* 6th ed. London.

Mandelbaum, M. 1958. "Darwin's Religious Views," *Journal of the History of Ideas* 19: 363–78.

Meteyard, E. 1871. *A Group of Englishmen 1795–1815.* London: Longmans, Green.

Paley, W. 1819. *Natural Theology. In Collected Works.* London: Rivington.

Ruse, M. 1971. "Natural Selection in *The Origin of Species,*" *Studies in History and Philosophy of Science* 1: 311–51.

———. 1973. "The Nature of Scientific Models: Formal v. Material Analogy," *Philosophy of the Social Sciences* 3: 63–80.

———. 1975a. "Darwin's Debt to Philosophy: An Examination of the Influence of the Philosophical Ideas of John F. W. Herschel and William Whewell on the Development of Charles Darwin's Theory of Evolution," *Studies in History and Philosophy of Science* 6: 159–81.

———. 1975b. "Charles Darwin and Artificial Selection," *Journal of the History of Ideas* 36: 339–50.,

———. 1975c. "Charles Darwin's Theory of Evolution: An Analysis," *Journal of the History of Biology* 8: 219–41.

Schweber, S. S. 1977. "The Origin of the *Origin* Revisited," *Journal of the History of Biology* 10: 229–316.

Sebright, J. 1809. "The Art of Improving the Breeds of Domestic Animals," in a letter addressed to the Right Hon. Sir Joseph Banks, K. B. London.

Stauffer, R. 1975. *Charles Darwin's Natural Selection.* Cambridge: Cambridge University Press.

Thagard, P. 1977. "Darwin and Whewell," *Studies in History and Philosophy of Science* 8: 353–56.

Vorzimmer, P. 1963. "Charles Darwin and Blending Inheritance," *Isis* 54: 371–90.

———. 1969. "Darwin, Malthus, and the Theory of Natural Selection," *Journal of the History of Ideas* 30: 527–42.

Whewell, W. 1837. *History of the Inductive Sciences.* London: Parker.

———. 1840. *Philosophy of the Inductive Sciences.* London: Parker.

Wilkinson, J. 1820. "Remarks on the Improvement of Cattle, etc.," in a letter to Sir John Saunders Sebright, Bart. M. P. Nottingham.

Youatt, W. 1831. *The Horse, with a Treatise on Draught.* London: Library of Useful Knowledge.

———. 1834. *Cattle, Their Breeds, Management, and Diseases.* London: Library of Useful Knowledge.

———. 1837. *Sheep, Their Breeds, Management, and Diseases.* London: Library of Useful Knowledge.

5

On the Origin of Species

Charles Darwin

If during the long course of ages and under varying conditions of life, organic beings vary at all in the several parts of their organization, and I think this cannot be disputed; if there be, owing to the high geometrical powers of increase of each species, at some age, season, or year, a severe struggle for life, and this certainly cannot be disputed; then, considering the infinite complexity of the relations of all organic beings to each other and to their conditions of existence, causing an infinite diversity in structure, constitution, and habits, to be advantageous to them, I think it would be a most extraordinary fact if no variation ever had occurred useful to each being's own welfare, in the same way as so many variations have occurred useful to man. But if variations useful to any organic being do occur, assuredly individuals thus characterized will have the best chance of being preserved in the struggle for life; and from the strong principle of inheritance they will tend to produce offspring similarly characterized. This principle of preservation, I have called, for the sake of brevity, Natural Selection. Natural selection, on the principle of qualities being inherited at corresponding ages, can modify the egg, seed, or young, as easily as the adult. Among many animals, sexual selection will give its aid to ordinary selection, by assuring to the most vigorous and best adapted males the greatest number of offspring. Sexual selection will also give characters useful to the males alone, in their struggles with other males.

Whether natural selection has really thus acted in nature, in modifying and adapting the various forms of life to their several conditions and stations, must be judged of by the general tenor and balance of evidence given in

Extract from *Origin of Species,* first published in 1859.

the following chapters. But we already see how it entails extinction; and how largely extinction has acted in the world's history, geology plainly declares. Natural selection, also, leads to divergence of character; for more living beings can be supported on the same area the more they diverge in structure, habits, and constitution, of which we see proof by looking at the inhabitants of any small spot or at naturalized productions. Therefore during the modification of the descendants of any one species, and during the incessant struggle of all species to increase in numbers, the more diversified these descendants become, the better will be their chance of succeeding in the battle of life. Thus the small differences distinguishing varieties of the same species, will steadily tend to increase till they come to equal the greater differences between species of the same genus, or even of distinct genera.

We have seen that it is the common, the widely diffused, and widely ranging species, belonging to the larger genera, which vary most; and these will tend to transmit to their modified offspring that superiority which now makes them dominant in their own countries. Natural selection, as has just been remarked, leads to divergence of character and to much extinction of the less improved and intermediate forms of life. On these principles, I believe, the nature of the affinities of all organic beings may be explained. It is a truly wonderful fact—the wonder of which we are apt to overlook from familiarity—that all animals and all plants throughout all time and space should be related to each other in group subordinate to group, in the manner which we everywhere behold—namely, varieties of the same species most closely related together, species of the same genus less closely and unequally related together, forming sections and sub-genera, species of distinct genera much less closely related, and genera related in different degrees, forming sub-families, families, orders, sub-classes, and classes. The several subordinate groups in any class cannot be ranked in a single file, but seem rather to be clustered round points, and these round other points, and so on in almost endless cycles. On the view that each species has been independently created, I can see no explanation of this great fact in the classification of all organic beings; but, to the best of my judgment, it is explained through inheritance and the complex action of natural selection, entailing extinction and divergence of character, as we have seen illustrated in the diagram. [See fig. 5, p. 93.]

The affinities of all the beings of the same class have sometimes been represented by a great tree. I believe this simile largely speaks the truth. The green and budding twigs may represent existing species; and those produced during each former year may represent the long succession of extinct species. At each period of growth all the growing twigs have tried to branch out on all sides, and to overtop and kill the surrounding twigs and branches, in the same manner as species and groups of species have tried to overmaster other species in the great battle for life. The limbs divided into great branches, and these into lesser and lesser branches, were themselves once, when the tree was small, budding twigs; and this connection of the former and present buds by ramifying branches may well represent the classification of all extinct and living species in groups subordinate to groups. Of the many twigs which flourished when the tree was a mere bush, only two or three, now grown into great branches, yet survive and bear all the other branches; so with the species which lived during long-past geological periods, very few now have living and modified descendants. From the first growth of the tree, many a limb and branch has decayed and dropped off; and these lost branches of various sizes may represent those whole orders, families, and genera which have now no living representatives, and which are known to us only from having been found in a fossil state. As we here and there see a thin straggling branch springing from a fork low down in a tree, and which by some chance has been favored and is still alive on its summit, so we occasionally see an animal like the Ornithorhynchus or Lepidosiren, which in some small degree connects by its affinities two large branches of life, and which has apparently been saved from fatal competition by having inhabited a protected station. As buds give rise by growth to fresh buds, and these, if vigorous, branch out and overtop on all sides many a feebler branch, so by generation I believe it has been with the great Tree of Life, which fills with its dead and broken branches the crust of the earth, and covers the surface with its ever branching and beautiful ramifications.

6

Objections to Mr. Darwin's Theory of the Origin of Species

Adam Sedgwick

Before writing about the transmutation theory, I must give you a skeleton of what the theory is:—

1st. *Species* are *not permanent; varieties* are the beginning of new species.

2nd. Nature began from the simplest forms—probably from one form— the primeval *monad,* the parent of all organic life.

3d. There has been a continual ascent on the organic scale, till organic nature became what it is, by one continued and unbroken stream of onward movement.

4th. The organic ascent is secured by a Malthusian principle through nature,—by a battle of life, in which the best in organization (the best varieties of plants and animals) encroach upon and drive off the less perfect. This is called the theory of *natural selection.*

It is admirably worked up, and contains a great body of important truth; and it is eminently amusing. But it gives no element of strength to the fundamental theory of transmutation; and without specific transmutations natural selection can do nothing for the general theory.[1] The flora and fauna of North America are very different from what they were when the Pilgrim Fathers were driven out from old England; but changed as they are, they do not one jot change the collective fauna and flora of the actual world.

5th. We do not mark any great organic changes *now,* because they are so slow that even a few thousand years may produce no changes that have fixed the notice of naturalists.

From *The Spectator,* April 7, 1860.

6th. But *time is the agent,* and we can mark the effects of time by the organic changes on the great geological scale. And on every part of that scale, where the organic changes are great in two contiguous deposits of the scale, there must have been a corresponding lapse of time between the periods of their deposition—perhaps millions of years.

I think the foregoing heads give the substance of Darwin's theory; and I think that the great broad facts of geology are directly opposed to it.

Some of these facts I shall presently refer to. But I must in the first place observe that Darwin's theory is not *inductive,*—not based on a series of acknowledged facts pointing to a *general conclusion,*—not a proposition evolved out of the facts, logically, and of course including them. To use an old figure, I look on the theory as a vast pyramid resting on its apex, and that apex a mathematical point. The only facts he pretends to adduce, as true elements of proof, are the *varieties* produced by domestication, or the *human artifice* of cross-breeding. We all admit the varieties, and the very wide limits of variation, among domestic animals. How very unlike are poodles and greyhounds! Yet they are of one species. And how nearly alike are many animals,—allowed to be of distinct species, on any acknowledged views of species. Hence there may have been very many blunders among naturalists in the discrimination and enumeration of species. But this does not undermine the grand truth of nature and the continuity of true species. Again, the varieties, built upon by Mr. Darwin, are varieties of domestication and human *design.* Such varieties could have no existence in the old world. Something may be done by cross-breeding; but mules are generally sterile, or the progeny (in some rare instances) passes into one of the original crossed forms. The Author of Nature will not permit His work to be spoiled by the wanton curiosity of Man. And in a state of nature (such as that of the old world before Man came upon it) wild animals of different species do not desire to cross and unite.

Species have been constant for thousands of years; and time (so far as I see my way) though multiplied by millions and billions would never change them, so long as the conditions remained constant. Change the conditions, and old species would disappear; and new species *might* have room to come in and flourish. But how, and by what causation? I say by *creation.* But, what do I mean by creation? I reply, the operation of a power quite beyond the powers of a pigeon-fancier, a cross-breeder, or hybridizer; a power I cannot imitate or comprehend; but in which I can believe, by a legitimate conclusion of sound reason draw from the laws and harmonies of Nature. For I can see in all around me a design and purpose, and a mutual adaptation of parts which I *can* comprehend,—and which prove that there is exterior to, and above, the mere phenomena of Nature a great prescient and designing cause. Believing this, I have no difficulty in the repetition of new species during successive epochs in the history of the earth.

But Darwin would say I am introducing a miracle by the supposition. In one sense, I am; in another, I am not. The hypothesis does not suspend or interrupt an established law of Nature. It does suppose the introduction of a new phenomenon unaccounted for by the operation of any *known* law of Nature; and it appeals to a power above established laws, and yet acting in harmony and conformity with them.

The pretended physical philosophy of modern days strips Man of all his moral attributes, or holds them of no account in the estimate of his origin and place in the created world. A cold atheistical materialism is the tendency of the so-called material philosophy of the present day. Not that I believe that Darwin is an atheist; though I cannot but regard his materialism as atheistical; because it ignores all rational conception of a final cause. I think it untrue because opposed to the obvious course of Nature, and the very opposite of inductive truth. I therefore think it intensely mischievous.

Let no one say that it is held together by a cumulative argument. Each series of facts is laced together by a series of assumptions, which are mere repetitions of the one false principle. You cannot make a good rope out of a string of air-bubbles.

I proceed now to notice the manner in which Darwin tries to fit his principles to the facts of geology.

I will take for granted that the known series of fossil-bearing rocks or deposits may be divided into the Palaeozoic; the Mesozoic; the Tertiary or Neozoic; and the Modern—the Fens, Deltas, &c., &c., with the spoils of the actual flora and fauna of the world, and with wrecks of the works of Man.

To begin then, with the Palaeozoic rocks. Surely we ought on the trans-mutation theory, to find near their base great deposits with *none but the lowest forms of organic life.* I know of no such deposits. Oken contends that life began with the infusorial forms. They are at any rate well fitted for fossil preservation; but we do not find them. Neither do we find beds exclusively of hard corals and other humble organisms, which ought, on the theory, to mark a period of vast duration while the primeval monads were working up into the higher types of life. Our evidence is, no doubt, very scanty; but let not our opponents dare to say that it makes *for them.* So far as it is positive, it seems to me pointblank *against them.* If *we* build upon imperfect evidence, they commence without any evidence whatsoever, and against the evidence of actual nature. As we ascend in the great stages of the Palaeozoic series (through Cambrian, Silurian, Devonian, and Carboni-ferous rocks) we have in each a *characteristic* fauna; we have no wavering of species,—we have the noblest cephalopods and brachiopods that ever existed; and they preserve their typical forms till they disappear. And a few of the types have endured, with specific modifications, through all succeeding ages of the earth. It is during these old periods that we have some of the noblest ichthyic forms that ever were created. The same may be said, I think, of the carboniferous flora. As a whole, indeed, it is lower than the living flora of our own period; but many of the old types were grander and of higher organization than the corresponding families of the living flora; and there is no wavering, no wanting of organic definition, in the old types. We have some land reptiles (batrachians), in the higher Palaeozoic periods, but not of a very low type; and the reptiles of the permian groups (at the very top of the Palaezoic rocks), are of a high type. If all this be true (and I think it is), it gives but a sturdy grist for the transmutation-mill, and may soon break its cogs.[2]

We know the complicated organic phenomena of the Mesozoic (or Oolitic) period. It defies the transmutationist at every step. Oh! but the document, says Darwin, is a fragment. I will interpolate long periods to account for all the changes. I say, in reply, if you deny my conclusion grounded on positive evidence, I toss back your conclusions, derived from negative evidence—the inflated cushion on which you try to bolster up the defects of your hypothesis. The reptile fauna of the Mesozoic period is the grandest and highest that ever lived. How came these reptiles to die off, or to degenerate? And how came the dinosaurs to disappear from the face of Nature, and leave no descendants like themselves, or of a corresponding nobility? By what process of *natural selection* did they disappear? Did they tire of the land, and become Whales, casting off their hind-legs? And, after they had lasted millions of years as whales, did they tire of the water, and leap out again as Pachyderms? I have heard of both hypotheses; and I cannot put them into words without seeming to use the terms of mockery. This I do affirm, that if the transmutation theory were proved true in the actual world, and we could hatch rats out of eggs of geese, it would still be difficult to account for the successive forms of organic life in the old world. They appear to me to give the lie to the theory of transmutation at every turn of the pages of Dame Nature's old book.

The limits of this letter compel me to omit any long discussion of the Tertiary Mammals, of course including man at their head. On physical grounds, the transmutation theory is untrue, if we reason (as we ought to do) from the known to the unknown. To this rule, the Tertiary Mammals offer us no exception. Nor is there any proof, either ethnographical or physical, of the bestial origin of man.

And now for a few words upon Darwin's long inter*polated periods* of geological ages. He has an eternity of past time to draw upon; and I am willing to give him ample measure; only let him use it logically, and in some probable accordance with facts and phenomena.

1st. I place the theory against facts viewed collectively. I see no proofs of enormous *gaps* of geological time (I say nothing of years or centuries), in those cases where there is a sudden change in the ancient fauna and flora. I am willing, out of the stock of past time, to lavish millions or billions upon each epoch, if thereby we can gain rational results from the operation of *true causes*. But time and "natural selection" can do nothing if there be not a *vera causa* working with them.[3] I must confine myself to a very small number of the collective instances.

2d. Towards the end of the carboniferous period, there was a vast extinction of animal and vegetable life. We can, I think, account for this extinction mechanically. The old crust was broken up. The sea bottom underwent a great change. The old flora and fauna went out; and a new flora and fauna appeared, in the ground, now called permian, at the base of the new red sandstone, which overlies the carboniferous rocks. I take the fact as it *is,* and I have no difficulty. The time in which all this was brought about *may* have been very long, even upon a geological scale of time. But where do the *intervening* and connecting types exist, which are to mark the *work of natural selection?* We do not find them. Therefore, the step onwards gives no true resting-place to a baseless theory; and is, in fact, a stumbling-block in its way.

3d. Before we rise through the new red sandstone, we find the muschel-kalk (wanting in England, though its place on the scale is well-known) with an *entirely new* fauna: where have we a proof of any enormous lapse of geological time to account for the change? We have no proof in the deposits themselves: the presumption they offer to our senses is of a contrary kind.

4th. If we rise from the muschel-kalk to the Lias, we find again a new fauna. All the anterior species are gone. Yet the passage through the upper members of the new red sandstone to the Lias is by insensible gradations, and it is no easy matter to fix the physical line of their demarcation. I think it would be a very rash assertion to affirm that a great geological interval took place between the formation of the upper part of the new red sandstone and the Lias. Physical evidence is against it. To support a baseless theory, Darwin would require a countless lapse of ages of which we have no commensurate physical monuments; and he is unable to supply any of the connecting organic links that ought to bind together the older fauna with that of the Lias.

I cannot go on any further with these objections. But I will not conclude without expressing my deep aversion to the theory; because of its unflinching materialism;—because it has deserted the inductive track,—the only track that leads to physical truth;—because it utterly repudiates final causes, and thereby indicates a demoralized understanding on the part of its advocates. By the word, demoralized, I mean a want of capacity for comprehending the force of moral evidence, which is dependent on the highest faculties of our nature. What is it that gives us the sense of right and wrong, of law, of duty, of cause and effect? What is it that enables us to construct true theories on good inductive evidence? Theories which enable us, whether in the material or the moral world, to link together the past and the present. What is it that enables us to anticipate the future, to act wisely with reference to future good, to believe in a future state, to acknowledge the being of a God? These faculties, and many others of like kind, are a part of ourselves quite as much so as our organs of sense. All nature is subordinate to law. Every organ of every sentient being has its purpose bound up in the very law of its existence. Are the highest conceptions of man, to which he is led by the necessities of his moral nature, to have no counterpart or fruition? I say *no,* to all such questions; and fearlessly affirm that we cannot speculate on man's position in the actual world of nature, on his destinies, or on *his origin,* while we keep his highest faculties out of our sight. Strip him of these faculties, and he becomes entirely bestial; and he may well be (under such a false and narrow view) nothing better than the natural progeny of a beast, which has to live, to beget its likeness, and then die forever.

By gazing only on material nature, a man may easily have his very senses bewildered (like one under the cheatery of an electro-biologist); he may become so frozen up, by a too long continued and exclusively material study, as to lose his relish for moral truth, and his vivacity in apprehending it. I think I can see traces of this effect, both in the origin and in the details of certain portions of Darwin's theory; and, in confirmation of what I now write, I would appeal to all that he states about those marvelous structures,—the comb of a common honey-bee, and the eye of a mammal. His explanations make demands on our credulity, that are utterly beyond endurance, and do not give us one true natural step towards an explanation of the phenomena— viz., the perfection of the structures, and their adaptation to their office. There *is* a light by which a man may see and comprehend facts and truths such as these. But Darwin wilfully shuts it out from our senses; either because he does not apprehend its power, or because he disbelieves in its existence. This is the grand blemish of his work. Separated from his sterile and contracted theory, it contains very admirable details and beautiful views of nature,— especially in those chapters which relate to the battle of life, the variations of species, and their diffusion through wide regions of the earth.

In some rare instances, Darwin shows a wonderful credulity. He seems to believe that a white bear, by being confined to the slops floating in the Polar basin, might in time be turned into a whale; that a lemur might easily be turned into a bat; that a three-toed tapir might be the great-grandfather of a horse; or that the progeny of a horse may (in America) have gone back into the tapir. . . . all interpreted hypothetically—produces, in some minds, a kind of pleasing excitement, which predisposes them in its favor and if they are unused to careful reflection, and averse to the labor of accurate investigation, they will be likely to conclude that what is (apparently) *original,* must be a production of original *genius,* and that anything very much opposed to prevailing notions must be a grand *discovery,*—in short, that whatever comes from "the bottom of a well" must be the "truth" which has been long hidden there.

NOTES

1. It is worth remarking that though no species of the *horse* genus was found in America when discovered, two or three *fossil* species have been found there. Now, if these horses had (through some influence of climate) been transmuted into tapirs or buffaloes, one might expect to see the *tendency* at least towards such a change in the numerous herds of wild horses— the descendants of those brought from Europe—which are now found in both South and North America.

2. I forebear to mention the Stagonolepis, a very highly organized reptile, the remains of which were found, by Sir R. I. Murchison, in a rock near Elgin, supposed to belong to the

old red sandstone. Some doubts have been expressed about the age of the deposit. Should the first opinion prove true (and I think it will), we shall then have one of the oldest reptiles of the world exhibiting, not a very low, but a very high organic type.

3. See reference on *Time,* in the *Annotations of Bacon's Essays.*

7

The Origin of Species

T. H. Huxley

The Darwinian hypothesis has the merit of being eminently simple and comprehensible in principle, and its essential positions may be stated in a very few words: all species have been produced by the development of varieties from common stocks; by the conversion of these, first into permanent races and then into new species, by the process of *natural selection,* which process is essentially identical with that artificial selection by which man has originated the races of domestic animals—the *struggle for existence* taking the place of man, and exerting, in the case of natural selection, that selective action which he performs in artificial selection.

The evidence brought forward by Mr. Darwin in support of this hypothesis is of three kinds. First, he endeavors to prove that species may be originated by selection; secondly, he attempts to show that natural causes are competent to exert selection; and thirdly, he tries to prove that the most remarkable and apparently anomalous phenomena exhibited by the distribution, development, and mutual relations of species, can be shown to be deducible from the general doctrine of their origin, which he propounds, combined with the known facts of geological change; and that, even if all these phenomena are not at present explicable by it, none are necessarily inconsistent with it.

There cannot be a doubt that the method of inquiry which Mr. Darwin has adopted is not only rigorously in accordance with the canons of scientific logic, but that it is the only adequate method. Critics exclusively trained in classics or in mathematics, who have never determined a scientific fact in

From *Collected Essays,* vol. 2, *Darwiniana* (London: Macmillan, 1893 [originally published in 1860]) pp. 71–79.

their lives by induction from experiment or observation, prate learnedly about Mr. Darwin's method, which is not inductive enough, not Baconian enough, forsooth, for them. But even if practical acquaintance with the process of scientific investigation is denied them, they may learn, by the perusal of Mr. Mill's admirable chapter "On the Deductive Method," that there are multitudes of scientific inquiries in which the method of pure induction helps the investigator but a very little way.

> "The mode of investigation," says Mr. Mill, "which, from the proved inapplicability of direct methods of observation and experiment, remains to us as the main source of the knowledge we possess, or can acquire, respecting the conditions and laws of recurrence of the more complex phenomena, is called, in its most general expression, the deductive method, and consists of three operations: the first, one of direct induction; the second, of ratiocination; and the third, of verification."

Now, the conditions which have determined the existence of species are not only exceedingly complex, but, so far as the great majority of them are concerned, are necessarily beyond our cognizance. But what Mr. Darwin has attempted to do is in exact accordance with the rule laid down by Mr. Mill; he has endeavored to determine certain great facts inductively, by observation and experiment; he has then reasoned from the data thus furnished; and lastly, he has tested the validity of his ratiocination by comparing his deductions with the observed facts of Nature. Inductively, Mr. Darwin endeavors to prove that species arise in a given way. Deductively, he desires to show that, if they arise in that way, the facts of distribution, development, classification, etc., may be accounted for, *i.e.* may be deduced from their mode of origin, combined with admitted changes in physical geography and climate, during an indefinite period. And this explanation, or coincidence of observed with deduced facts, is, so far as it extends, a verification of the Darwinian view.

There is no fault to be found with Mr. Darwin's method, then; but it is another question whether he has fulfilled all the conditions imposed by that method. Is it satisfactorily proved, in fact, that species may be originated by selection? that there is such a thing as natural selection? that none of the phenomena exhibited by species is inconsistent with the origin of species in this way? If these questions can be answered in the affirmative, Mr. Darwin's view steps out of the rank of hypotheses into those of proved theories; but, so long as the evidence at present adduced falls short of enforcing that affirmation, so long, to our minds, must the new doctrine be content to remain among the former—an extremely valuable, and in the highest degree probable, doctrine, indeed the only extant hypothesis which is worth anything in a scientific point of view; but still a hypothesis, and not yet the theory of species.

After much consideration, and with assuredly no bias against Mr. Darwin's views, it is our clear conviction that, as the evidence stands, it is not absolutely proven that a group of animals, having all the characters exhibited by species in Nature, has ever been originated by selection, whether artificial or natural. Groups having the morphological character of species—distinct and permanent races in fact—have been so produced over and over again; but there is no positive evidence, at present, that any group of animals has, by variation and selective breeding, given rise to another group which was, even in the least degree, infertile with the first. Mr. Darwin is perfectly aware of this weak point, and brings forward a multitude of ingenious and important arguments to diminish the force of the objection. We admit the value of these arguments to their fullest extent; nay, we will go so far as to express our belief that experiments, conducted by a skillful physiologist, would very probably obtain the desired production of mutually more or less infertile breeds from a common stock, in a comparatively few years; but still, as the case stands at present, this "little rift within the lute" is not to be disguised nor overlooked.

In the remainder of Mr. Darwin's argument our own private ingenuity has not hitherto enabled us to pick holes of any great importance; and judging by what we hear and read, other adventurers in the same field do not seem to have been much more fortunate. It has been urged, for instance, that in his chapters on the struggle for existence and on natural selection, Mr. Darwin does not so much prove that natural selection does occur, as that it must occur; but, in fact, no other sort of demonstration is attainable. A race does not attract our attention in Nature until it has, in all probability, existed for a considerable time, and then it is too late to inquire into the conditions of its origin. Again, it is said that there is no real analogy between the selection which takes place under domestication, by human influence, and any operation which can be effected by Nature, for man interferes intelligently. Reduced to its elements, this argument implies that an effect produced with trouble by an intelligent agent must, à fortiori, be more troublesome, if not impossible, to an unintelligent agent. Even putting aside the question whether Nature, acting as she does according to definite and invariable laws, can be rightly called an unintelligent agent, such a position as this is wholly untenable. Mix salt and sand, and it shall puzzle the wisest of men, with his mere natural appliances, to separate all the grains of sand from all the grains of salt; but a shower of rain will effect the same object in ten minutes. And so, while man may find it tax all his intelligence to separate any variety which arises, and to breed selectively from it, the destructive agencies incessantly at work in Nature, if they find one variety to be more soluble in circumstances than the other, will inevitably, in the long run, eliminate it.

A frequent and a just objection to the Lamarckian hypothesis of the transmutation of species is based upon the absence of transitional forms between many species. But against the Darwinian hypothesis this argument has no force. Indeed, one of the most valuable and suggestive parts of Mr. Darwin's work is that in which he proves, that the frequent absence of transitions is a necessary consequence of his doctrine, and that the stock whence two or more species have sprung, need in no respect be intermediate between these species. If any two species have arisen from a common stock in the same way as the carrier and the pouter, say, have arisen from the rock-pigeon, then the common stock of these two species need be no more intermediate between the two than the rock-pigeon is between the carrier and pouter. Clearly appreciate the force of this analogy, and all the arguments against the origin of species by selection, based on the absence of transitional forms, fall to the ground. And Mr. Darwin's position might, we think, have been even stronger than it is if he had not embarrassed himself with the aphorism, *"Natura non facit saltum,"* which turns up so often in his pages. We believe, as we have said above, that Nature does make jumps now and then, and a recognition of the fact is of no small importance in disposing of many minor objections to the doctrine of transmutation.

But we must pause. The discussion of Mr. Darwin's arguments in detail would lead us far beyond the limits within which we proposed, at starting, to confine this article. Our object has been attained if we have given an intelligible, however brief, account of the established facts connected with species, and of the relation of the explanation of those facts offered by Mr. Darwin to the theoretical views held by his predecessors and his contemporaries, and, above all, to the requirements of scientific logic. We have ventured to point out that it does not, as yet, satisfy all those experiments; but we do not hesitate to assert that it is as superior to any preceding or contemporary hypothesis, in the extent of observational and experimental basis on which it rests, in its rigorously scientific method, and in its power of explaining biological phenomena, as was the hypothesis of Copernicus to the speculations of Ptolemy. But the planetary orbits turned out to be not quite circular after all, and, grand as was the service Copernicus rendered to science, Kepler and Newton had to come after him. What if the orbit of Darwinism should be a little too circular? What if species should offer residual phenomena, here and there, not explicable by natural selection? Twenty years hence naturalists may be in a position to say whether this is, or is not, the case; but in either event they will owe the author of "The Origin of Species" an immense debt of gratitude. We should leave a very wrong impression on the reader's mind if we permitted him to suppose that the value of that work depends wholly on the ultimate justification of the theoretical views which it contains. On the contrary, if they were disproved tomorrow, the book would still be the best of its kind—the most compendious statement of well-sifted facts bearing on the doctrine of species that has ever appeared. The chapters on Variation, on the Struggle for Existence, on Instinct, on Hybridism, on the Imperfection of the Geological Record, on Geographical Distribution, have not only no equals, but, so far as our knowledge goes, no competitors, within the range of biological literature. And viewed as a whole, we do not believe that, since the publication of Von Baer's "Researches on Development," thirty years ago, any work has appeared calculated to exert so large an influence, not only on the future of Biology, but in extending the domination of Science over regions of thought into which she has, as yet, hardly penetrated.

Part Two

Evolution Today

Introduction

A great deal has happened to evolutionary theory in the hundred years since Darwin died, and yet it is amazing just how many of the same questions and problems keep surfacing. In particular, again and again we get queries about the legitimacy of evolution as an approach and of natural selection as an adequate cause. This section raises some of these questions. Through the writings of myself and others, some answers are offered. Whether the responses are adequate is for you to judge.

As a kind of prolegomenon, I offer a small piece of my own: "Is There a Limit to Our Knowledge of Evolution?" I make a threefold division (not particularly original with me) between evolution as a *fact*, evolution as a *path*, and evolution as a *theory* or mechanism. If nothing else, I do hope I scotch the depressingly familiar lament (or gloat) that "evolution is just a theory, not a fact." As so often is the case, what this claim implies and whether it has any critical force depends on how you are using your terms. What do you mean by "theory"? What do you mean by "fact"? With these questions answered, the slogan crumbles. Taken one way, the claim is a truism. Taken another way, it is false.

We move next to one of the really meaty articles in the collection, Francisco Ayala's "The Mechanisms of Evolution." Mainly, this discusses issues from the third part of my division (evolution as theory), but toward the end Ayala does have some remarks about the second part (evolution as path). As was mentioned in the introduction to Part One, Darwin's big problem stemmed from the lack of an adequate theory of heredity. Even if natural selection does work, what does it work on? In this area major biological advances have been made for, as Ayala explains, we now know much about the nature and causes of variation in populations of organisms, and about how selection might work on such variation.

Let me draw your attention particularly to one very important point that Ayala makes. Frequently, it is thought that selection could never be really effective because there is insufficient favorable material on which it can work. However, contrary to this supposition (which is expressed by, among others, Karl Popper in the discussion following Ayala's) selection does not have to wait for the occasional new, needed variation. Rather there are always masses of variation existing in natural populations, ready for new needs and new challenges.

I will express this truth in a different way by means of a metaphor. Suppose, for some college course, you were asked to write an essay on dictatorship, and the only source material you had was that which arrived through the post from the Book-of-the-Month Club. I forecast that the course would be finished and you would have failed long before an appropriate volume came your way. Suppose, however, that you had a good library to draw on. The essay could be finished before the weekend is over. If there be nothing suitable on the shelves on Hitler, then there will be material on Stalin; and if not Stalin, then someone else. Too frequently, it is thought that selection waits (and fails) because it relies on variation through the book club equivalent— something that arrives rarely and randomly. But, as Ayala shows, there is a library to draw on. If a group of organisms faces a new challenge—a predator or drought, say—then already it has all sorts of material to draw on. Selection can start working and evolution can occur.

We move next to Popper's critique of orthodox modern theory (generally called "neo-Darwinism" or the "synthetic theory"). Among other charges he levels one of the most familiar of all, namely, that the theory is not genuine science because selection is not a genuine mechanism. Darwin's argument reduces to the shallow tautology that those that survive are those that survive. You must decide whether I answer Popper adequately. Note that my critique of Popper does not challenge his "criterion of demarcation" between science and nonscience. I do not deny that a mark of genuine science is that it be, in principle, falsifiable—that is to say, open to test and refutation.

I should say incidentally that, apparently independently of my critique, Popper has modified his thinking somewhat, agreeing that perhaps he was somewhat harsh on natural selection. (See, especially, "Natural Selection and the Emergence of Mind," *Dialectica* 32 (1978): 339–55.) However, the critical Popper seems much better known than the conciliatory Popper, and his critical views are worth having here because it is these that have been influential, persuading not a few that there is something rotten about Darwinism.

Popper himself, like Huxley, favors an evolutionary theory that allows for jumps or "saltations" of a kind. In this, he anticipated a view of evolution that has gathered much publicity in recent years, the so-called theory of "punctuated equilibria." One of its chief exponents is Stephen Jay Gould, represented by the essay: "Is a New and General Theory of Evolution Emerging?" Gould's ideas are clear, although sometimes the language is a little complex. I trust you will find the glossary at the end of this book helpful in this respect.

Together with Gould's contribution, I include a lively response by the English evolutionist, John Maynard Smith ("Did Darwin Get It Right?"). Let me draw your attention to two points concerning this exchange. First, using the trichotomy introduced in my own essay at the beginning of the section, do realize that this is a debate about theories or mechanisms. Gould is certainly not at all questioning the fact of evolution. (He was a witness against creationism in Arkansas!) I have heard it said that we evolutionists should not disagree, even about mechanisms, because this only opens the door to our critics. I believe that such caution would be cowardly and false. Precisely because claims about evolution are scientific, there is room for— and expectation of—debate. Unanimity is not the mark of a healthy science.

Second, as with Darwin and his own critics, I am sure that at least part of the Gould/Maynard Smith exchange transcends brute fact. The simple matter is that when Maynard Smith looks at organisms, he thinks first and primarily "adaptation"; Gould does not. There is a matter of attitudes at stake here, notwithstanding the fact that Gould (like Huxley, I am sure) certainly does not want to deny that adaptation is widespread in the organic world, or that Maynard Smith admits to a weakness for "hopeful monsters." (Note, however, that Maynard Smith does not give way to his weakness precisely because of worries about adaptedness.)

Concluding Part Two, Richard Dawkins in "Universal Darwinism" backs up what I have just been saying about attitudes. Dawkins centers on adaptive complexity as the key aspect of organic nature, admits to being a neo-Paleyist, and plumps solidly for natural selection. (He also very usefully points out that there is a lot of ambiguity in much talk about Darwinism's rivals, especially about saltationism and punctuated equilibria.) I should mention that Dawkins's article is controversial, having been dismissed by one eminent evolutionist of my acquaintance as "banal." Perhaps, taken as science, it does not say much new; but, this is to read it the wrong way. Dawkins is doing philosophy, trying to set up the very conditions for an adequate evolutionary theory. This study is far from banal.

Yet, let me leave you here with one disturbing thought. Suppose Dawkins's arguments are well taken? Does he not prove too much? Is not Darwinism, with all rivals shown conceptually impossible, once again nigh tautological? If nothing else, Dawkins's success would suggest what I hint at in my critique of Popper, namely, that selection is deeply embedded in and connected to beliefs about the general lawlike nature of the world. Denial of selection may not be contradictory, but given all of selection's implications about its systematic effect, such denial may push you close to a disregard for the general uniformity of biological nature.

8

Is There a Limit to
Our Knowledge of Evolution?

Michael Ruse

Charles Darwin died just over one hundred years ago, in 1882, some twenty-three years after he published his major evolutionary work, *On the Origin of Species by Means of Natural Selection* (1859). Darwin was not the first to argue for evolution—gradual descent of all organisms from "one or a few forms"—but it was he more than anyone who made the doctrine respectable and plausible (Ruse 1979).

Today, however, evolution stands under great threat from biblical fundamentalists. (For full discussions, see Eldredge 1982, Futuyma 1982, Kitcher 1982, Newell 1982.) In a different way, the adequacy of the mechanism that Darwin himself proposed for evolution—natural selection—is strongly attacked from many different quarters (Gould 1980, Popper 1974, and in virtually every issue of *Systematic Zoology*). It seems, therefore, peculiarly appropriate to stand back for a moment, and to assess the extent that we have knowledge of evolution, and the extent to which we can hope for further knowledge.

In thinking of evolution, it is convenient to make a threefold division. First, there is what we might call the putative *fact* or happening of evolution, the claim that organisms did not arrive here on earth miraculously, but by a process of descent. Second, there is the question of the *paths* taken in the process, what evolutionists call "phylogenies." Did birds evolve via the dinosaurs, or directly from more primitive organisms? Third, there is the matter of the *mechanism* of evolution: the causes behind the process. Given Darwin's

From *BioScience* 34, no. 2 (1984): 100–104. Reprinted by permission of the publisher.

seminal contributions, this question often becomes one of deciding just how far selection extends, how it operates, and the nature of the alternatives.

Fact. Paths. Mechanisms. These are not three entirely separate issues. Paths and mechanisms presuppose the fact of evolution. But, for the purposes of exposition, it is convenient to separate out the three issues. Accordingly, let us now go on to ask about the limits of knowledge, as they apply to the three aspects of evolution.

IS THE FACT OF EVOLUTION SECURELY KNOWN?

In the development of scientific theories, there seems always to be the prospect of more work and more discoveries. For instance, in physics one deals with ever-smaller entities, constrained only by the rising costs (Rescher 1978). But, as far as historical facts are concerned, including the fact of evolution, this infinite prospect does not seem to apply. Either evolution occurred, or it did not. Therefore, in at least one sense, one could surely establish the fact, once and for all.

However, one does encounter difficulty at another level, because of the temporal dimension to evolution. No one was around to see the evolution of organisms, from the beginning to the sophisticated forms we see about us today. There are those—particularly the so-called "Scientific Creationists"— who argue that this is the end of the matter. Without eyewitnesses, they claim that any talk of origins is but a matter of faith. Scientific knowledge is impossible. Claims about evolution are no less religious than claims about Genesis (Morris 1947, Gish 1973).

Outside the Creationist movement, few would go so far. Nevertheless, many people feel that there is something a bit "iffy" about evolution. A common complaint is that "evolution is only a theory, not a fact." It is felt that claims about evolution are somewhat on a par with speculations about Kennedy's assassination. Indeed, even some professional biologists feel a little this way. For instance, Colin Patterson, a deservedly well-known ichthyologist at the British Museum, allows only that evolution is "neither fully scientific, like physics, for example, nor unscientific, like history" (Patterson 1978, p. 146). And this is a sentiment shared by others. (For more details, see Ruse 1982.)

In response, two things need to be said. First, in some ultimate sense it is true that logical certainty can never be achieved in science. There is always the possibility, at some level, that one might be mistaken. In this sense, if you like, all science is tentative. If you doubt this, remember how certain the early Victorians were about the essential truth of Newtonian physics. Some even went so far as to "prove" that Newton's laws are a priori necessary.

But, second, let it also be noted that in science one can achieve as much certainty as any reasonable person could demand. By this, I mean as much certainty as any of us ask in everyday life. The metaphor I myself like is one borrowed from the law. A jury is told to convict, if the guilt seems established "beyond reasonable doubt." There is no demand that the guilt be put beyond all logical possibility. Similar standards do (and should) apply in science. It is "beyond reasonable doubt" that the earth goes round the sun, and that water is H_2O (Ruse 1973). In a very strong sense of "know," we know that these facts are true. (Note that this sense of knowledge is compatible with tentativeness, of the kind mentioned above. Even though we find a person guilty "beyond reasonable doubt," it could still be that new evidence would force a reopening of the case.)

Now, let us push the metaphor a little further. Why would one find someone guilty "beyond reasonable doubt"? Obviously eyewitness testimony would count very heavily. But, suppose one didn't have this. Then one would have to work by circumstantial evidence: Lord Rake was found dead in his library, with a dagger through his heart. He was stabbed by a left-handed man, and the butler is left-handed. He was stabbed with an oriental knife, and the butler's parents spent years in China as missionaries. He was a notorious libertine, and the butler's daughter is pregnant. And so forth, and so forth. The evidence taken as a whole points overwhelmingly to the butler's guilt.

To philosophers, this form of argumentation is very familiar. It is what the nineteenth-century man of science William Whewell (1840) called a "consilience of inductions." And, possibly the best consilience one can find in science is that pointing to evolution. Biological phenomenon after biological phenomenon converges on evolution: Why are homologies so common? Why do we have such similar bone-structures between the functionally different arm of man, forelimb of horse, wing of bird, flipper of whale? Because of evolution! Why are embryos (e.g., man and dog) so similar, when adults are so different? Because of evolution! Why are the facts of biogeographical distribution so distinctive? Why would a group of islands like the Galapagos Archipelago have no less than fourteen different species of Darwin's finch? Because of evolution! These and many other phenomena put the actual fact or happening of evolution "beyond reasonable doubt." (See the excellent discussions in Futuyma 1982, and Kitcher 1982.)

I should add that Darwin's consilience in the *Origin* did not come by chance. Darwin learnt methodology from Whewell, even though later Whewell refused to allow the *Origin* into the library of the Cambridge College of which he was by then Master (Ruse 1979).

CAN ONE KNOW OF PHYLOGENIES?

We move on to the next set of questions to do with evolution. What about the actual paths taken by evolution? Can we know any of these? Can we know all of these?

The overall pattern of evolution seems fairly well established. It shows the transition from relatively simple early life forms to the overwhelming organic diversity of today's world. Moreover, it shows fully the incredibly branching nature of the evolutionary process. Indeed, the main outlines of the history of life were discovered by anti-evolutionists before Darwin and the *Origin!* In recent years much progress has been made in uncovering the early history of life, and we can now trace it back about 3.5 billion years, that is, for about three-fourths of the earth's 4.6-billion-year span. (For details see Ayala and Valentine 1979, Eldredge 1982, Luria et al. 1981, Schopf 1978.)

Again, certain specific items of evolution seem now to have been established, as firmly as any reasonably minded person could demand or wish. The evolution of birds and mammals, from reptiles, springs to mind. The fossil record showing the transitions is rock solid, to use an appropriate metaphor! Archaeopteryx, for instance, is a fossil organism which is—quite literally—a reptile with feathers. Indeed, just before the *Origin,* Archaeopteryx feathers had been discovered, which were simply thought to be rather humdrum bird feathers. It was only when the complete organism was discovered, just after the *Origin,* that the full significance of the earlier discoveries was appreciated (Feduccia 1980).

Coming closer to home, the fossil evidence of our own simian ancestry is overwhelming. Nineteenth-century critics of Darwin demanded the "missing link." In *Australopithecus* we have it. These organisms, which lived up to 4 million years ago, had brains the size of apes, but walked upright like humans. It is no longer reasonable to pretend that humans stand apart from chimps and gorillas (Futuyma 1982, Johanson and White 1979).

But, having said this much, one must acknowledge that there are many, many gaps in the fossil record. Moreover, given the high improbability of fossilization, there is no reason to think that all or most of these gaps will be bridged. In short, we will probably reach a limit of fossil evidence for phylogenies, with many things still unknown. Pertinent information will simply have been lost, irretrievably. Also, it would be disingenuous not to recognize that even with good fossil evidence, additional theorizing is needed to establish exactly what did happen in the past (Schaeffer et al. 1972).

Nevertheless, we must not end this section on a negative note, suggesting that evolutionists can extract only a finite fixed amount of information from their fossils, and then that is it. There are ever-developing techniques of getting more information from what we have already got. For instance, in the human sphere, much new information is being gathered from examining not simply the size of the brain, but also the shape and marks that the brain left behind. This is a most valuable new tool for tracing phylogenies. The same holds also of techniques for examining microscopic scratches on teeth. One can discover all sorts of things about diet, thus tracing lineages and gleaning new information about the evolving animals. (Johanson and Edey 1981).

Also—and this is terribly important—although fossils perform most visibly and crucially in the quest for phylogenies, they are far from providing the only source of pertinent information. Comparative studies on today's organisms tell a great deal. Long before any fossil evidence was discovered, it was obvious that humans are most closely related to apes. Similarities of every kind— anatomical, embryological, behavioral—cried out about common ancestry.

Moreover, powerful new techniques are also being developed in this direction. For instance, systematists now use the computer in their quest for the understanding of phylogenetic affinities. And molecular biology has been brought into play, most directly with the concept of a "molecular clock." If there is a molecular dimension to organisms—one which lies below the level at which most evolutionary processes operate—then change would be random and reasonably constant. In other words, comparison of molecular similarities and differences between organisms would tell of relationships, and of how long ago it was that different organisms took different evolutionary paths (Fitch 1976, Goodman 1982a).

Suffice it to say that this notion of a clock is already yielding much new information. Interest now centers on the human case. Molecular biology shows phenomenal similarities between men and apes. But when did the break come? Conventional paleoanthropology claims that we broke from the apes more than 10 million years ago. The clock suggests that the break occurred less than 5 million years ago. (We are closer to the apes, than dogs are to foxes.) Which position is right? It is to be hoped that the answers will come fairly soon, as more is learned about the record and about the accuracy of the clock.

I must emphasize that the clock is still a controversial notion. One should not assume automatically that when molecules (physics) and fossils (biology) fall out, it is the former that has to be correct. A hundred years ago, relying on thermodynamical calculations about the Earth's rate of cooling, Lord Kelvin criticized Darwin for demanding too much time for evolution. But it was Darwin who was right! Perhaps history repeats itself, and again the fossils are a surer guide to the past than physics. Nevertheless, whatever the final judgment on the molecular clock, no one doubts that it is stimulating a successful drive to learn more about the course of human prehistory. (For a fascinating discussion of this whole topic, see Gribbin and Cherfas 1982. See also Goodman 1982b, King and Wilson 1975, Templeton 1983.)

I conclude, therefore, that some (many/most?) phylogenies will never be known—certainly not in full detail. But, heuristically, there is no reason to stop the quest. Much is being learned, and will be learned, about phylogenies. Where fossils fail, comparative studies often succeed.

WHAT ABOUT MECHANISMS?

We come now to the most difficult question of all. What do we know of evolutionary mechanisms? What can we hope to know?

As noted earlier, for historical reasons if for no other, these questions usually begin with Darwin's mechanism of natural selection. In the *Origin*, Darwin argued first to a universal struggle for existence and reproduction. Then he took this fact, that not all organsims do indeed survive and reproduce, and he argued that the differential success will be a function of distinctive "helpful" characteristics ("adaptations"). This process Darwin labeled "natural selection," because of the analogy with the artificial selection practiced by animal and plant breeders. Since the time of Darwin we have had the coming of Mendelian and of molecular genetics. Hence, selection today is usually thought of as a process involving gene ratios in populations. Ultimately, however, as with Darwin, the link is made to those organic features which help in the battle for life (Ayala and Valentine 1979, Ruse 1982).

That natural selection can be effective, and that it has been effective, is very solidly established. Many selection experiments have been performed, showing how selection can mold organisms as theory predicts. Whole new species (reproductively isolated breeding groups) have been formed, through the power of selection (Jones 1981). Moreover, selection operating in the wild has been documented, again and again. The best-known (although far-from-unique) case is that of the industrial melanism of moths in Britain. As trees got dirtier, because of pollution from the industrial revolution, many moth species got darker, thus remaining camouflaged against their main predators: robins. That this occurred through selection is granted by all (Kettlewell 1978). Indeed, even Creationists agree that selection is at work here (Gish 1978).

Additionally, there is much indirect evidence for the power of selection. Certainly, no one has a convincing alternative for the many sophisticated adaptations that we find in nature. Why, for example, is the trilobite eye exactly the form predicted by Descartes and Huygens, as that needed to avoid distortion due to different colors of light (Clarkson and Levi-Setti 1975)? Why the eye at all, indeed, if natural selection is not at work? Alternative explanations are occasionally suggested, like Lamarckism (the inheritance of acquired characteristics). Invariably, they have come tumbling down, despite the initial enthusiasm of proponents (Ruse 1982).

It is true that adaptations are indirect evidence of selection; but, invoking the legal analogy again, do not think the worse of them for that. Fingerprints are indirect evidence that the culprit held the murder weapon. Are they inferior even to eyewitness reports? I suggest that adaptations are natural selection's fingerprints.

Allow, then, that natural selection is an effective component in the evolutionary process. The question now becomes one of just how effective a component selection is. There are three important matters of fact which bear on this most difficult question.

First, all agree that natural selection could not have been responsible for absolutely everything. Even the most enthusiastic of Darwinians (like myself!) accept this much. As we saw in the last section, if there is any merit to the idea of a molecular clock, then there is at least non-selection-powered evolution at the molecular level (King and Jukes 1969). But, even at more "organic" levels, selection is not all-powerful. For instance, we know from studies of heredity that many genes are "pleiotropic," that is to say, they affect more than one characteristic at once. Both theory and experiment therefore suggest that, if one feature is very adaptive (and thus strongly selected), another feature, which may have no selective advantage at all, could thus "piggyback" in. Because the second was pleiotropically linked to the first, such a non-useful feature could evolve right along with useful features.

Second, notwithstanding the point just made—that all agree that selection doesn't do absolutely everything—evolutionists today are split right down the middle on the full extent of the operation of non-selective forces and processes. There are strong neo-Darwinians, like Ernest Mayr (1982), Edward O. Wilson (1975), and John Maynard Smith (1981), who think selection very, very important. And, there are critics, like Stephen Jay Gould (1980), Niles Eldredge and Joel Cracraft (1980), and Steven M. Stanley (1979), who think there are many other processes at work.

These latter evolutionists endorse the theory of "punctuated equilibria," arguing that evolution goes in fits and starts, as opposed to the gradual change supposed by orthodox Darwinians. Because of constraints like this on organisms, it has been suggested that the "engineering constraints" involved in putting a functioning organism together may dictate many features, which are not therefore themselves produced directly by selection. For instance, strong Darwinians think the four-limbedness of vertebrates to have been the result of selection. Some punctuated equilibria supporters think it may simply be a condition of getting vertebrates to work at all. One thus gets rapid switches in evolution, from one well-integrated functioning form to another. (See Lewontin and Gould 1979 for some thoughts in this direction, and Maynard Smith 1981 for a Darwinian response.)

Yet other critics of conventional Darwinism favor yet other causal factors. Some think randomness, at the non-molecular level, has been very important (e.g., Kimura and Ohta 1971). Recently, others have argued that thermodynamics offers insights into the evolutionary process (Wiley and Brooks 1982).

The third point about the effectiveness of natural selection is crucially relevant to our quest for the limits of our knowledge of selection. There seems little hope of some sort of algorithm (as it were) which would decide effectively and definitively between Darwinians (selectionists) and critics. If there were such a decision procedure, we could all start working together! Unfortunately, rows about the effectiveness of selection have existed since the day of the *Origin's* publication, and, as a matter of empirical fact, I see no end to them. Nor do I see theoretical hope of ending the rows. However often the Darwinian shows some feature to be controlled by selection, there will always be yet more features for the critic who would deny that selection applies generally.

Overall, therefore, I am somewhat pessimistic about the possibility of our ever knowing fully the causes of evolutionary change. Like Kafka's castle, it is a goal we shall never achieve.

SHOULD WE GIVE UP?

But, although this is a somewhat negative note which I have just struck, I don't intend it as a dismal note. Let me conclude by making three more points, qualifying and amplifying my claim about the unlikelihood of ever finding the full story to evolutionary mechanisms.

First, even though we may never know every causal detail behind evolutionary change, we can certainly anticipate learning a lot more about them. Indeed, in the past two decades, we have learned more about the causes of evolution than at any previous time—at least, since the coming of Mendelism, if not since the coming of Darwinism itself. There is no reason why this rapid advance should not continue.

To justify such a feeling, consider the recent advances made in sociobiology, particularly those pertaining to the Hymenoptera (ants, bees, wasps). Two questions about this group have puzzled evolutionists since Darwin. Why do we find sterile workers in this group? Why are the workers always female and never male? At one stroke, William Hamilton (1964) solved these problems through the concept of "kin selection." The Hymenoptera are haplodiploid. Females have mother and father, and hence have a full (diploid) chromosome set. Males have only mothers, and have a half (haploid) chromosome set. This means that the sister-sister relationship is closer than the mother-daughter relationship. Hence, sterile workers better their evolutionary prospects when they raise fertile sisters rather than when they raise daughters. Evolution by proxy, as it were! Sons have no such close relationships, and hence for them there is no advantage to such sterile altruism. Thus, we get sterile females, but not sterile males. (For details, see Maynard Smith 1978, or Oster and Wilson 1978.)

Although details of this explanation are still queried, the outline is now generally accepted. A major advance in our understanding of the evolutionary process has occurred. I see no reason why similar such advances should not occur—at least, in the foreseeable future.

Second, despite the prospect of lack of complete success, I deny that evolutionists today are necessarily on the wrong track in what they are doing. Sometimes, it is suggested that we need a whole new "paradigm," as it were, a model which would cut through all our present difficulties and insecurities, throw dazzling new light on evolution and its processes. (See Lewontin 1974, Rose 1982.) Of course, part of the trouble with paradigms is that you don't know that you need them until you're right in the middle of them. However, given the continuing success of evolutionists in discovering mechanisms, I very much doubt that a totally new approach is needed. Certainly, there is no reason to think that a new approach would speed up discoveries or open the prospect of total knowledge.

I am not saying that no new breakthroughs will occur or are needed. I suspect, in fact, that we will learn more and more as we discover and apply molecular biology. (Already the molecular world has paid great dividends, showing that widespread nature of genetic variation.) But, there is no reason to conclude that evolutionary studies today are radically misdirected. And given the very great advances in evolutionary work, there are good reasons to think the contrary: that such studies are well directed indeed.

Third and finally, let me end with a warning. A virtual axiom of research in the physical world is that, as one delves ever deeper and deeper and smaller and smaller, costs escalate until they become prohibitive. We face a similar problem in evolutionary studies, although (as hinted before) the limits here are temporal, not spatial. Once one has gotten down to molecules (or, at most, to atoms), I doubt that going much smaller pays many evolutionary dividends. However, evolutionary studies—both in experiment and in nature—do demand lots of time. Even with relatively fast-breeding organisms, like fruit flies, one usually needs years to complete experiments. And, one-a-year breeders (or slower) require very-long-term commitments.

This all does lead to problems, particularly given the way research is structured today. Evolutionists want ready results, particularly if they are up for tenure! And, funding bodies want reassurance that their largess is not wasted. Hence, there is a natural tendency to go for quick answers, irrespective of their theoretical and practical significance and centrality. Consequently, important questions about the long-term effects of selection—particularly those operating in the wild—go unanswered. (See the warning by A. J. Cain 1979.)

We need to think about this matter carefully. At the very least, the National Science Foundation needs to break from its short-term funding policies, and to earmark some funds for long-term studies. Perhaps this is a vain hope, given the present fiscal climate. But, those of us who care about evolutionary studies must push for more enlightened research policies and attitudes. Evolution is one of the most glorious ideas of all time. Even though we may never achieve absolute knowledge, at least let us make sure that the limits are not self-imposed, for reasons of academic expediency.

NOTE

1. Douglas Futuyma gave a typically thoughtful reading of the first draft of this essay.

REFERENCES

Ayala, F. J., and J. W. Valentine. 1979. *Evolving: The Theory and Process of Organic Evolution.* Benjamin Cummings, Menlo Park, Calif.

Cain, A. J. 1978. "Reply to Gould and Lewontin," *Proceedings of the Royal Society 8,* 205: 599–604.

Clarkson, E. N. K., and R. Levi-Setti. 1975. "Trilobite Eyes and the Optics of Descartes and Huygens," *Nature* 254: 663–667.

Darwin, C. 1859. *On the Origin of Species.* Murray, London.

Eldredge, N. 1982. *The Monkey Business: A Scientist Looks at Creationism.* Washington Square Press, New York.

Eldredge, N., and J. Cracraft. 1980. *Phylogenetic Patterns and the Evolutionary Process.* Columbia University Press, New York.

Feduccia, A. 1980. *The Age of Birds.* Harvard University Press, Cambridge, Mass.

Fitch, W. M. 1976. "Molecular Evolutionary Clocks." Pp. 160–178 in F. J. Ayala, ed., *Molecular Evolution*. Sinauer, Sunderland, Mass.

Futuyma, D. 1982. *Science on Trial: The Case for Evolution*. Pantheon, New York.

Gish, D. T. 1973. *Evolution: The Fossils Say No!* Creation-Life Publishers, San Diego.

Goodman, M. 1982a. "Decoding the Pattern of Protein Evolution," *Progress in Biophysics and Molecular Biology 38:* 105.

———. 1982b. "Biomolecular Evidence on Human Origins from the Standpoint of Darwinian Theory," *Human Biology 54:* 247–264.

Gould, S. J. 1980. "Is a New and General Theory of Evolution Emerging?" *Paleobiology. 6:* 119–130.

Gribbin, J, and J. Cherfas. 1982. *The Monkey Puzzle*. Bodleyhead, London.

Hamilton, W. D. 1964. "The Genetical Evolution of Social Behavior," *Journal of Theoretical Biology 7:* 1–16, 17–32.

Johanson, D., and M. Edey. 1981. *Lucy: The Beginnings of Humankind*. Simon and Schuster, New York.

Johanson, D., and T. D. White, 1979. "A Systematic Assessment of Early African Hominids," *Science 203:* 321–330.

Jones, J. S. 1981. "Models of Speciation—The Evidence from *Drosophilia,*" *Nature 289:* 743 744.

Kettlewell, H. B. D. 1978. *The Evolution of Melanism*. Clarendon, Oxford.

Kimura, M., and T. Ohta, 1971. *Theoretical Aspects of Population Genetics*. Princeton University Press, Princeton, N.J.

King, J. L., and T. H. Jukes. 1969. "Non-Darwinian Evolution," *Science 164:* 788-798.

King, M. C., and A. C. Wilson. 1975. "Evolution at Two Levels: Molecular Similarities and Biological Differences between Humans and Chimpanzees." *Science 188:* 107–116.

Kitcher, P. 1982. *Abusing Science*. Massachusetts Institute of Technology Press, Cambridge, Mass.

Lewontin, R. 1974. *The Genetic Basis of Evolutionary Change*. Columbia University Press, New York.

Luria, S. E., S. J. Gould, and S. Singer. 1981. *A View of Life*. Benjamin Cummings, Menlo Park, Calif.

Maynard Smith, J. 1978. "The Evolution of Behavior," *Scientific American 239:* 176–193.

———. 1981. "Did Darwin Get it Right?" *London Review of Books 3* (11): 10–11.

Mayr, E. 1982. *The Growth of Biological Thought*. Harvard University Press, Cambridge, Mass.

Morris, H. M., ed. 1974. *Scientific Creationism*. Creation-Life, San Diego.

Newell, N. 1982. *Creation and Evolution*. Columbia University Press, New York.

Oster, G., and E. O. Wilson. 1978. *Caste and Ecology in the Social Insects*. Princeton University Press, Princeton, N.J.

Patterson, C. 1978. *Evolution*. British Museum (Natural History), London.

Popper, K. 1974. "Darwinism as a Metaphysical Reserarch Programme." Pp. 133–143 in P. A. Schilpp, ed., *The Philosophy of Karl Popper*. Open Court, LaSalle, Ill.

Rescher, N. 1978. *Scientific Progress*. University of Pittsburgh Press, Pittsburgh.

Rose, S. 1982. *Towards a Liberatory Biology*. Allison and Busby, London.

Ruse, M. 1973. *The Philosophy of Biology*. Hutchinson, London.

Ruse, M. 1979. *The Darwinian Revolution: Science Red in Tooth and Claw.* University of Chicago Press, Chicago.

———. 1982. *Darwinism Defended: A Guide to the Evolution Controversies.* Addison-Wesley, Reading, Mass.

Schaeffer, B., M. Hecht, and N. Eldredge. 1972. "Phylogeny and Paleontology." Pp. 31–46 in T. Bodzhansky, M. Hecht, and W. Steere, eds., *Evolutionary Biology.* Appleton-Century-Crofts.

Schopf, J. W. 1978. "The Evolution of the Earliest Cells," *Scientific American,* September: 110–138.

Stanley, S. M. 1979. *Macroevolution: Pattern and Process.* W. H. Freeman, San Francisco.

Templeton, A. R. 1983. "Phylogenetic Inference from Restriction Endonuclease Clearage Site Maps with Particular Reference to the Evolution of Humans and Apes," *Evolution: 37:* 221–244.

Whewell, W. 1840. *The Philosophy of the Inductive Sciences.* Parker, London.

Wiley, E. O., and D. R. Brooks, 1982. "Victims of History—a Non-equilibrium Approach to Evolution," *Systematic Zoology 31:* 1–24.

Wilson, E. O. 1975. *Sociobiology: The New Synthesis.* Harvard University Press, Cambridge, Mass.

9

The Mechanisms of Evolution

Francisco J. Ayala

In the 119 years since the publication of *On the Origin of Species* Darwin's basic principles have been progressively refined. According to Darwin, the basis of evolution is the occurrence of random heritable modifications in the individuals of a population. The advantageous modifications are then adopted and the disadvantageous ones are discarded through natural selection: the differential survival and reproduction of genetically variant individuals. In this way evolutionary adaptation involves a mixture of variation and selection, of chance and necessity.

Darwin thought of variation as a transient phenomenon. Because a population of organisms is closely adapted to its environment, he argued, the vast majority of modifications will be disadvantageous and the modified individuals will accordingly be eliminated by natural selection. In the rare event that a modification is advantageous it will render the individual more likely to survive and reproduce. As a result the advantageous modification will gradually spread to all the members of the population over the generations, ultimately replacing the type that was formerly dominant.

Darwin's theory implies that natural populations are made up of a more or less common genetic type with a few rare variants. In recent years this assumption has been contradicted by evidence that natural populations possess an enormous reservoir of genetic variation, suggesting that the role of chance in the evolutionary process is subtler than Darwin supposed. The advances in molecular biology, together with the statistical approach to evolution provided by population genetics, have enabled biologists to better understand

From *Scientific American* (September 1978): pp. 56-69. Reprinted by permission of W. H. Freeman and Company.

where genetic variation comes from, how it is maintained in populations and how it contributes to evolutionary change.

In Darwin's day the science of genetics had not yet been born. The discrete units of heredity called genes were first identified by Gregor Mendel in Darwin's lifetime but did not become widely known until the 20th century. Darwin's vague but prescient notion of random fluctuations in the hereditary material nonetheless turned out to be an approximation of Mendel's more precise concept of genetic variation, and so Mendelian genetics could be incorporated into the theory of natural selection without too much difficulty. The fusion of the two disciplines from the early 1920's through the late 1950's is often referred to as Neo-Darwinism or the modern synthesis.

The dramatic discoveries of molecular genetics over the past 20 years have led to yet another synthesis, encompassing an understanding of evolutionary processes at the molecular level. A gene is now known to be a segment of one of the extremely long DNA molecules in the cell that store the organism's genetic information in their structure. The sequence of four kinds of nucleotide base (adenine, cytosine, guanine and thymine) along each strand of the DNA double helix represents a linear code. The information contained in that code directs the synthesis of specific proteins; the development of an organism depends on the particular proteins it manufactures. Proteins are made up of long chains of amino acids, and the specific properties of each protein are determined by the sequence of amino acids in the chain. This sequence is in turn specified by the sequence of nucleotide bases in the DNA of the genes.

The genetic information stored in the DNA molecule is expressed in two steps. In the first process, called transcription, the sequence of nucleotide bases along one of the DNA strands is copied onto a complementary strand of RNA (which is made up of the same nucleotide bases as DNA except that thymine is replaced by the closely related uracil). In the second process, called translation, the genetic program of the organism is "read" from the RNA in codons, or successive groups of three nucleotide bases. The four RNA bases form 64 different codons that specify the 20 common amino acids in proteins. (The discrepancy between the 64 codons and the 20 amino acids is due to the redundancy of the genetic code and the fact that certain codons represent instructions such as "Start" and "Stop.")

In protein synthesis the amino acids specified by the sequence of codons along the gene are added one by one to the growing chain. Once the protein has been assembled it spontaneously assumes a specific three-dimensional form and begins to function as an enzyme, as a structural component or in some other biological role. The characteristics and behavior of organisms depend ultimately on the sequences of amino acids in their proteins, and evolution consists largely in the progressive substitution of one amino acid for another.

The new understanding of the chemical nature of the gene has provided a view of mutation at the molecular level. A mutation can be considered an error in the replication of DNA prior to its translation into protein. Such an error is often confined to the replacement of one nucleotide-base pair by another (a point mutation), and it may lead to the replacement of one amino acid by another in the protein specified for by that gene. Point mutations that result in the substitution of an amino acid are called missense mutations; those that convert the codon for an amino acid into a "stop" codon are called nonsense mutations. Other mutations may involve the insertion of a nucleotide into the DNA molecule or the deletion of a nucleotide from it; such mutations may have more pervasive effects by shifting the "frame" in which the nucleotide sequence is read, and they may lead to several missense or nonsense substitutions. If these DNA mutations occur in the germ cells of the organism, they will be passed on to the next generation.

In addition to changes in the structure of the genes by mutation, evolution involves changes in the amount and organization of the genes. A human being has in each cell many times more DNA than our single-cell ancestors of a billion years ago had. Evolutionary increments (or decrements) in the hereditary material occur largely by means of duplications (or deletions) of DNA segments; the duplicated segments can then evolve toward serving new functions while the prexisting segments retain the original function.

The forces that give rise to gene mutations operate at random in the sense that genetic mutations occur without reference to their future adaptiveness in the environment. In other words, a mutant individual is no more likely to appear in an environment in which it would be favored than in one in which it would be selected against. If a favored mutation does appear, it can be viewed as exhibiting a "preadaptation" to that particular environment: it did not arise as an adaptive response but rather proved to be adaptive after it appeared.

A population consisting of several million individuals is likely to have a few mutations per generation in virtually every gene carried by the population. Mutations that give rise to substantial changes in the physical characteristics of the organism, however, are unlikely to be advantageous. Since a population is usually well adapted to its environment, major changes are usually maladaptive, just as a large random change in the construction of a clock (the removal of a spring or the addition of a gear) is not likely to improve its functioning. Most evolutionary changes seem to occur by the gradual accumulation of minor mutations (analogous to the tightening of a screw) accompanied by slow transitions in the physical characteristics of individuals in the population.

The DNA molecules in the nucleus of higher cells are associated with protein and packed into the dense bodies called chromosomes. The number of chromosomes in the cell nucleus differs from species to species: eight in the fruit fly *Drosophila*, 20 in corn, 24 in the tomato, 40 in the house mouse, 46 in man, 48 in the potato. A substantial reorganization of the hereditary material can result from transpositions of chromosomal segments, each of which comprises hundreds or thousands of nucleotide bases. The total number of chromosomes can be increased by duplication or reduced by fusion. A segment of a chromosome can be deleted, an extra piece can be inserted or a segment can be removed, inverted and put back. A segment from one chromosome can be transferred to another, or noncorresponding pieces can be exchanged. All these chromosomal aberrations alter the organization of the genes and contribute new raw material for evolutionary change.

Of the 46 chromosomes in every human cell, 23 are copies of those originating in the sperm of the father and the other 23 are copies of those originating in the egg of the mother. The genes thus occur in pairs, one on a maternal chromosome and the other on the homologous, or corresponding, paternal chromosome. The two genes in a pair are said to occupy a certain locus, or position, on each of the homologous chromosomes. For example, there is a locus on one pair of homologous chromosomes that codes for eye color. Each chromosome may comprise many thousands of gene loci.

A gene at a given locus may have variant forms known as alleles. There may be several alleles at a locus in a large population, although there can be only two in any one individual. Each allele arises by mutation from a preexisting gene and may differ from it at one or several parts of its nucleotide-base sequence. When the two alleles at a certain locus on the homologous chromosomes in an individual are identical, the individual is said to be homozygous at that locus; when the two alleles are different, the individual is said to be heterozygous at that locus.

Hereditary variation, as reflected in the existence of multiple alleles in a population, is clearly a prerequisite for evolutionary change. If all the individuals in a population are homozygous for the same allele at a given locus, there can be no evolution at that locus until a new allele arises by mutation. If, on the other hand, two or more alleles are present in a population, the frequency of one allele can increase at the expense of the other or others as a consequence of natural selection. Of course, the selective value of an allele is not fixed. The environment is variable in space and time; under certain conditions one allele is favored and under different conditions another allele is favored. Hence a population that has considerable amounts of genetic variation may be hedged against future changes in the environment.

Laboratory experiments have demonstrated that the greater the amount of genetic variation in a population, the faster its rate of evolution. In one experiment two populations of the fruit fly *Drosophila* were bred so that one population had initially about twice as much genetic variation as the other. The populations were then allowed to evolve in the laboratory for 25 generations with intense competition for food and living space, conditions that tend to stimulate rapid evolutionary change. Although both types of population evolved, gradually becoming better adapted to the laboratory environment, the rate of evolution was substantially higher in the population that had the greater initial variation.

The question of how much variation exists in natural populations is therefore of central interest to biologists, since it determines to a large extent the evolutionary plasticity of a species. The task of estimating genetic variation, however, is a difficult one since much genetic variation is hidden in each generation and is not expressed as manifest traits. The reason is that at a given locus in a heterozygous individual one allele is usually dominant and the other is recessive, that is, only the dominant allele is expressed in the heterozygous state. If a human being has a dominant allele for brown eye color and a recessive one for blue eye color, his eyes will be brown and the fact that he carries a gene for blue eye color will be concealed.

Such hidden variation can be revealed by breeding experimental organisms with their close relatives. When this inbreeding is done, some of the recessive alleles that have been concealed in the heterozygous state will become homozygous and will then be expressed. For example, intensive inbreeding of fruit flies has revealed they possess several recessive alleles that when the locus is homozygous results in the expression of grossly abnormal traits such as extremely short wings, deformed bristles, blindness and other serious defects.

Another indication of the magnitude of genetic variation in natural populations has been provided by artificial-selection experiments. In such experiments those individuals of a population that exhibit the greatest expression of a particular commercially desirable trait are chosen to breed the next generation. If a plant breeder wants to increase the yield of a variety of wheat, he will select those plants with the greatest yield at each generation and utilize their seed to grow new progeny. If the selected population changes over the generations in the direction of the applied selection, then it is clear the original plants possessed a reservoir of genetic variation with respect to the selected trait.

Indeed, the changes obtained by artificial selection are often enormous. In one flock of White Leghorn chickens egg production increased from 125.6 eggs per hen per year in 1933 to 249.6 eggs per hen per year in 1965: an increase of nearly 100 percent in 32 years! Selection can also be successfully practiced in opposite directions. For example, selection for high protein content in a variety of corn increased the protein content from 10.9 to 19.4 percent, whereas selection for low protein content reduced the protein content from 10.9 to 4.9 percent. Artificial selection has been successful in creating large numbers of commercially desirable traits in domesticated species such as cattle, swine, sheep, poultry, corn, rice and wheat, as well as in experimental animals such as fruit flies, in which artificial selection of more than 50 different traits has been accomplished. The fact that artificial selection works almost every time it is attempted indicates there is genetic variation in populations for virtually every characteristic of the organism.

This kind of evidence suggested to biologists that natural populations do have large stores of genetic variation. Yet until quite recently the limitations of traditional genetic analysis prevented investigators from determining precisely how much variation there is. Consider what would be required to find out what proportion of the genes of an individual are heterozygous. It is almost impossible to study every gene locus because of the scale of the task, but if one could obtain an unbiased sample of all the genes of an organism, it would be possible to extrapolate the values observed in that sample to the population as a whole. Indeed, opinion pollsters are able to predict with fair accuracy how millions of people will vote in a U.S. Presidential election on the basis of a representative sample of about 2,000 people: .001 percent of the population. The fact remains that with Mendelian techniques it is impossible to obtain an unbiased sample of all the genes in an individual because classical genetic analysis (involving crossbreeding of individuals exhibiting different traits) detects only those loci that are variable (that have different alleles). Since there is no way to detect invariant loci, it was impossible to obtain a truly random sample of all the genes.

The way out of this dilemma was provided by the molecular biological revolution of the past two decades. Since many genes code for proteins, one can infer variation in the genetic material from variation in the proteins manufactured by individuals. If a certain protein is invariant among the individuals of a population, the gene coding for that protein is probably also invariant; if the protein is variable, then the gene too is variable. By selecting a number of proteins that represent an unbiased sample of the genes in an organism it is therefore possible to estimate the number of alleles in a population and the frequencies at which they occur.

Biochemists have known since the early 1950's how to determine the amino acid sequence of proteins, but several months or years are usually required to sequence one protein, let alone the thousands that would be needed to obtain a statistically valid sample. Fortunately there is a simple technique, gel electrophoresis, that makes it possible to study protein variation with only a moderate investment of time and resources. Since the late 1960's this technique has been exploited to estimate the genetic variation in several natural populations.

In gel electrophoresis ground tissue or blood from several individuals is inserted into a gel consisting of starch, the synthetic polymer acrylamide or some other substance providing a homogeneous matrix. When an electric current is passed through the gel, the proteins in the tissue migrate at a rate that is determined primarily by the electric charge on their constituent amino acids (although the size and conformation of the protein may also influence the migration). Electrophoresis is so sensitive that it can detect proteins that differ by a single amino acid out of a total of some hundreds—provided that the substitution of one amino acid for another results in a change in the total electric charge on the molecule.

The proteins manufactured by different individuals in a population are compared by running them side by side in a gel for a certain time interval. The positions of the proteins after they have migrated are determined by applying a stain specific for the protein under study, which is usually an enzyme. Because each amino acid chain in a protein (some proteins have more than one chain) is the product of a single gene this approach enables the investigator to estimate how many loci in the population have multiple alleles and how many are invariant. To obtain a rough survey of variation in natural populations about 20 loci are usually examined. One useful measure of variation is hetero-zygosity: the average proportion of loci at which an individual in the population possesses two alleles.

Electrophoretic techniques were first applied to estimating genetic variation in natural populations in 1966, when three studies were published, one dealing with man and the other two with *Drosophila.* Since then numerous populations have been surveyed and many more are studied every year. One recent survey concerned the krill *Euphausia superba,* a shrimplike organism that thrives in the waters near Antarctica and is a major food source of whales. A total of 36 gene loci coding for different enzymes were examined in 126 krill individuals. No variation was detected at 15 loci, but at each of the other 21 loci two, three or four allelic gene products were found in the population. In other words, approximately 58 percent of the loci in this krill population had two or more alleles. On the average each krill individual was heterozygous at 5.8 percent of its loci.

Large amounts of genetic variation have been found in most natural populations studied, including 125 animal species and eight plant species. Among animals, invertebrates generally show more genetic variation than vertebrates, although there are some exceptions. The average heterozygosity for invertebrates is 13.4 percent; the average for vertebrates is 6.6 percent. The heterozygosity for man is 6.7 percent, close to the vertebrate average. Plants have a great deal of genetic variation: the average heterozygosity for eight species is 17 percent.

These estimates become even more dramatic when it is taken into account that electrophoresis underestimates genetic variation. One reason is the redundancy of the genetic code: not all mutations or substitutions in the DNA result in changes in the amino acid sequence of proteins. Moreover, since electrophoresis distinguishes proteins that have different amino acid compositions by their differential migration in an electric field, if a mutation does not alter the electrical properties of the molecule, it will not be detected. For example, if a positively charged amino acid (say glutamic acid) is replaced in a variant protein by another positively charged amino acid (say aspartic acid), the two proteins may be indistinguishable by electrophoretic criteria. Although it is clear that the estimates of variation in natural populations obtained by electrophoresis are underestimates, the degree of underestimation is not yet known. Several laboratories are now attempting to solve this problem so that genetic variation can be more precisely estimated.

In any case the extent of the variation observed in natural populations is vastly greater than that predicted by classical Darwinian theory. Instead of being homozygous for a dominant allele at most loci, individuals are hetero-zygous at a large proportion of loci. This fact has important consequences, particularly for animals that reproduce sexually.

Sexual reproduction involves the fusion of two germ cells (the sperm and the egg in animals), each of which possesses only one set of chromosomes instead of the two homologous sets possessed by each tissue cell. The germ cells are formed by the process of meiosis, or reduction division, in which the normal complement of chromosomes is reduced by half. In the first stage of meiosis the chromosomes duplicate themselves and the homologous chromosomes then pair up. At this stage the paired chromosomes may break in several places and exchange pieces, the process called recombination. The resulting chromosomes are a mosaic of the homologous paternal and maternal chromosomes and hence have a new combination of alleles. In the second stage of meiosis each cell divides twice to yield four germ cells. During the second division the homologous chromosomes are randomly assorted, so that there is a mixture of maternal and paternal chromosomes in each germ cell.

The scrambling of the genes by recombination (which generates new combinations of alleles on the same chromosome) and random assortment (which results in new combinations of chromosomes in the germ cells) does not in itself alter gene frequencies or cause evolution. Indeed, as it was first independently postulated by the mathematician G. H. Hardy and the biologist W. Weinberg in 1908, recombination and random assortment cause no net change in the frequencies of alleles in a population. In the absence of selection gene frequencies will remain constant from generation to generation, a hypothetical situation that has been named the Hardy-Weinberg equilibrium. The effect of recombination and random assortment is merely to reshuffle the existing genes in a population so that new combinations of alleles are exposed to selection at each generation. Sexual reproduction therefore generates a large amount of genetic diversity, greatly increasing the possibilities for evolution and providing the population with an adaptability to a changing environment far beyond the reach of an asexual species. It may be for this reason that sexuality is virtually universal in the living world, except for organisms such as bacteria, which reproduce rapidly and exist in large numbers and so may incorporate mutations in short periods of time.

Clearly the greater the heterozygosity of individuals in a sexually reproducing population is, the larger will be the number of possible combinations of alleles in the germ cells and hence in the potential progeny. Consider man, with an average heterozygosity of 6.7 percent. If we assume that there are 100,000 gene loci in man, a human individual would be heterozygous for about 6,700 loci. Such an individual could potentially produce $2^{6,700}$ ($10^{2,017}$) different germ cells, a number vastly greater than the number of atoms in the known universe (roughly estimated as being 10^{80}). Of course, such a number of germ cells will never be produced by any human individual, not even all of mankind. It follows that no two human beings ever have been or ever will be genetically identical (with the exception of identical twins and other multiple births from the same zygote, or fertilized egg). Such is the genetic basis of human individuality. The same can be said of any other organism that reproduces sexually.

It therefore seems clear that, contrary to Darwin's conception, most of the genetic variation in populations arises not from new mutations at each generation but from the reshuffling of previously accumulated mutations by recombination. Although mutation is the ultimate source of all genetic variation, it is a relatively rare event, providing a mere trickle of new alleles into the much larger reservoir of stored genetic variation. Indeed, recombination alone is sufficient to enable a population to expose its hidden variation for many generations without the need for new genetic input by mutation.

One can conclude that large numbers of alleles are stored in populations even though they are not maximally adaptive for that time or place; instead they are maintained at low frequency in the heterozygous state until the environment changes and they suddenly become adaptive, at which point their frequency gradually increases under the influence of natural selection until they become the dominant genetic type. But how do natural populations maintain the large reservoirs of genetic variation needed to respond to a changing environment? When one allele is locally more adaptive than another, one would expect that natural selection would gradually eliminate the less advantageous alleles in favor of the more advantageous ones until every locus is homozygous. Hence the persistence of locally disadvantageous alleles in a population can be explained only by postulating mechanisms that actively maintain diversity in spite of the selective forces tending to eliminate it.

On such mechanism is heterozygote superiority. If the heterozygote Aa survives or reproduces better than either homozygote AA or aa, then neither allele can eliminate the other. The most striking example of the mechanism is sickle-cell anemia. This human disease, which is prevalent in tropical Africa and the Middle East, results from an allele that gives rise to a variant form of hemoglobin, the oxygen-transporting protein in red blood cells. Biochemical studies have shown that the trait is due ultimately to the substitution of one amino acid (valine) for another (glutamic acid) at one position along two of the four constituent chains (with a total of nearly 600 amino acids) in the hemoglobin molecule; the abnormal hemoglobin can be distinguished from the normal form by electrophoresis. The slight change in the structure of the variant hemoglobin has catastrophic effects: it causes the hemoglobin molecules inside the red blood cells to form long strands. As a result the cells collapse to the shape of a sickle, resulting in a severe form of anemia that is usually fatal before reproductive age.

Since the sickle-cell allele is obviously disadvantageous, why does it persist in the human population of tropical Africa at frequencies of as high as 30 percent? It turns out that individuals who are heterozygous for the sickle-cell trait are protected against the most lethal form of malaria, whereas normal homozygotes are not. Hence the heterozygote individual is clearly superior over either homozygote; he is protected from malaria and does not suffer from sickle-cell anemia. As a result the heterozygotes preferentially survive and reproduce and the sickle-cell allele is maintained at high frequency in the population.

Selection may also act directly to maintain multiple alleles in a population. If the range of a species encompasses several different environments, natural selection will diversify the gene pool in such a way that several alleles will be optimally adapted to the different subenvironments. Indeed, recent investigations have shown that variant enzymes (coded for by different alleles) may differ in their catalytic efficiency, in their sensitivity to temperature, acidity or alkalinity and in their response to other environmental factors, thereby rendering them subject to natural selection. For example, some variants of the enzyme alcohol dehydrogenase in populations of the fruit fly *Drosophila melanogaster* have been found to be more resistant to heat than other variants; the heat-resistant variants are commoner in the fruit-fly populations of warmer environments than they are in those of cooler environments. This finding provides strong evidence that multiple alleles may be maintained at some loci by "diversifying selection" in populations that live in heterogeneous environments. Individuals that are heterozygous at a number of loci are also usually stronger and reproductively more successful than individuals homozygous at a large number of loci; the phenomenon is known as hybrid vigor. Perhaps the manufacture of slightly variant proteins and enzymes by the heterozygote enables it to adapt to a broader range of environmental conditions or to exploit marginal habitats.

A fourth mechanism by which multiple alleles can be maintained in a population is frequency-dependent selection, in which the fitnesses of two alleles are not constant but change with their frequency. If one allele is less advantageous than the other when it is at a high frequency but gains the advantage when its frequency declines to a certain level, then the frequency of that allele will tend to stabilize at about that level.

It is also possible that some of the variation observed in proteins represents insignificant changes at the functional level that do not alter the survival or reproductive success of the organism; such mutations would then be selectively neutral. For example, although some variant enzymes (such as the variants of alcohol dehydrogenase) have been found to have different functional characteristics, others may not. If this is the case, the few variant genes that are subject to natural selection might be scattered along a chromosome, together with other variant genes that are selectively neutral. Although some of the alleles would be selected for, the majority would merely be carried along without being tested. The extent to which evolution, particularly at the molecular level, is not subject to selection is a matter of continuing debate among evolutionary biologists.

Another controversy that has been aroused by the finding of large amounts of variation in populations is the problem of genetic load. If large numbers of less fit alleles are maintained in populations by heterozygote superiority, there will be a very high probability that at each generation a zygote will be homozygous at one or more loci for a disadvantageous allele. As a result a large number of less fit zygotes might be expected, which could be a burden of mortality and infertility too great for the population to bear. Yet it must be remembered that each locus is not subject to selection separate from the others, so that thousands of selective processes would be summed as if they were individual events. The entire individual organism, not the chromosomal locus, is the unit of selection, and alleles at different loci interact in complex ways to yield the final product. Since alleles are more likely to be tested as members of groups than as isolated units, the cost of maintaining variation in a population is actually far lower than was originally believed.

In any case there can be no doubt that the staggering amount of genetic variation in natural populations provides ample opportunities for evolution to occur. Hence it is not surprising that whenever a new environmental challenge materializes—a change of climate, the introduction of a new predator or competitor, man-made pollution—populations are usually able to adapt to it.

A dramatic recent example of such adaptation is the evolution by insect species of resistance to pesticides. The story is always the same: when a new insecticide is introduced, a relatively small amount is enough to achieve satisfactory control of the insect pest. Over a period of time, however, the concentration of the insecticide must be increased until it becomes totally inefficient or economically impractical. Insect resistance to a pesticide was first reported in 1947 for the housefly (*Musca domestica*) with respect to DDT. Since then resistance to one or more pesticides has been reported in at least 225 species of insects and other arthropods. The genetic variants required for resistance to the most diverse kinds of pesticides were apparently present in every one of the populations exposed to these man-made compounds.

The process of evolution has two dimensions: phyletic evolution and speciation. Phyletic evolution is the gradual changes that occur with time in a single lineage of descent; as a rule these changes result in greater adaptation to the environment and often reflect environmental changes. Speciation occurs when a lineage of descent splits into two or more new lineages and is the process that accounts for the great diversity of the living world.

In sexually reproducing organisms a species is a group of interbreeding natural populations that are reproductively isolated from any other such groups. The inability to interbreed is important because it establishes each species as a discrete and independent evolutionary unit; favorable alleles can be exchanged between populations of a species but cannot be passed on to individuals of other species. Since species are unable to exchange genes, they must evolve independently of one another.

The reproductive isolation of species is maintained by means of biological barriers known as reproductive isolating mechanisms. These mechanisms are of two types: prezygotic mechanisms, which impede the mating between members of different populations and so prevent the formation of hybrid offspring, and postzygotic mechanisms, which reduce the viability or fertility of hybrid offspring. Both types of isolation mechanisms serve to forestall the exchange of genes between populations.

The prezygotic reproductive isolating mechanisms are of five major types: (1) ecological isolation, where populations occupy the same territory but live in different habitats and so do not meet; (2) temporal isolation, where mating in animals and flowering in plants occur in different seasons or at different times of day; (3) ethological isolation, where sexual attraction between males and females is weak or absent; (4) mechanical isolation, where copulation in animals or pollen transfer in plants is prevented because of the different size or shape of the genitalia or the different structure of flowers, and (5) gametic isolation, where the gametes, or male and female germ cells, fail to attract each other. The spermatozoa of male animals may also be inviable in the sexual tract of females or pollen inviable in the stigma of flowers.

The postzygotic isolating mechanisms are of three major types: (1) hybrid inviability, where the hybrid zygotes fail to develop or at least to reach sexual maturity; (2) hybrid sterility, where hybrids fail to produce functional gametes, and (3) hybrid breakdown, where the offspring of hybrids have reduced viability or fertility.

All these reproductive isolating mechanisms do not act simultaneously between two species, but two or more are usually operating. Temporal isolation tends to be commoner in plants and ethological isolation commoner in animals, but even among closely related species different sets of isolating mechanisms are often operating when different pairs of species are compared. The evolutionary function of reproductive isolating mechanisms is to prevent inbreeding, but how this end is achieved depends on the opportunism of natural selection acting in the context of the specific environmental circumstances and the available genetic variation.

Clearly the waste of reproductive effort is far greater for postzygotic isolating mechanisms than it is for prezygotic ones. If a hybrid zygote is produced that is inviable, two gametes have been wasted that could have been used in non-hybrid reproduction. Even worse, if the hybrid is viable but sterile, the waste includes not only the gametes but also the resources utilized by the hybrid breakdown, which involves the resources utilized by both the hybrids and their offspring. Although gametic isolation also wastes gametes, and some other prezygotic mechanisms waste energy in unsuccessful courtship or failed copulation, in general prezygotic mechanisms are much less wasteful than postzygotic ones. For this reason whenever two populations that have already been reproductively isolated by postzygotic mechanisms come in contact natural selection rapidly promotes the development of prezygotic isolating mechanisms.

Since species are reproductively isolated groups of populations, the question of how species arise is equivalent to that of how reproductive isolating mechanisms arise. Speciation commonly has two stages: a first stage in which reproductive isolation starts as an incidental by-product of the genetic divergence between two populations, and a second stage in which reproductive isolation is completed when it is directly promoted by natural selection.

The first stage of speciation requires that the exchange of genes between two populations of a species be interrupted, usually by means of a geographical separation (say the formation of a mountain range between them or the emigration of one of the populations to an island). The absence of gene exchange between the two populations makes it possible for them to diverge genetically, at least in part as a consequence of their adaptation to local conditions or ways of life. As the isolated populations become increasingly different genetically, postzygotic isolating mechanisms may appear between them because hybrid offspring would have disharmonious genetic constitutions and hence a reduced viability or fertility.

The first stage of speciation is usually a gradual process, and it is often difficult to decide whether or not two populations have entered it. Moreover, the first stage may be reversible: if two populations that have been geographically isolated for some time come to have overlapping ranges, it is possible for the two populations to fuse back into a single one if the loss of fitness in the hybrids is not too great. If, on the other hand, crossbreeding yields offspring with significantly reduced viability or fertility, the populations will undergo the second stage of speciation.

The second stage involves the development of prezygotic isolating mechanisms, a process that is directly promoted by natural selection. Consider the following simplified situation. Assume that at a certain locus there are two alleles: A, which favors matings within the population, and a, which favors crossbreeding with other populations. If postzygotic isolating mechanisms operate between two populations, A will be common in offspring of normal fitness and a will be common in hybrid offspring of low fitness. As a result the a allele will decrease in frequency from generation to generation. Natural selection therefore favors the development of prezygotic isolating mechanisms that avoid the formation of hybrid zygotes.

Speciation can take place without the second stage if gene exchange between two populations is prevented long enough for them to diverge genetically to a significant extent. For example, the ancestors of many plants and animals now indigenous to the Hawaiian Islands arrived there from the mainland several million years ago. There they evolved and became adapted to the local conditions. Although natural selection did not directly promote reproductive isolation between the species evolving in Hawaii and the species on the mainland, the reproductive isolation of many species has nonetheless become complete.

The two stages of speciation are apparent in a group of closely related species of *Drosophila* that live in the American Tropics. The group consists of 15 species, six of which are morphologically very similar and so are termed sibling species. One of the sibling species, *D. willistoni*, consists of two subspecies (races of a species that inhabit different geographical areas): *D. willistoni quechua*, which lives in continental South America west of the Andes, and *D. willistoni willistoni*, which lives east of the Andes and also in Central America, Mexico and the island of the Caribbean. These two subspecies do not meet in nature; they are separated by the Andes because the flies cannot survive at high altitudes. Tests have shown that there is incipient reproductive isolation between the subspecies, particularly in the form of hybrid sterility, although the result depends on the direction of the matings. When a female *willistoni* is crossed with a male *quechua,* the male and female offspring are fertile. If, however, a male *willistoni* is crossed with a female *quechua,* the female offspring will be fertile and the males will be sterile. If these two subspecies came in geographical contact and crossbred, natural selection would favor the development of prezygotic reproductive isolating mechanisms because of the subspecies' partial hybrid sterility. The two subspecies are therefore considered to be in the first stage of speciation.

Drosophila equinoxialis is another species that consists of two geographically separated subspecies: *D. equinoxialis equinoxialis,* which inhabits continental South America, and *D. equinoxialis caribbensis,* which lives in Central America and the Caribbean islands. Laboratory crosses between the two subspecies always yield fertile female offspring and sterile male offspring, independent of the direction of the cross. Thus there is somewhat greater reproductive isolation between the two subspecies of *D. equinoxialis* than there is between the two subspecies of *D. willistoni.* Natural selection in favor of prezygotic reproductive isolating mechanisms would accordingly be stronger for *D. equinoxialis* because all the hybrid males are sterile. There is no evidence, however, of prezygotic isolating mechanisms among the subspecies of either *D. willistoni* or *D. equinoxialis,* and therefore they are not yet considered different species.

The second stage of the speciation process can also be found within the *D. willistoni* group. *Drosophila paulistorum* is a species consisting of six semispecies, or incipient species. As in *D. equinoxialis,* crosses between males and females of these semispecies yield fertile females and sterile males. In places where two or three semispecies have come in geographical contact, however, the second stage of speciation has advanced to the point where ethological isolation—the most effective prezygotic isolating mechanism in *Drosophila* and many other animals—is nearly complete. Semispecies from the same locality will not crossbreed in the laboratory but semispecies from different localities will; the reason is that the genes involved in ethological isolation have not fully spread throughout the populations. The semispecies of *D. paulistorum* therefore provide a striking example of the action of natural selection in the second stage of speciation. When ethological isolation is complete, the six semispecies will have become fully distinct species.

The final result of the process of geographical speciation can be observed in the species of the *D. willistoni* group. *D. willistoni, D. equinoxialis, D. tropicalis* and other species of this group coexist over wide territories without ever interbreeding. Hybrids are never found in nature, are extremely difficult to obtain in the laboratory and are always completely sterile.

Speciation is only one step, albeit the most fundamental one, in the diversification of the living world. Once reproductive isolation has been completed each newly formed species will take an independent evolutionary course; inevitably the species will become increasingly different as time passes. Since evolution is a gradual process, organisms that share a recent common ancestor are likely to be more similar to one another than organisms that share a remoter ancestor. This simple assumption is the logical basis of efforts to reconstruct evolutionary history by comparative studies of living organisms, which traditionally have been based on comparative morphology, embryology, cell biology, ethology, biogeography and other biological disciplines.

The task of reconstructing evolutionary history is far from simple: rates of evolutionary change may vary at different times, in different groups of organisms or with respect to different morphological features. Moreover, resemblances due to common descent must be distinguished from those due to similar ways of life, to the occupation of similar habitats or to accidental convergence. Sometimes the study of the fossil remains of extinct organisms provides clues to the evolutionary history of a group of species, but the fossil record is always incomplete and often altogether lacking.

In recent years the comparative study of nucleic acids (DNA and RNA) and proteins has become a powerful tool for the reconstruction of evolutionary history. These informational molecules retain a considerable amount of evolutionary information in their sequence of nucleotides or amino acids. Since at the molecular level evolution proceeds by the substitution of one nucleotide or amino acid for another, the number of differences in the sequence of an equivalent nucleic acid or protein in two species provides some indication of the recency of their common ancestry. One widely studied protein is cytochrome *c,* a protein involved in cell respiration; another is hemoglobin.

Investigations of evolutionary history at the molecular level have notable advantages over comparative anatomy and other classical disciplines. One is that the information is more readily quantifiable: the number of amino acids or nucleotides that are different is readily established when the sequence of units in a protein or a nucleic acid is known for several organisms. The second advantage is that very different types of organisms can be compared. There is little that comparative anatomy can tell us about organisms as diverse as yeast, a pine tree and a fish, but there are proteins common to all three that can be compared readily.

For example, the amino acid sequence of cytochrome *c* has been determined for several organisms, from bacteria and yeast to insects and human beings. Since each amino acid substitution can involve one, two or three nucleotide substitutions in the corresponding DNA codon, one can calculate the maximum or minimum number of nucleotide changes that could have given rise to the observed amino acid substitutions. Taking the minimum number of possible nucleotide differences between the genes coding for cytochrome *c* as a basis of comparison for 20 different organisms, Walter M. Fitch and Emanuel Margoliash at Northwestern University were able to construct a phylogeny of these animals. The overall relations agree fairly well with those inferred from the fossil record and other traditional sources. The cytochrome *c* phylogeny disagrees with the traditional one in several instances, including the following: the chicken appears to be related more closely to the penguin than to ducks and pigeons; the turtle, a reptile, appears to be related more closely to birds than to the rattlesnake, and man and monkeys diverge from the mammals before the marsupial kangaroo separates from the placental mammals.

In spite of these erroneous relations, it is remarkable that the study of a single protein yields a fairly accurate representation of the evolutionary history of 20 diverse organisms. A more accurate molecular phylogeny of these species and others should be obtained when the sequences of additional proteins and nucleic acids have been determined. The study of informational molecules from an evolutionary standpoint is a young science that was founded only about a decade ago. It is a powerful approach that should make increasingly important contributions to our understanding of biological evolution.

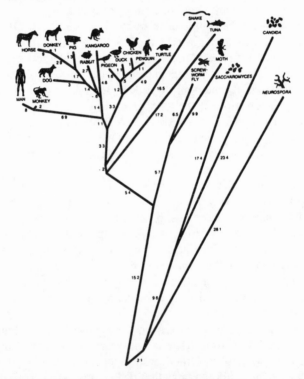

Figure 1. COMPUTER-GENERATED PHYLOGENY of 20 diverse organisms, based on differences in the amino acid sequence of cytochrome *c* from each species, was prepared by Walter M. Fitch and Emanuel Margoliash at Northwestern University. The phylogeny agrees fairly well with evolutionary relations inferred form the fossil record and other sources. The numbers of the branches are the minimum number of nucleotide substitutions in the DNA of the genes that could have given rise to observed differences in amino acid sequence.

REFERENCES

Ayala, F. J. (ed.) 1976. *Molecular Evolution.* Sinauer Associates, Inc.
Dobzhansky, T., F. J. Ayala, G. Ledyard Stebins, and J. W. Valentine. 1977. *Evolution.* W. H. Freeman and Company.
Fitch, W. H., and E. Margoliash. 1967. "Construction of Phylogenetic Trees," *Science* 155, no. 3760, (January 20): 279–84.
Goodman, M., G. W. Moore, and G. Matsuda. 1975. "Darwinian Evolution in the Genealogy of Haemoglobin," *Nature* 253, no. 5493, (February 20): 603–608.
Lewontin, R. C. 1974. *The Genetic Basis of Evolutionary Change.* Columbia University Press, N.Y.

10

Darwinism as a
Metaphysical Research Program

Sir Karl Popper

I have always been extremely interested in the theory of evolution, and very ready to accept evolution as a fact. I have also been fascinated by Darwin as well as by Darwinism—though somewhat unimpressed by most of the evolutionary philosophers; with the one great exception, that is, of Samuel Butler.[1]

My *Logik der Forschung* contained a theory of the growth of knowledge by trial and error-elimination, that is, by Darwinian *selection* rather than Lamarckian *instruction;* this point (at which I hinted in that book) increased, of course, my interest in the theory of evolution. Some of the things I shall have to say spring from an attempt to utilize my methodology and its resemblance to Darwinism to throw light on Darwin's theory of evolution.

The Poverty of Historicism[2] contains my first brief attempt to deal with some epistemological questions connected with the theory of evolution. I continued to work on such problems, and I was greatly encouraged when I later found that I had come to results very similar to some of Schrödinger's.[3]

In 1961 I gave the Herbert Spencer Memorial Lecture in Oxford, under the title "Evolution and the Tree of Knowledge."[4] In this lecture I went, I believe, a little beyond Schrödinger's ideas; and I have since developed further what I regard as a slight improvement on Darwinian theory,"[5] while keeping strictly within the bounds of Darwinism as opposed to Lamarckism—within natural selection, as opposed to instruction.

I tried also in my Compton lecture (1966)[6] to clarify several connected

From *Unended Quest* (LaSalle, Ill.: Open Court, 1976), pp. 167–79, 234–35. Reprinted by permission of the author.

questions; for example, the question of the *scientific status* of Darwinism. It seems to me that Darwinism stands in just the same relation to Lamarckism as does:

Deductivism	*to*	Inductivism,
Selection	*to*	Instruction by Reception,
Critical Error Elimination	*to*	Justification.

The logical untenability of the ideas on the right-hand side of this table establishes a kind of logical explanation of Darwinism (i.e. of the left-hand side). Thus it could be described as "almost tautological"; or it could be described as applied logic—at any rate, as applied *situational logic* (as we shall see).

From this point of view the question of the scientific status of Darwinian theory—in the widest sense, the theory of trial and error-elimination—becomes an interesting one. I have come to the conclusion that Darwinism is not a testable scientific theory, but a *metaphysical research program*—a possible framework for testable scientific theories.[7]

Yet there is more to it: I also regard Darwinism as an application of what I call "situational logic." Darwinism as situational logic can be understood as follows.

Let there be a world, a framework of limited constancy, in which there are entities of limited variability. Then some of the entities produced by variation (those which "fit" into the conditions of the framework) may "survive," while others (those which clash with the conditions) may be eliminated.

Add to this the assumption of the existence of a special framework—a set of perhaps rare and highly individual conditions—in which there can be life or, more especially, self-reproducing but nevertheless variable bodies. Then a situation is given in which the idea of trial and error-elimination, or of Darwinism, becomes not merely applicable, but almost logically necessary. This does not mean that either the framework or the origin of life is necessary. There may be a framework in which life would be possible, but in which the trial which leads to life has not occurred, or in which all those trials which led to life were eliminated. (The latter is not a mere possibility but may happen at any moment: there is more than one way in which all life on earth might be destroyed.) What is meant is that if a life-permitting situation occurs, and if life originates, than this total situation makes the Darwinian idea one of situational logic.

To avoid any misunderstanding: it is not in every possible situation that Darwinian theory would be successful; rather, it is a very special, perhaps even a unique situation. But even in a situation without life Darwinian selection can apply to some extent: atomic nuclei which are relatively stable (in the situation in question) will tend to be more abundant than unstable ones; and the same may hold for chemical compounds.

I do not think that Darwinism can explain the origin of life. I think it quite possible that life is so extremely improbable that nothing can "explain" why it originated; for statistical explanation must operate, *in the last instance,* with very high probabilities. But if our high probabilities are merely low probabilities which have become high because of the immensity of the available time (as in Boltzmann's "explanation"), then we must not forget that in this way it is possible to "explain" almost everything.[8] Even so, we have little enough reason to conjecture that any explanation of this sort is applicable to the origin of life. But this does not affect the view of Darwinism as situational logic, once life and its framework are assumed to constitute our "situation."

I think that there is more to say for Darwinism than that it is just one metaphysical research program among others. Indeed, its close resemblance to situational logic may account for its great success, in spite of the almost tautological character inherent in the Darwinian formulation of it, and for the fact that so far no serious competitor has come forward.

Should the view of Darwinian theory as situational logic be acceptable, then we could explain the strange similarity between my theory of the growth of knowledge and Darwinism: both would be cases of situational logic. The new and special element in the *conscious scientific approach to knowledge*— conscious criticism of tentative conjectures, and a conscious building up of selection pressure on these conjectures (by criticizing them)—would be a consequence of the emergence of a descriptive and argumentative language; that is, of a descriptive language whose descriptions can be criticized.

The emergence of such a language would face us here again with a highly improbable and possibly unique situation, perhaps as improbable as life itself. But given this situation, the theory of the growth of exosomatic knowledge through a conscious procedure of conjecture and refutation follows "almost" logically: it becomes part of the situation as well as part of Darwinism.

As for Darwinian theory itself, I must now explain that I am using the term "Darwinism" for the modern forms of this theory, called by various names, such as "neo-Darwinism" or (by Julian Huxley) "The New Synthesis." It consists essentially of the following assumptions or conjectures, to which I will refer later.

(1) The great variety of the forms of life on earth originate from very few forms, perhaps even from a single organism: there is an evolutionary tree, an evolutionary history.

(2) There is an evolutionary theory which explains this. It consists in the main of the following hypotheses.

(a) Heredity: the offspring reproduce the parent organisms fairly faithfully.

(b) Variation: there are (perhaps among others) "small" variations. The most important of these are the "accidental" and hereditary mutations.

(c) Natural selection: there are various mechanisms by which not only the variations but the whole hereditary material is controlled by elimination. Among them are mechanisms which allow only "small" mutations to spread; "big" mutations ("hopeful monsters") are as a rule lethal, and thus eliminated.

(d) Variability: although *variations* in some sense—the presence of different competitors—are for obvious reasons prior to selection, it may well be the case that *variability*—the scope of variation—is controlled by natural selection; for example, with respect to the frequency as well as the size of variations. A gene theory of heredity and variation may even admit special genes controlling the variability of other genes. Thus we may arrive at a hierarchy, or perhaps at even more complicated interaction structures. (We must not be afraid of complications; for they are known to be there. For example, from a selectionist point of view we are bound to assume that something like the genetic code method of controlling heredity is itself an early product of selection, and that it is a highly sophisticated product.)

Assumptions (1) and (2) are, I think, essential to Darwinism (together with some assumptions about a changing environment endowed with some regularities). The following point (3) is a reflection of mine on point (2).

(3) It will be seen that there is a close analogy between the "conservative" principles (a) and (d) and what I have called dogmatic thinking; and likewise between (b) and (c), and what I have called critical thinking.

I now wish to give some reasons why I regard Darwinism as metaphysical, and as a research program.

It is metaphysical because it is not testable. One might think that it is. It seems to assert that, if ever on some planet we find life which satisfies conditions (a) and (b), then (c) will come into play and bring about in time a rich variety of distinct forms. Darwinism, however, does not assert as much as this. For assume that we find life on Mars consisting of exactly three species of bacteria with a genetic outfit similar to that of three terrestrial species. Is Darwinism refuted? By no means. We shall say that these three species were the only forms among the many mutants which were sufficiently well adjusted to survive. And we shall say the same if there is only one species (or none). Thus Darwinism does not really *predict* the evolution of variety. It therefore cannot really *explain* it. At best, it can predict the evolution of variety under "favorable conditions." But it is hardly possible to describe in general terms what favorable conditions are—except that, in their presence, a variety of forms will emerge.

And yet I believe I have taken the theory almost at its best—almost in its most testable form. One might say that it "almost predicts" a great variety of forms of life.[9] At first sight natural selection appears to explain it, and in a way it does; but hardly in a scientific way. To say that a species now living is adapted to its environment is, in fact, almost tautological. Indeed we use the terms "adaptation" and "selection" in such a way that we can say that, if the species were not adapted, it would have been eliminated by natural selection. Similarly, if a species has been eliminated it must have been ill adapted to the conditions. Adaptation or fitness is *defined* by modern evolutionists as survival value, and can be measured by actual success in survival: there is hardly any possibility of testing a theory as feeble as this.[10]

And yet, the theory is invaluable. I do not see how, without it, our knowledge could have grown as it has done since Darwin. In trying to explain experiments with bacteria which become adapted to, say, penicillin, it is quite clear that we are greatly helped by the theory of natural selection. Although it is metaphysical, it sheds much light upon very concrete and very practical researches. It allows us to study adaptation to a new environment (such as a penicillin-infested environment) in a rational way: it suggests the existence of a mechanism of adaptation, and it allows us even to study in detail the mechanism at work. And it is the only theory so far which does all that.

This is, of course, the reason why Darwinism has been almost universally accepted. Its theory of adaptation was the first nontheistic one that was convincing; and theism was worse than an open admission of failure, for it created the impression that an ultimate explanation had been reached.

Now to the degree that Darwinism creates the same impression, it is not so very much better than the theistic view of adaptation; it is therefore important to show that Darwinism is not a scientific theory, but metaphysical. But its value for science as a metaphysical research program is very great, especially if it is admitted that it may be criticized, and improved upon.

Let us now look a little more deeply into the research program of Darwinism, as formulated above under points (1) and (2).

First, though (2), that is, Darwin's theory of evolution, does not have sufficient explanatory power to *explain* the terrestrial evolution of a great variety of forms of life, it certainly *suggests* it, and thereby draws attention to it. And it certainly does *predict* that *if* such an evolution takes place, it will be *gradual*.

The nontrivial *prediction of gradualness* is important, and it follows immediately from (2)(a)-(2)(c); and (a) and (b) and at least the smallness of the mutations predicted by (c) are not only experimentally well supported, but known to us in great detail.

Gradualness is thus, from a logical point of view, the central prediction of the theory. (It seems to me that it is its only prediction.) Moreover, as long as changes in the genetic base of the living forms are gradual, they are— at least "in principle"—explained by the theory; for the theory does predict the occurrence of small changes, each due to mutation. However, "explanation in principle"[11] is something very different from the type of explanation which we demand in physics. While we can explain a particular eclipse by predicting it, we cannot predict or explain any particular evolutionary change (except perhaps certain changes in the gene population *within* one species); all we can say is that if it is not a small change, there must have been some intermediate steps—an important suggestion for research: a research program.

Moreover, the theory predicts *accidental* mutations, and thus *accidental* changes. If any "direction" is indicated by the theory, it is that throwback mutations will be comparatively frequent. Thus we should expect evolutionary sequences of the random-walk type. (A random walk is, for example, the track described by a man who at every step consults a roulette wheel to determine the direction of his next step.)

Here an important question arises. How is it that random walks do not seem to be prominent in the evolutionary tree? The question would be answered if Darwinism could explain "orthogenetic trends," as they are sometimes called; that is, sequences of evolutionary changes in the same "direction" (nonrandom walks). Various thinkers such as Schrödinger and Waddington, and especially Sir Alister Hardy, have tried to give a Darwinian explanation of orthogenetic trends, and I also have tried to do so, for example, in my Spencer lecture.

My suggestions for an enrichment of Darwinism which might explain orthogenesis are briefly as follows.

(A) I distinguish external or environmental selection pressure from internal selection pressure. Internal selection pressure comes from the organism itself and, I conjecture, ultimately from its *preferences* (or "aims") though these may of course change in response to external changes.

(B) I assume that there are different classes of genes: those which mainly control the *anatomy*, which I will call *a*-genes; those which mainly control *behavior*, which I will call *b*-genes. Intermediate genes (including those with mixed functions) I will here leave out of account (though it seems that they exist). The *b*-genes in their turn may be similarly subdivided into *p*-genes (controlling *preferences* or "aims") and *s*-genes (controlling *skills*).

I further assume that some organisms, under external selection pressure, have developed genes, and especially *b*-genes, which allow the organism a certain variability. The *scope* of behavioral variation will somehow be controlled by the genetic *b*-structure. But since external circumstances vary, a not too rigid determination of the behavior by the *b*-structure may turn out to be as successful as a not too rigid genetic determination of heredity, that is to say of the scope of gene variability. (See (2)(d) above.) Thus we may speak of "purely behavioral" changes of behavior, or variations of behavior, meaning nonhereditary changes within the genetically determined scope or repertoire; and we may contrast them with genetically fixed or determined behavioral changes.

We can now say that certain environmental changes may lead to new problems and so to the adoption of new preferences or aims (for example, because certain types of food have disappeared). The new preferences or aims may at first appear in the form of new tentative behavior (permitted but not fixed by the *b*-genes). In this way the animal may tentatively adjust itself to the new situation without genetic change. But this *purely behavioral* and tentative change, if successful, will amount to the adoption, or discovery, of a new ecological niche. Thus it will favor individuals whose *genetic p*-structure (that is, their instinctive preferences or "aims") more or less anticipates or fixes the new behavioral pattern of preferences. This step will prove decisive; for now those changes in the skill structure (*s*-structure) will be favored which conform to the new preferences: skills for getting the preferred food, for example.

I now suggest that *only after the s-structure has been changed will certain changes in the a-structure be favored; that is, those changes in the anatomical structure which favor the new skills*. The internal selection pressure in these cases will be "directed," and so lead to a kind of orthogenesis.

My suggestion for this internal selection mechanism can be put schematically as follows:

$$p \rightarrow s \rightarrow a.$$

That is, the preference structure and its variations control the selection of the skill structure and its variations; and this in turn controls the selection of the purely anatomical structure and its variations.

This sequence, however, may be cyclical: the new anatomy may in its turn favor changes of preference, and so on.

What Darwin called "sexual selection" would, from the point of view expounded here, be a special case of the internal selection pressure which I have described; that is, of a cycle starting with new *preferences*. It is characteristic that internal selection pressure may lead to comparatively bad adjustment to the environment. Since Darwin this has often been noted, and the hope of explaining certain striking maladjustments (maladjustments from a survival point of view, such as the display of the peacock's tail) was one of the main motives for Darwin's introduction of his theory of "sexual selection." The original preference may have been well adjusted, but the internal selection pressure and the feedback from the changed anatomy to changed preferences (*a* to *p*) may lead to exaggerated forms, both behavioral forms (rites) and anatomical ones.

As an example of nonsexual selection I may mention the woodpecker. A reasonable assumption seems to be that this specialization started with a *change in taste* (preferences) for new foods which led to genetic behavioral changes, and then to new skills, in accordance with the schema

$$p \rightarrow s;$$

and that the anatomical changes came last.[12] A bird undergoing anatomical changes in its beak and tongue without undergoing changes in its taste and skill can be expected to be eliminated quickly by natural selection, *but not the other way round*. (Similarly, and not less obviously: a bird with a new skill but without the new preferences which the new skill can serve would have no advantages.)

Of course there will be a lot of feedback at every stage: $p \rightarrow s$ will lead to feedback (that is, s will favor further changes, including genetic changes, in the same direction as p), just as a will act back on both s and p, as indicated. It is, one may conjecture, this feedback which is mainly responsible for the more exaggerated forms and rituals.[13]

To explain the matter with another example, assume that in a certain situation external selection pressure favors bigness. Then the same pressure will also favor sexual *preference* for bigness: preferences can be, as in the case of food, the result of external pressure. But once there are new p-genes a whole new cycle will be set up: it is the p-mutations which trigger off the orthogenesis.

This leads to a general principle of mutual reinforcement: we have on the one hand a primary *hierarchical control* in the preference or aim structure, over the skill structure, and further over the anatomical structure; but we also have a kind of secondary interaction or feedback between those structures. I suggest that this hierarchical system of mutual reinforcement works in such a way that in most cases the control in the preference or aim structure largely dominates the lower controls throughout the entire hierarchy.[14]

Examples may illustrate both these ideas. If we distinguish genetic changes (mutations) in what I call the "preference structure" or the "aim structure" from genetic changes in the "skill structure" and genetic changes in the "anatomical structure," then as regards the interplay between the aim structure and the anatomical structure there will be the following possibilities:

(a) Action of mutations of the aim structure on the anatomical structure: when a change takes place in taste, as in the case of the woodpecker, then the anatomical structure relevant for food acquisition may remain unchanged, in which case the species is most likely to be eliminated by natural selection (unless extraordinary skills are used); or the species may adjust itself by developing a new anatomical specialization, similar to an organ like the eye: a stronger interest in seeing (aim structure) in a species may lead to the selection of a favorable mutation for an improvement of the anatomy of the eye.

(b) Action of mutations of the anatomical structure on the aim structure: when the anatomy relevant for food acquisition changes, then the aim structure concerning food is in danger of becoming fixed or ossified by natural selection, which in its turn may lead to further anatomical specialization. It is similar in the case of the eye: a favorable mutation for an improvement of the anatomy will increase keenness of interest in seeing (this is similar to the opposite effect).

The theory sketched suggests something like a solution to the problem of how evolution leads towards what may be called "higher" forms of life. Darwinism as usually presented fails to give such an explanation. It can at best explain something like an improvement in the degree of adaptation. But bacteria must be adapted at least as well as men. At any rate, they have existed longer, and there is reason to fear that they will survive men. But what may perhaps be identified with the higher forms of life is a behaviorally richer preference structure—one of greater scope; and if the preference structure should have (by and large) the leading role I ascribe to it, then evolution towards higher forms may become understandable.[15] My theory may also be presented like this: higher forms arise through the primary hierarchy of $p \rightarrow s \rightarrow a$, that is, whenever and as long as the preference structure is in the lead. Stagnation and reversion, including overspecialization, are the result of an inversion due to feedback within this primary hierarchy.

The theory also suggests a possible solution (perhaps one among many) to the problem of the separation of species. The problem is this: mutations on their own may be expected to lead only to a change in the gene pool of the species, not to a new species. Thus local separation has to be called in to explain the emergence of new species. Usually one thinks of geographic separation.[16] But I suggest that geographic separation is merely a special case of separation due to the adoption of new behavior and consequently of a new ecological niche; if a *preference* for an ecological niche—a certain *type* of location—becomes hereditary, then this could lead to sufficient local separation for interbreeding to discontinue, even though it was still physiologically possible. Thus two species might separate while living in the same geographical region—even if this region is only of the size of a mangrove tree, as seems to be the case with certain African molluscs. Sexual selection may have similar consequences.

The description of the possible genetic mechanisms behind orthogenetic trends, as outlined above, is a typical situational analysis. That is to say, only if the developed structures are of the sort that can simulate the methods of situational logic will they have any survival value.

Another suggestion concerning evolutionary theory which may be worth mentioning is connected with the idea of "survival value," and also with teleology. I think that these ideas may be made a lot clearer in terms of problem solving.

Every organism and every species is faced constantly by the threat of extinction; but this threat takes the form of concrete problems which it has to solve. Many of these concrete problems are not as such survival problems. The problem of finding a good nesting place may be a concrete problem for a pair of birds without being a survival problem for these birds, although it may turn into one for their offspring; and the species may be very little affected by the success of these particular birds in solving the problem here and now. Thus I conjecture that most problems are posed not so much by survival, but by *preferences*, especially *instinctive preferences*, and even if the instincts in question (*p*-genes) should have evolved under external selection pressure, the problems posed by them are not as a rule survival problems.

It is for reasons such as these that I think it is better to look upon organisms as problem-solving rather than as end-pursuing: as I have tried to show in "Of Clouds and Clocks,"[17] we may in this way give a rational account— "in principle," of course—of *emergent evolution*.

I conjecture that the origin of *life* and the origin of *problems* coincide. This is not irrelevant to the question whether we can expect biology to turn out to be reducible to chemistry and further to physics. I think it not only possible but likely that we shall one day be able to recreate living things from nonliving ones. Although this would, of course, be extremely exciting in itself[18] (as well as from the reductionist point of view), it would not *establish* that biology can be "reduced" to physics or chemistry. For it would not establish a physical explanation of the emergence of problems—any more than our ability to produce chemical compounds by physical means establishes a physical theory of the chemical bond or even the existence of such a theory.

My position may thus be described as one that upholds a theory of *irreducibility and emergence*, and it can perhaps best be summarized in this way:

(1) I conjecture that there is no biological process which cannot be regarded as correlated in detail with a physical process or cannot be progessively analyzed in physicochemical terms. But no physicochemical theory can explain the emergence of a new problem, and no physicochemical process can as such solve a *problem*. (Variational principles in physics, like the principle of least action or Fermat's principle, are perhaps similar but they are not solutions to problems. Einstein's theistic method tries to use God for similar purposes.)

(2) If this conjecture is tenable it leads to a number of distinctions. We must distinguish from each other:

a physical problem = a physicist's problem;

a biological problem = a biologist's problem;

an organism's problem = a problem like: How am I to survive? How
    am I to propagate? How am I to change? How am I to adapt?

a man-made problem = a problem like: How do we control waste?

From these distinctions we are led to the following thesis: *the problems of organisms are not physical: they are neither physical things, nor physical laws, nor physical facts. They are specific biological realities; they are "real" in the sense that their existence may be the cause of biological effects.*

(3) Assume that certain physical bodies have "solved" their problem of reproduction: that they can reproduce themselves; either exactly, or, like crystals, with minor faults which may be chemically (or even functionally) *inessential*. Still, they might not be "living" (in the full sense) if they cannot adjust themselves: they need reproduction *plus* genuine variability to achieve this.

(4) The "essence" of the matter is, I propose, *problem solving*. (But we should not talk about "essence"; and the term is not used here seriously.) Life as we know it consists of physical "bodies" (more precisely, structures) which are problem solving. This the various species have "learned" by natural selection, that is to say by the method of reproduction plus variation, which itself has been learned by the same method. This regress is not necessarily infinite—indeed, it may go back to some fairly definite moment of emergence.

Thus men like Butler and Bergson, though I suppose utterly wrong in their theories, were right in their intuition. Vital force ("cunning") does, of course, exist—but it is in its turn a product of life, *of selection,* rather than anything like the "essence" of life. It is indeed the preferences *which lead the way.* Yet the way is not Lamarckian but Darwinian.

This emphasis on *preferences* (which, being dispositions, are not so very far removed from propensities) in my theory is, clearly, a purely "objective" affair: we *need not* assume that these preferences are conscious. But they *may* become conscious; at first, I conjecture, in the form of states of well-being and of suffering (pleasure and pain).

My approach, therefore, leads almost necessarily to a research program that asks for an explanation, in objective biological terms, of the emergence of states of consciousness.

NOTES

1. Samuel Butler has suffered many wrongs from the evolutionists, including a serious wrong from Charles Darwin himself who, though greatly upset by it, never put things right. They were put right, as far as possible, by Charles's son Francis, after Butler's death. The story, which is a bit involved, deserves to be retold. See pp. 167–219 of Nora Barlow, ed., *The Autobiography of Charles Darwin* (London: Collins, 1958), esp. p. 219, where references to most of the other relevant material will be found.

2. See [1945(a)], section 27; cp. [1957(g)] and later editions, esp. pp. 106–8.

3. "I am alluding to Schrödinger's remarks on evolutionary theory in *Mind and Matter*, especially those indicated by his phrase "Feigned Lamarckism"; see *Mind and Matter*, p. 26; and p. 118 of the combined reprint cited in n. 214 above.

4. The lecture [1961(j)] was delivered on October 31, 1961, and the manuscript was deposited on the same day in the Bodleian Library. It now appears in a revised version, with an addendum, as Chap. 7 of my [1972(a)].

5. See [1966(f)]; now Chap. 6 of [1972(a)].

6. See [1966(f)].

7. The term "metaphysical research programme" was used in my lectures from about 1949 on, if not earlier; but it did not get into print until 1958, though it is the main topic of the last chapter of the *Postscript* (in galley proofs since 1957). I made the *Postscript* available to my colleagues, and Professor Lakatos acknowledges that what he calls "scientific research programmes" are in the tradition of what I described as "metaphysical research programmes" ("metaphysical" because nonfalsifiable). See p. 183 of his paper "Falsification and the Methodology of Scientific Research Programmes," in *Criticism and the Growth of Knowledge*, edited by Imré Lakatos and Alan Musgrave (Cambridge: Cambridge University Press, 1970).

8. See *L.Sc.D.*, section 67.

9. For the problem of "degrees of prediction" see F. A. Hayek, "Degrees of Explanation," first published in 1955 and now Chap. 1 of his *Studies in Philosophy, Politics and Economics* (London: Routledge & Kegan Paul, 1967); see esp. n. 4 on p. 9. For Darwinism and the production of "a great variety of structures," and for its irrefutability, see esp. p. 32.

10. Darwin's theory of sexual selection is partly an attempt to explain falsifying instances of this theory; such things, for example, as the peacock's tail, or the stag's antlers. See the text before n. 12.

11. For the problem of "explanation in principle" (or "of the principle") in contrast to "explanation in detail," see Hayek, *Philosophy, Politics and Economics*, Chap. 1, esp. section VI, pp. 11–14.

12. David Lack makes this point in his fascinating book, *Darwin's Finches* (Cambridge: Cambridge University Press, 1947), p. 72: ". . . in Darwin's finches all the main beak differences between the species may be regarded as adaptations to difference in diet." (Footnote references to the behavior of birds I owe to Arne Petersen.)

13. As Lack so vividly describes it, *ibid.*, pp. 58 f., the *absence* of a long tongue in the beak of a woodpeckerlike species of Darwin's finches does not prevent this bird from excavating in trunks and branches for insects—that is, it sticks to its taste; however, due to its particular anatomical disability it has developed a skill to meet this difficulty: "Having excavated, it picks up a cactus spine or twig, one or two inches long, and holding it lengthwise in its beak, pokes it up the crack, dropping the twig to seize the insect as it emerges." This striking behavioral trend may be a nongenetical "tradition" which has developed in that species with or without teaching among its members; it may also be a genetically entrenched behavior pattern. That is to say, a genuine behavioral invention can take the place of an anatomic change. However this may be, this example shows how the behavior of organisms can be a "spearhead" of evolution: a type of biological problem solving which may lead to the emergence of new forms and species.

14. See now my 1971 Addendum, "A Hopeful Behavioural Monster," to my Spencer Lecture,

Chap. 7 of [1972(a)], and Alister Hardy, *The Living Stream: A Restatement of Evolution Theory and Its Relation to the Spirit of Man* (London: Collins, 1965), Lecture VI.

15. This is one of the main ideas of my Spencer Lecture, now Chap. 7 of [1972(a)].

16. The theory of geographic separation or geographic speciation was first developed by Moritz Wagner in *Die Darwin'sche Theorie und das Migrationsgesetz der Organismen* (Leipzig: Duncker and Humblot, 1868); English translation by J. L. Laird, *The Darwinian Theory and the Law of Migration of Organisms* (London: Edward Stanford, 1873). See also Theodosius Dobzhansky, *Genetics and the Origin of Species*, 3d rev. ed. (New York: Columbia University Press, 1951), pp. 179-211.

17. See [1966(f)], pp. 20-26, esp. pp. 24 f., point (11). Now [1972(a)], p. 244.

18. See [1970(1)], esp. pp. 5-10; [1972(a)], pp. 289-95.

REFERENCES

This reference list follows in its numbering (such as "[1945 (a)]") the "Bibliography of the Writings of Karl Popper," compiled by Troels Eggers Hansen for *The Philosophy of Karl Popper,* vols. 14/I and 14/II of *The Library of Living Philosophers,* ed. by Paul A. Schilpp (La Salle, Ill.: Open Court Publishing Co., 1974), pp. 1199-1287.

1945 (a) "The Poverty of Historicism, III," *Economica,* 12, pp. 69-89.

1957 (g) *The Poverty of Historicism,* Routledge & Kegan Paul, London, and the Beacon Press, Boston, Mass.

1961 (j) *Evolution and the Tree of Knowledge,* Herbert Spencer Lecture, delivered on October 30th, 1961, in Oxford. (Now Chapter 7 of [1972 (a)].)

1966 (f) *Of Clouds and Clocks: An Approach to the Problem of Rationality and the Freedom of Man,* Washington University Press, St. Louis, Missouri. (Now in [1972 (a)].)

1970 (l) "A Realist View of Logic, Physics, and History," *Physics, Logic and History,* edited by Wolfgang Yourgrau and Allen D. Breck, Plenum Press, New YOrk and London, pp. 1-30, and 35-37.

1972 (a) *Objective Knowledge: An Evolutionary Approach,* Clarendon Press, Oxford.

11

Karl Popper's Philosophy of Biology

Michael Ruse

Although Sir Karl Popper has not yet given us a full length philosophical treatment of evolutionary biology, enough of his general position has been sketched to make possible a preliminary evaluation. There are at least two reasons why such an evaluation seems worthwhile. First, a number of biologists are taking seriously Popper's views on science generally and biology in particular. (See, for example, [1], [17].) Secondly, Popper has labeled his overall epistemology "evolutionary," and has drawn a very strong analogy between what he takes to be biological evolution and the evolution of scientific knowledge (if indeed they are not for him part and parcel of the same thing). "The theory of knowledge which I wish to propose is a largely Darwinian theory of the growth of knowledge" ([19], p. 261). Hence, such an evaluation of Popper's work, one which concentrates primarily on his views on biology rather than on his wider position about the growth of knowledge, is the aim of this paper. In the concluding section, however, I shall make some brief remarks about the implications of the biological discussion for Popper's more general position.

I

Popper's first detailed and significant comments about evolutionary biology are to be found in the *Poverty of Historicism* ([18]). A primary aim of that work is to deny that there are any human laws of progress or destiny—"there

From *Philosophy of Science* 44, no. 4 (1977):638–61. Reprinted by permission of the Philosophy of Science Association.

can be no prediction of the course of human history by scientific or any other rational methods" ([18], p. iv.)—and in the course of his argument Popper poses the question: "Is there a law of evolution?" Although, obviously, Popper's major concern is with evolutionary laws as applied to humans, he keeps his discussion sufficiently general that some of his views about biological evolution become apparent.

Broadly speaking, as might be expected from one concerned to deny laws of human destiny, Popper's position is that there are no such laws of evolution. "There are neither laws of succession, nor laws of evolution" ([18], p. 117). Moreover, Popper's reason for taking such a position is simple and clear. Laws require repeatability. However: "The evolution of life on earth, or of human society, is a unique historical process" ([18], p. 108). Hence, there can be no evolutionary laws.

At this point the scientist or philosopher concerned with the welfare of evolutionary biology may perhaps be undergoing simultaneous emotions of *déjà vu* and depression. If indeed there are no evolutionary laws, then, as the many critics of evolutionary biology who have made reference to the uniqueness argument have been happy to point out, this surely says little for and much against evolutionary biology as a science. (See, for example, [14], p. 31.) However, as equally many commentators have noted, such an approach to evolutionary biology is even more surely misguided. (See, for example, [25], pp. 89–91.) Although indeed the evolution of (say) elephants is unique, this is no more of a bar to evolutionary laws than is the uniqueness of Earth a bar to astronomical laws. The evolutionary biology sympathizer will point out that one must draw a distinction between the unique history of life on earth (involving "phylogenies") and the biologists' theory of evolution through natural selection: a theory which speaking generally argues in a lawlike way that, given groups of organisms, one gets a differential reproduction which, combined with the constant injection of new variation (through random mutation), leads to a "selecting" and eventually to an evolution of forms.

Nevertheless, careful study of what Popper has to say shows that, were one to fault him for confusing unique phylogenies with general evolutionary claims, one would be doing him an injustice. His position is more subtle than a casual reading implies, and he is indeed making important distinctions and claims that, regretfully, have not always been made by biological commentators writing since he did. Indeed, even more regretfully, at least one such commentator has mistakenly cited Popper's authority in support of the above given rather crude argument to the nonlawlike nature of evolutionary claims from the uniqueness of evolutionary phenomena. (See [6], p. 124.)

For a start, Popper is fully and explicitly aware of the distinction to be drawn between the unique history of the evolution of organisms and a theory, perhaps giving a mechanism, of evolutionary change. Moreover, although he replies in the negative to his question about a law of evolution, he makes it clear that he does not deny the possibility of laws being involved in a theory of evolutionary change. Far from it: "Such a process [as the evolution of life on earth], we may assume, proceeds in accordance with all kinds of causal laws, for example, the laws of mechanics, of chemistry, of heredity and segregation, of natural selection, etc." ([18], p. 108). What Popper is concerned to deny is overall extrapolations from the course of evolution—extrapolations pointing to a general progression in the course of evolutionary history, and the like. And this is a denial in which the great majority of evolutionists would no doubt join with Popper. Certainly, modern evolutionists seem to find little room in their theorizing for progressive speculations. Apart from anything else, the theory of evolution through natural selection implies that change is opportunistic. Even though a change from a human viewpoint might be "degenerative," say the loss of some complex organ, it is quite possible that such might happen if the conditions favor a degenerative change. One thinks here, for example, of the loss of sight of cave dwelling mammals, or, a case to be mentioned later, the loss of flying ability of oceanic island dwelling insects.

But even so, one might feel that Popper is still being unnecessarily firm in his stand against evolutionary laws (excluding from consideration now laws to do with selection, and the like). After all, one might argue, although overall laws of progress and such things may not hold, may indeed be just not the thing that a phylogenetic description is about, it is well-known that evolutionary history shows many *trends*. ([30].) For instance, one frequently sees (in the fossil record) a trend towards larger bodily size—so frequently, in fact, that it is referred to (following its discoverer) as "Cope's Rule." Could one not argue that such as Cope's Rule or "Dollo's Law" (that evolution never reverses itself) give the lie to Popper's denial of evolutionary laws—that here we go beyond the unique?

There are two points to be made in reply, both of which are in Popper's favor. First, such "rules" or "laws" of evolution tend to have an awful lot of exceptions. A recent discussant of Cope's rule, S. M. Stanley, admits candidly that: "Because numerous exceptions are known, recognition of the concept as a law has been rejected by most workers" ([32], p. 1). Similarly, in a spirited attack which regretfully is not always as careful in its treatment of uniqueness as is Popper, the paleontologist G. G. Simpson points out that some evolutionary "laws" have up to 36% exceptions! ([31], p. 29). Even Boyle's law has a better batting average than this, and one might well feel that such frail reeds need raise no qualms for a Popperian denial of evolutionary laws.

But this brings me to the second point. Suppose one insisted in regarding some of these claims about evolutionary trends as being at least quasilaws or near laws. There is some justification for doing this. For a start, some of the claims do not have that many exceptions. Having denied that Cope's rule is a law, Stanley nevertheless continues: "Still, it [*i.e.,* Cope's rule] has been widely upheld as a valid empirical generalization, and of the definitions for 'rule' listed by Webster, 'a generally prevailing condition,' describes it accurately" ([32], p. 1). Second, by a process of excluding exceptions specifically, so common a practice with laws, one can make such claims even more accurate. Cope's rule seems to break down primarily when a new type of environment is being invaded, as when the first amphibians and the first birds evolved. Hence, by restating Cope's rule as excluding such instances, one has even more accurate general claims. Third, and most important, many evolutionary "laws" have the kind of theoretical justificatory backing which tends to distinguish laws from mere accidental generalizations. Stanley, for example, argues that there are good adaptive reasons why major evolutionary breakthroughs usually involve comparatively small body sizes, which sizes after the breakthrough can then be increased. Moreover, he argues also that there are reasons why this might not hold in the exceptional cases. (See [25], for more argumentation in this vein.)

However, despite these arguments, which seem to have some strength, Popper's position is unscathed—because he allows for and points to the position they are supporting! First, Popper points out that merely showing that there are evolutionary trends does not prove the existence of evolutionary laws. Laws do not assert existence. *"But trends are not laws.* A statement asserting the existence of a trend is existential, not universal" ([18], p. 115, his italics). However, secondly, in reply to the obvious counter, Popper then goes on to say: "If we succeed in determining the complete or sufficient singular conditions *c* of a singular trend *t,* then we can formulate the universal law: 'Whenever there are conditions of the kind *c* there will be a trend of the kind *t*'" ([18], p. 129n). In other words, if evolutionists really can set up the conditions for a trend, then Popper will allow them a law—although he does have some doubts about testing such a law. Hence, the way still seems open for the possibility of something like Cope's rule being made into a law, and Popper points out this way.

In short, although one might feel—given what he does allow—that Popper is a little paradoxical in his firm denial of evolutionary laws,[1] there is much of real value in his first excursion into the philosophy of evolutionary biology.

II

We come now to Popper's more recent comments about evolutionary biology. These are to be found in his *Objective Knowledge: An Evolutionary Approach* ([19]), and in his autobiographical contribution to the *Philosophy of Karl Popper* ([20]). Since, not surprisingly, many of the points made in the two works coincide, I shall here base my discussion on the second, fuller contribution, subtitled "Darwinism as a Metaphysical Research Program," feeling free to refer where necessary to the first contribution for points of clarification and expansion. (There are also some comments in Popper's contribution to the 1973 Herbert Spencer Lecture Series [21]. However these seem not to add anything else.)

Put simply, it is Popper's claim that in an important sense neo-Darwinian evolutionary theory, the modern theory of evolution, is not a genuine scientific theory. He argues that the theory is not properly testable, and then, true to his most fundamental philosophical tenets, he concludes that the theory is metaphysical. "I have come to the conclusion that Darwinism is not a testable scientific theory but a *metaphysical research program*—a possible framework for testable scientific theories" ([20], p. 134, his italics). One should add that in calling the theory "metaphysical," unlike a logical positivist, Popper is making a philosophical point and not thereby implying condemnation. Indeed, he puts Darwinism in the same column as Deductivism (as opposed to the column with Lamarckism and Inductivism), and there can surely be no higher Popperian praise than that—although perhaps revealingly Popper does later refer to evolutionary theory as a "feeble" theory ([20], p. 137), and perhaps even more revealingly he makes suggestions for "an enrichment of Darwinism" ([20], p. 138). In what follows I shall therefore first consider Popper's reasons for calling neo-Darwinism "metaphysical," and then I shall look at his suggested improvements.

III

Popper begins with an argument about possible life on Mars. Popper argues (correctly I think) that the Darwinian evolutionist would make at least the following three claims. First, organisms reproduce in kind fairly faithfully; second, there are small, accidental, hereditary mutations (causing change); third, there is a process of natural selection.[2] Now, argues Popper, the evolutionist would seem to be committed to the view that if ever on some planet we find life satisfying the first two claims, selection will come into play and cause a wide variety of organic forms. Hence, evolutionary theory would seem to make predictions which are testable. Therefore, evolutionary theory would seem to have genuine scientific content and would seem to offer the possibility of genuine scientific explanations. However, argues Popper, Darwinian evolutionary theory does not really make such a claim about a variety of forms.

> For assume that we find life on Mars consisting of exactly three species of bacteria with a genetic outfit similar to that of terrestrial species. Is Darwinism refuted? By no means. We shall say that these three species were the only forms among the many mutants which were sufficiently well adjusted to survive. And we shall say the same if there is only one species (or none). Thus Darwinism does not really *predict* the evolution of variety. It therefore cannot really *explain* it. ([20], p. 136, his italics)

I make two comments about this argument. First, although evolutionists do believe that natural selection working on random variation will normally lead to variety, they do not think that this is something which must follow necessarily. Selection can act to keep a population or species absolutely stable by eliminating all new mutations. (See [13], especially chapter 4.) But this leads on to the second point. Variety will come about when and only when there is, as it were, some advantage to or cause for such variety—when different ecological niches for example can be used. Now this claim, it seems, does lead to predictions testable at least in principle. Popper's Mars example is perhaps a little unfair, because we know already that Mars is not going to be very hospitable to life,[3] so one is hardly going to expect so much organic variety as here on earth (remember, here on earth one gets a lot more variety in the jungle than in the desert). But let us consider for a moment some of the reasons or conditions for variety, in particular (following Popper) let us consider some of the reasons why one might get different species created. (A species is a group of organisms, breeding between themselves, but potentially or actually isolated reproductively from all others. The differences between species are a reflection of and dependent upon differences in genetic constitution, "gene pools." Different species therefore have different gene pools, although how different may vary and is indeed a matter of some debate. We can sidestep this issue here. See [16] for more details.)

First, there is the question of isolation or separation. There is, in fact, some controversy between evolutionary biologists about exactly how or why speciation occurs, and the role played by isolation—in particular, some feel that speciation (between two groups originally of the same species) always requires a period of geographical separation. Others believe that although most speciation may be of this kind, "allopatric" speciation, it can occur between groups not separated ("sympatric" speciation). But even those who allow sympatric speciation often demand some kind of ecological separation—say, speciating groups being on different parts of the same host. So, separation or isolation seems most important for speciation. (See [13] and [16] for details.)

Obviously however, whether speciation is allopatric or sympatric, more is needed. What is needed is some reason or reasons to push apart the genes of the two speciating groups. Selection is clearly going to be the main thing operative here—for instance, the ecological and geographical conditions of the two groups may be very different, and these in turn might well lead to different selective pressures. And finally, let us mention something which may well be of great importance in speciation, namely the so-called "founder principle" ([2], [3], [16]). It is generally believed today that species of organisms contain a great deal of genetic variability—there is no such thing as a standard typical member, for nearly all members will have some genes which are not common and perhaps not have some genes which are fairly common. The reasons why such variability is supposed are several—one reason for instance is the probable widespread occurrence of superior heterozygote fitness (*i.e.*, the heterozygote for two alleles is fitter than either homozygote), which in turn leads to different genes being "balanced" and maintained in a population. But the consequence of this variability is that if a small group (of "founders") is isolated from the main population (say on an island) they will not be typical, because nothing is typical. Hence, not only will they *a fortiori* be on the way to being a different gene group, but as they breed between themselves and "shake down" into a cohesive groups they may alter drastically the various selective fitnesses of the genes they do possess, thus leading to rapid change. For instance, a gene A_1

may have selective value when it is rare (as in the total population) but little at all when it is common (as in the founder population).

Now, let us start putting some of these points together in possible models.[4] Suppose for a start one came across a planet where the chances of allopatric speciation seemed difficult rather then otherwise—suppose, for instance, the planet were fairly small and uniformly covered with water (without freakish currents, and so on). One suspects that were but a few aquatic species discovered on such a planet, no evolutionist would be desperately perturbed. On the other hand, suppose one came across a planet with conditions which seemed tailor-made for speciation. Suppose, for instance, one had an area with fairly large populations, which investigation showed to be variable genetically (which in turn manifested itself as phenotypic variation). Suppose also on such a planet one had other isolated areas, with differing conditions—cold, warm, dry, wet, and the like. And suppose finally there seemed possible rare (but only rare) ways in which organisms might go from the main area to the isolated areas (which isolated areas were now inhabited). Had one reason to believe that life on the planet was fairly old (*e.g.*, through the fossil record or general complexity of structure), yet were one to find that absolutely no speciation at all had occurred, then I suggest that, *contra* Popper, modern evolutionists would be worried. Their theory, parts of it at least, would have been falsified. The claims that they make about speciation would seem not to hold.

Of course, talk of hypothetical planets at best makes evolutionary theory testable in principle, but this seems all that is necessary to counter Popper here, since his argument is at the hypothetical level. One would add, however, that there is empirical evidence from this world which seems both to support and test evolutionists' claims about speciation. Repeatedly through the world have been found and are found cases where populations, isolated from the main group under the kinds of conditions described above, have evolved into new species. The classic case is probably that of the finches on the Galapagos Archipelago off the coast of South America, those very finches which made Darwin an evolutionist. ([8].) Moreover, there is experimental evidence based on populations of captive fruit flies which supports the founder principle hypothesis in particular, apart from more general evidence supporting claims about the genetic variability always in populations, which claims are so crucial to modern thinking about speciation. (See [3] and [13] for more details.)

IV

Next, in his campaign to show that evolutionary theory is metaphysical, Popper brings up that popular suggestion that adaptation and selection are just about vacuous.

> Take "Adaptation." At first sight natural selection appears to explain it, and in a way it does, but it is hardly a scientific way. To say that a species now living

is adapted to its environment is, in fact, almost tautological. Indeed we use the terms "adaptation" and "selection" in such a way that we can say that, if the species were not adapted, it would have been eliminated by natural selection. Similarly, if a species has been eliminated it must have been ill adapted to the conditions. Adaptation or fitness is *defined* by modern evolutionists as survival value, and can be measured by actual success in survival: there is hardly any possibility of testing a theory as feeble as this. ([20]), p. 137, his italics)

Before attempting a criticism of this passage, one must in fairness to Popper point out that in making a claim of this kind he is in very distinguished *biological* company. Indeed, one of today's leading evolutionists, R. C. Lewontin, at one point used virtually the same language as Popper, although in later writings he has backtracked somewhat. (See [11], p. 309 and [12], pp. 41-42.) Because of such ambiguity, what one starts to suspect is that Popper (and Lewontin) are right in thinking that there is something tautological or analytic surrounding selection—a definition perhaps—but that Popper is wrong in thinking that this is all there is to the matter. There are probably some very solid empirical claims being made. And this, I would suggest, is in fact the case.

First, take natural selection itself. This is a systematic differential reproduction between organisms, ultimately brought about on the one hand by organisms' tendency to increase in number in geometric fashion, and on the other hand by the inevitable limitations of space and food supply. Now, in pointing to the fact that there is a differential reproduction—that not all organisms which are born survive and reproduce (offspring which are in turn viable)—we hardly have something which is tautological. The differential reproduction may be as "obvious" as the roundness of the earth, but neither is empirically empty—certainly the differential reproduction makes evolutionary theory testable. If we all just budded off one and only one offspring asexually, evolutionary theory would be false. At least, it would be false inasmuch as one tried to apply it to this world of ours, which is precisely what evolutionists do want and try to do. (In the short term, it must be allowed that one could have a differential reproduction, even though all organisms reproduce. It would just be a question of one organism or kind having more [viable] offspring than another organism or kind.)

Secondly, we have the point about selection which seems to cause so much trouble. Now, it certainly seems to be the case that evolutionists do link up adaptive value and fitness in terms of survival value (or, more precisely, in terms of reproductive value), and that what we have here are definitions— analytic or tautological statements. But even here we have the evolutionists doing more than just making straight analytic definitions. Evolutionists always emphasize that natural selection is *systematic*—the differential reproduction is not a random matter. Overall success is believed to be on average a function of organisms' peculiar characteristics, and so not only do we have the very nonanalytic matter of which characteristics aid survival and reproduction—

if we were all identical there could be no selection—but also we have the claim that things of adaptive value in one situation, will also be of value in similar situations. This may be difficult to test, but it is an empirical claim and could be false. Suppose one found on an island a group of men who had lost their sight, and who were not troglodytes or in any other way peculiar. Once again evolutionists would have to rethink their theory. That no such men have been found does not prove the theory analytic—it may just be that the theory is true. Falsifiability should not be confused with being falsified.

Of course, once again in practice it may be difficult actually to test this aspect of evolutionary theory. No doubt it would not be easy to be able to rule out possible adaptive reasons for men losing their sight. But evolutionists do have positive evidence, from experiments and nature, that there is a kind of uniformity about adaptive value in similar situations. For example, wing-lessness seems to be of value to insects and other small animals on oceanic islands (because they stand less chance of being blown away), and experiments bear this out ([2])—and after all, animals do not lose their sight (in a systematic way) except for good reason, as with moles.

Moreover, even granting something analytic about the way in which adaptation is defined, we still have empirical questions surrounding the whole problem of evolutionary change. Why, for example, a characteristic is adaptive at one point in time or space, but not at a later time or different place? Suppose an animal color in a species changes from light to dark, as in the famous case of moths in industrial Britain. ([29].) In this particular instance, it has been shown that years ago light color was an adaptation, whereas today dark color is adaptive. The reasons supposed however, that because of pollution trees have become darker and that avian moth predators more easily see moths which do not match their backgrounds, clearly take one beyond the analytic.[5]

Thirdly, Popper ignores entirely the fact that evolutionists allow that it can be the less well-adapted which can survive and the more well-adapted which fails to survive. For a start, it is percentages which count not individuals—does one group on average have a better record than another, not does one individual survive rather than another. Secondly, Popper ignores entirely the hypothesis of genetic drift, which supposes that in certain special situations fortuitously the less well-adapted (or neutral) can succeed where more well-adapted gets eliminated. Genetic drift is still a highly contentious issue ([13]), but the way Popper argues it would be ruled out as contradictory, whereas even its strongest critics seem to feel the need to mount empirical counter-arguments. It should be added, lest it be thought that the possible existence of drift makes evolutionary theory unfalsifiable in the sense that all characteristics necessarily have an explanation—selection or drift—that no one today denies that selection fashioned major characteristics like the hand, the eye, and so on. Conversely, the above given hypothetical islanders could not have lost through drift so important a characteristic as their sight. Hence, drift is not a ubiquitous escape clause against falsification. (See [25], p. 115.)

Finally, let us mention that in evolutionary theory, linked with selection although perhaps not really part of it, we have the claim that the selected characteristics will be passed on from one generation to the next. This is obviously necessary for evolution, for were there no such transmission selection would have no effect. And the claim is clearly empirical—logically it is quite possible that the strong, sexy, or otherwise advantaged individuals always have puny, ugly, or otherwise disadvantaged offspring. Although an evolutionary theory based on selection does not necessarily have to use a theory of inheritance stemming from Mendelian or neo-Mendelian claims and findings—Darwin's did not, for example—it seems today that most evolutionists do work with an amalgam of Darwinian selection and Mendelian genetics. (See [25] for details.)

All things considered therefore, it seems ridiculous to keep claiming that evolutionary theory has at its heart a devastating tautology. The time has come to lay this misconception quietly to rest.[6]

V

Popper continues his analysis by taking up the gradualness of evolution as forecast by Darwinian evolutionary theory. He allows that the theory "certainly does *predict* that if such an evolution takes place, it will be *gradual*" [20], p. 137, his italics). However, Popper goes on to say:

> Gradualness is thus, from a logical point of view, the central prediction of the theory. (It seems to me that it is its only prediction.) Moreover, as long as changes in the genetic base of the living forms are gradual, they are—at least "in principle"— explained by the theory; for the theory does predict the occurrence of small changes, each due to mutation. However, "explanation in principle" is something very different from the type of explanation which we demand in physics. While we can explain a particular eclipse by predicting it, we cannot predict or explain any particular evolutionary change (except perhaps certain changes in the gene population *within* one species); all we can say is that if it is not a small change, there must have been some intermediate steps—an important suggestion for research: a research program. [29], pp. 137-8, his italics)

I am not quite sure what to make of this argument, because it seems to me to be so unfair. Either evolutionary theory predicts that change will be gradual, or it does not. In fact, for most cases it does, therefore we have a prediction, therefore it is testable, therefore it is not metaphysical. One may argue that the explanation is not of very much, but it is of something—although in fact if one looks at the matter historically one finds that the gradualists had a terrific battle to win over the nongradualists, the saltationists. Of course, the explanation is "in principle" in the sense that until one turns to an actual case, specific details are lacking, but this is the same as any theory until one turns to actual cases. The gradualness at least is no more in principle than are eclipses. One may agree with Popper that one has little more than a "research program," a start to explanations not an end, but by Popper's own philosophy the program is not metaphysical. (Incidentally, Popper seems unaware that evolutionists believe that in the plant world, evolution can occur nongradually in steps due to combining of complete sets of parental chromosomes in offspring. [33]. He seems also unaware of the great amount of explanatory information there is about some actual cases of gradual evolution, filling out the details— for example, in the case of the horse. [30].)[7]

VI

Finally in considering Popper's arguments about the nature of evolutionary theory, before turning to his suggestion for improving the theory, one might add that Popper's own position seems paradoxical—almost contradictory. Popper believes that neo-Darwinian evolutionary theory is metaphysical because it is unfalsifiable. Yet, Popper seems to believe not merely that it leads to falsifiable predictions, but that these are false! He writes:

> It is clear that a sufficient increase in fecundity depending fundamentally on genetical factors, or a shortening of the period of immaturity, may have the same survival value as, or even a greater survival value than, say, an increase in skill or in intelligence.
> From this point of view it may be a little hard to understand why natural selection should have produced anything beyond a general increase in rates of reproduction, and the elimination of all but the most fertile breeds. ([19], p. 271)

Then, in a footnote, Popper comments: "This is only one of the countless difficulties of Darwin's theory to which some Neo-Darwinists seem to be almost blind" ([19], p. 271n).

If we pursue this line of argument then Popper's point seems to be that selection theory is not merely not tautological but false, because there are in fact many breeds which are far less fertile with respect to numbers than others—elephants as opposed to herring, for instance. Of course, the truth of the matter is that selection theory does not necessarily have the implications that Popper seems to think it has, and that neo-Darwinists are far from blind to this whole question of reproductive rates. Indeed, there has been considerable attention paid to and controversy surrounding the problem of reproductive rates—why it is that organisms have the particular rates that they do.

All biologists seem to agree that controlled reproductive rates are a function of organisms being able to survive only with limited numbers in the face of limited resources, but that with such control these organisms can exploit those resources more successfully perhaps than faster breeders. However, some biologists have argued for a kind of "group selection," claiming that when one has (say) limited resources, there is selection on the group to keep reproductive rates down, or the whole group may perish. (See particularly

[38].) Other biologists argue that such group selection is chimerical and that individual selection is the cause. ([35].) Plovers, for instance, have a very stable clutch size, and they do not reproduce to their possible limit (if eggs are removed from the nest, then the female brings the number back up to the norm). It has been argued by David Lack simply that there is selection on the individual plover for such clutch size—more eggs and the individuals collectively are less fit, and a fewer egg number does not make for fitness which compensates the fewness in number. ([95], [10].) Incidentally, there is nothing unfalsifiable about the evolutionists' position here—that whatever the reproduction rates may be evolutionary theory will necessarily have an answer. If experiments showed (what they do not) that plovers given additional eggs have more fit offspring, then Lack would be wrong.

In short, at this point we can save Popper from the unwelcome implications of his own arguments. Evolutionists have paid attention to the problem of reproductive rates, and they do have arguments showing why there might be selective advantages to keeping such rates at a (comparatively) low level. Curiously however, despite his criticisms of evolutionists at this point, Popper seems to know the truth all along. Immediately after the quotation given above (from the main text), he writes "[There may be many different factors involved in the processes which determine the rates of reproduction and of mortality, for instance the ecological conditions of the species, its interplay with other species, and the balance of the two (or more) populations.]" ([19], p. 271, his square brackets). Even if Popper thinks these claims ill-founded, and he gives no arguments to such an effect, it is a little odd that he should accuse neo-Darwinians of blindness on this point. ([37] contains a discussion of recent work on the group versus individual selection controversy.)

VII

Let us turn now to Popper's own suggestions for "enriching" evolutionary theory. Popper argues that evolutionary theory "predicts *accidental* mutations, and thus *accidental* changes. If any 'direction' is indicated by the theory, it is that throwback mutations will be comparatively frequent. Thus we should expect evolutionary sequences of the random-walk type" ([20], p. 138, his italics). But then Popper asks: "How is it that random walks do not seem to be prominent in the evolutionary tree?" ([20], p. 138). Rather, we are told, we get trends, and to explain these Popper introduces his own theory, distinguishing between genes controlling anatomy ("*a*-genes") and those controlling behavior (subdivided into "*p*-genes," genes controlling preferences, and "*s*-genes," genes controlling skills), and arguing that changes in the latter sometimes prepare the way for changes in the former. He writes:

> We can now say that certain environmental changes may lead to new preferences or aims (for example, because certain types of food have disappeared). The new

> preferences or aims may at first appear in the form of new tentative behavior (permitted but not fixed by *b*-genes). In this way the animal may tentatively adjust itself to the new situation without genetic change. But this *purely behavioral* and tentative change, if successful, will amount to the adoption, or discovery, of a new ecological niche. Thus it will favor individuals whose *genetic* p-structure (that is, their instinctive preferences or "aims") more or less anticipates or fixes the new behavioral pattern of preferences. This step will prove decisive; for now those changes in the skill structure (*s*-structure) will be favored which conform to the new preferences: skills for getting the preferred food, for example. I now suggest that *only after the s-structure has been changed will certain changes in 'he a-structure be favored; that is, those changes in the anatomical structure which favor the new skills.* ([20], p. 139, his italics)

And by way of example, having schematically put his position as $p \rightarrow s \rightarrow a$, Popper invites us to consider the case of the woodpecker and his beak: "A reasonable assumption seems to be that this specialization started with a *change in taste* (preferences) for new foods which led to genetic behavioral changes, and then to new skills, . . . and that the anatomical changes came last" ([20], pp. 139-140, his italics).

As a preliminary towards evaluating Popper's suggestions let us note that he is perhaps a little unfair towards, or at least misleading in his presentation of, modern evolutionary theory. The whole point about the theory is that, given selection, the one thing we do not expect is, as Popper seems to imply we should get, something analogous to "the track described by a man who at every step consults a roulette wheel to determine the direction of his next step" ([20], p. 138). This is what we get with drift, not selection.

Selection implies that there will be systematic, sustained changes in (or maintenance of) gene ratios, and evolutionists look for such changes (or maintenance) for evidence of selection. One certainly gets the feeling that because he sees selection as meaning no more than that those which survive are those which survive, Popper has blocked out from his own view the fact that any old (or new) accidental variation will not necessarily have an equal chance with any other in the reproduction stakes—something which would indeed imply randomness. Given any situation which persists for even the shortest time we are going to get some kind of channeling by selection of certain variations into a kind of nonrandom order, and this channeling will show direction. One suspects that compounding Popper's problem here is first the fact that he fails to realize that mostly modern evolutionists believe that populations harbor at all times a terrific amount of genetic or phenotypic variation due to such things as balance from superior heterozygote fitness and selection for rareness, and possibly selective neutrality ([3], [13].) Hence, should a population start to feel selective pressure in a certain direction, it has already a great deal of genetic variation to draw on, and so can often respond at once in the required direction. Secondly, Popper fails to realize that often characteristics are controlled by many genes at once and hence although individual mutations may be rare and accidental, on average the chances of an appropriate mutation are much higher. Thus, were there a new selective advantage in size, for example, the species of organism would not have to wait for the one unique mutation for increased size but only for one of many possible mutations.

Of course, at this point Popper could rightly reply that the potential from all this genetic variation does not in itself imply that there will be trends—particularly some of the rather long-term trends of which we find evidence in the fossil record. Reasons must be given to show that selection will be sufficiently stable to lead to such trends. Again however, evolutionists are ahead of him. For instance, we have seen already some of the discussion which surrounds Cope's rules, which concerns trends towards larger body sizes. And there is in the literature much discussion of other trends. (See, for example, [30].)

But enough of this preliminary skirmishing. What about Popper's own suggestions? In order to get to the crux of the matter, let us grant Popper's way of speaking of genes just controlling anatomy and genes just controlling behavior, even though the widespread existence of pleiotropy (genes with more than one function) may put such a clean categorization in jeopardy.[8] Part of the difficulty with evaluating Popper's position is to know in what sense he is saying something original. Popper argues for behavior preceding structure. But this is generally granted by evolutionists. Thus, for example, Mayr (whom Popper cites in support) writes as follows:

A shift into a new niche or adaptive zone requires, almost without exception, a change in behavior. . . . It is very often the new habit which sets up the selection pressure that shifts the mean of the curve of structural variation. Let us assume, for instance, that a population of fish acquires the habit of eating small snails. In such a population any mutation or gene combination would be advantageous that would makes the teeth stronger and flatter, facilitating the crushing of snail shells. In view of the ever present genetic variation, it is virtually a foregone conclusion that the new selection pressures (owing to the changed habit) would soon have an effect on the facilitating structure. ([15], p. 371; see also [16])

So far, so good. But one assumes that Popper wants to say, indeed feels that he is saying, something more. The question is precisely what. One possibility is that Popper is hypothesizing a kind of behavioral mutation pressure, which as it were, following on a non-genetically caused behavior switch, takes organisms irreversibly from one behavior pattern to another (with anatomy hopefully following). Thus, Popper writes of the possibility that "when a change takes place in taste, as in the case of the woodpecker, then the anatomical structure relevant for food acquisition may remain unchanged . . . ([20], p. 140). This certainly seems different from Mayr's position, who seems to be envisaging just a reversible switch of preference, not one dependent on or necessarily involving a whole new set of genes.

If this is in fact Popper's position, then, for a start, it is hard to see precisely why Popper's hypothesis is even needed. Organisms switch behaviors and eventually anatomical changes follow. Why bother with sandwiching preference-and-skill behavior changing genes in the middle? Of course, one might perhaps get such genes, but it is difficult to see what they are doing (and nothing else is doing) that Popper feels needs doing, particularly since all they seem to be doing is cutting down an organism's options. Popper writes that (non-genetically fixed) behavior change "will favor individuals whose *genetic p*-structure (that is, their instinctive preferences or 'aims') more or less anticipates or fixes the new behavioral pattern of preferences" ([20], p. 139). But it is difficult to see why this is so, unless any back sliding would be dangerous—in which case there seems just as much likelihood of a need for anatomical genetic change as behavioral genetic change.

Of course, it might just be that Popper feels that these new preferences and skill behavior genes will, as it were, force the organism into action—they will push the organism into new preferences and thence into new skill behavior (or by being irreversible force the organism to stay with new preferences and skill behavior) when without the genes the organism would stay with or revert to the old preferences and skill behavior. If this is so, and talk of "anticipation" rather implies it, then first it is not easy to understand why (as Popper claims) natural selection would be less destructive on preference-cum-skill behavioral changes than anatomical changes. With respect to the woodpecker, Popper writes: "A bird undergoing anatomical changes in its beak and tongue without undergoing changes in its taste and skill can be expected to be eliminated quickly by natural selection, *but not the other way round*" ([20], p. 140, his italics). But why is this? If anything, I should have thought the opposite is the case. Certainly, a Canadian bird with a new, fixed, irreversible taste for maple syrup—a taste which it would not have or would avoid without its new genes—but with (as yet) no woodpecker-like changes in its beak, is going to be at a distinct disadvantage compared to its mates, who are satisfied with more humdrum fare and who are not wasting time beating their beaks futilely against maple trees. Secondly, one would truly like some empirical evidence to support Popper's position. If it is generally true, then it seems most odd that evolutionists, either working with wild or captive populations, have found no evidence of it—and they seem not to have done so.

Of course, the basic problem at this point is that Popper seems not to have the first idea about contemporary evolutionary thought about evolutionary change, speciation, movement into new ecological niches, and so on. For example, because he ties adaptive value so tightly to survival Popper seems not aware of the great genetic and phenotypic diversity evolutionists see in populations, which is thought to exist for the kinds of reasons mentioned earlier. But with this kind of diversity a population might well carry indefinitely (say) a bird-type with beak and tongue less well-adapted than most for the usual diet. Then, when a new niche opens up, because of dispersal, a change in the ecological balance, or some such thing, the previously less well-adapted type is ready to move in and take advantage in a way barred to the bird conforming to what hitherto had been the best-adapted type in the group. In short, it seems that evolutionary theory has no need of Popper's suggestions. (Incidentally, I am not denying that an initial preference and skill behavior switch might be caused in part by a new gene. No doubt this does happen sometimes. What worries me is Popper's succession: nongenetic switch → preference behavior gene switch → skill gene switch → anatomical switch. I see no reason or evidence for the necessity of a two or three part division at the beginning to the sequence—if anything, I see reasons against it.)

At this point, it might possibly be objected that (appearances to the contrary) I am attacking a straw Popper. In *Objective Knowledge* Popper writes of a "hopeful behavioral monster" ([19], pp. 281-284), and he portrays himself as providing a new form of the saltationary theory of R. B. Goldschmidt, who supposed that important evolutionary changes are a function of one step macromutations (rather than the many step microchanges of the Darwinian selectionist). Perhaps Popper favors major behavioral changes (to be followed by structural changes), which changes come in one generation from one or a combination of gene change(s)? The trouble here is that Popper counters none of the objections evolutionists have brought against saltationary theories positing such monsters—in particular against Goldschmidt's theory ([4], [5]). To take but one point—is this monster supposed to be of a species different from its parent? If it is not, then Popper is ignoring the major evolutionary problem of speciation. If it is, (and Popper hints that this is indeed so), then with whom it is going to reproduce, since its parents' species is ruled out? One is going to need several such monsters together in a single generation. But then, why do we seem to have absolutely no experimental or natural evidence of such clustering of new monsters? Why, in fact, does the empirical evidence point to gradual change? (See [2] for a detailed criticism of Goldschmidt's views.)

One could go on, perhaps indefinitely, trying to explicate Popper's precise position. But finally, in this attempt to understand what Popper is saying that is both new and important, let us mention that he makes friendly reference to the so-called "Baldwin effect," and although he speaks of his own hypothesis as having "considerably extended" the explanatory nature of the effect, there is indeed strong connection between the Baldwin and Popper effects. ([19], p. 268.) As Mayr paraphrases it: "The Baldwin effect designates the condition in which, owing to a suitable modification of the phenotype, an organism can stay in a favorable environment until selection has achieved the genetic fixation of this phenotype" ([16], p. 610). The kinds of things the Baldwin effect is (or was) invoked to explain are Lamarckian-type phenomena, apparently involving the inheritance of acquired characteristics. Typical would be the callus on the rump of the ostrich, supposedly first acquired in each generation through friction, but then becoming heritable.

In Popper's favor, let us note first that C. H. Waddington ([34]) (to whom

Popper makes reference) has experimentally produced something akin to the Baldwin effect, which he calls "genetic assimilation." Certain fruit flies show oddities if (and by and large only if) exposed to certain chemicals during development. Selection of these "freaks" eventually leads to heritable freaks (*i.e.*, freaks even without chemicals). Secondly, also in Popper's favor is the fact that some biologists believe that this genetic assimilation most crucially involves behavioral traits—one starts with new behavior which is not fixed, and eventually it becomes, through genetic assimilation, fixed. ([37].) However, before it be concluded that here at last we have vindication of Popper, other points must be noted.

First, let us consider Waddington's non-Lamarckian explanation of the Baldwin effect. He suggests that the chemicals do not cause freaks alone; rather they do it because the affected organisms *already* carry genes with a mild causal propensity towards freakiness. The chemicals reveal the carriers of these genes, and subsequent selection collects enough together in one individual to cause freakiness unaided. There is therefore no new mutation towards freakiness. Hence as before, inasmuch as Popper is relying on Waddingtonian genetic assimilation he is saying nothing new, and inasmuch as he is going beyond (and this by claim and fact he seems through his postulation of behavior genes to be doing), we have again the question raised earlier. In particular, what evidence, need, possibilities, are there for the new behavior fixing genes Popper postulates? But these questions bring me to my second point. I have suggested that, in fact, it seems normally not to an organism's advantage to have hitherto flexible behavior now fixed, and hence there seems little reason to suppose (as Popper supposes) that selection would work, even given new genes, to fix the behavior. And this worry is one which we find expressed also by many biologists about genetic assimilation, which does not even involve Popper's new genes. Thus, although Popper without comment ([19], p. 284) refers the reader to a discussion of the Baldwin effect in Mayr's classic *Animal Species and Evolution,* we find a hostile treatment of the concept and the following statement. "The Baldwin effect makes the tacit assumption that phenotypic rigidity is selectively superior to phenotypic flexibility. This is certainly often not true" ([16], p. 611). Unless this claim is completely false, and I doubt even those who allow some genetic assimilation would claim this (for their position seems to be that flexibility is lost as a side effect), the Popperian position fails. (See also [35].)

In summation Popper's "enrichments" of evolutionary theory seem at best unneeded and at worst wrong.

VIII

It is odd that one who, as we have seen, at one point dealt so sensitively with evolutionary biology, should now seem so determined to undervalue its nature and achievements. My suspicion is that, in part, Popper's attitude comes simply from the fact that he has not yet accepted the Darwinian revolution. Although at one point Popper says (truly) that Darwin's great move was to show how design and purpose in the organic world can be explained in purely physical terms, almost at once Popper then goes on to say that the difficulty with Darwinism is explaining something like the eye in terms of accidental mutations, and it is clear (by his own admission) that Popper is trying to get away from the accidental in evolution, and to give some direction to change ([19], p. 270). But this is the whole point—unless one can see that it is selection acting on "accidents" bringing about design-like effects, one misses entirely the force of Darwinism. Somehow one feels that Popper is in a tradition which started as soon as Darwin's *Origin* appeared—a tradition which includes such men as Charles Lyell and St. George Jackson Mivart, who were evolutionists but who felt that in order to account for the design-like effects of the organic world one must supplement selection with other mechanisms. Unlike them Popper may have no theological axe to grind; but, he seems a direct intellectual descendent. (Details of this tradition can be found in [27] and [28].)

Possibly also Popper's desire to see links between organic evolution and scientific theory growth leads him astray. This could be in at least two ways. First, fairly obviously, Popper's theory of theory growth is not itself scientific. It is, in a nonpejorative sense, metaphysical. Therefore, inasmuch as biological evolutionary theory can be shown metaphysical, links can be established between Popper's philosophical views and the beliefs of biologists. Second, the actual way in which Popper analyzes theory growth may be influencing his approach to biology. Using '*P*' for problem, '*TS*' for tentative solutions, '*EE*' for error-elimination, Popper sees all evolutionary sequences as following this pattern:

$$P_1 \rightarrow TS \rightarrow EE \rightarrow P_2 \qquad ([19], \text{p. } 243)$$

But the fact of the matter is that tentative knowledge solutions are frequently fairly large (saltationary) and often designed—think, for example, of Darwin's solution to the organic origins' problem. Could it be that (mistakenly) Popper is reading features of the evolution of knowledge into the evolution of organisms, and that it is for this reason that he wants to supplement biological evolutionary theory in the ways he suggests?

But even if my surmises are correct and my criticisms of Popper's views about biological evolutionary theory are well-taken, what then does this all imply? In particular, what does it imply about Popper's philosophical theory about scientific theory growth? In one sense, not a great deal. Darwinians do not have a monopoly on the word "evolution"; hence nothing I have argued can properly stop Popper characterizing his views as evolutionary. Nor has anything here proven his general philosophical theory mistaken, although this is not necessarily to say that it is true. However, I suggest my arguments do show one most important thing. No longer ought Popper claim close ties between his philosophical evolutionary theory and biological evolutionary theory, or feel that somehow some of the legitimacy of the latter rubs off on the former. The relationship between the two theories is at best one of weak analogy. In important respects, Popperian scientific theory evolution and neo-Darwinian biological evolution are different.

NOTES

1. In [19], Popper again somewhat paradoxically asserts: "There are no Darwinian laws of evolution" (p. 267). In [26] I consider the place in Darwin's theory of laws, which I think Popper would allow; and I do the same in [23], [24], [25], for modern evolutionary theory. A valuable discussion of these problems, with a slight different emphasis, is [7].

2. Popper sees the evolutionist making another claim, about variability, but this is irrelevant to his present argument.

3. However, recent space probes show that it is apparently not as inhospitable as was earlier thought.

4. Discussions of speciation usually cover the presumed events if and when isolated groups again come into contact. For brevity these will be ignored here.

5. Curiously, given his belief that selection is tautological, Popper does allow that finding out what are adaptations and why is an empirical matter.

6. In an above mentioned passage which influenced my analysis (and which is I trust captured by my analysis), Lewontin writes: "Evolution is the necessary consequence of three observations about the world. . . . They are: (1) There is phenotypic variation, the members of a species do not all look and act alike. (2) There is a correlation between parents and offspring. . . . (3) Differernt phenotypes leave different numbers of offspring in *remote* generations. . . . These are three contingent statements, all of which are true about at least some part of the biological world. . . . There is nothing tautological here" ([12], pp. 42–2, his italics).

Incidentally, in my reply to Popper I ignore the way Popper has blurred "adaptation" with being "adapted"—an organism can have an adaptation like the eye and still be ill-adapted to its environment. I ignore also the way Popper speaks only of inter specific selection and ignores intra specific selection.

7. In [22] and [25], I discuss in detail the problems of testing evolutionary theory. See also [36].

8. Popper allows for genes with mixed functions. I wonder how few these are. See [3].

REFERENCES

[1] Ayala, F. J., M. L. Tracey, L. G. Barr, J. F. McDonald, and S. Pevez-Salas, "Genetic Variation in Natural Populations of Five Drosophila Species and the Hypothesis of the Selective Neutrality of Protein Polymorphisms." *Genetics* 77 (1974): 343-384.

[2] Dobzhansky, T. *Genetics and the Origin of Species* (3rd ed.). New York: Columbia University Press, 1951.

[3] ———. *Genetics of the Evolutionary Process*. New York: Columbia University Press, 1970.

[4] Goldschmidt, R. *The Material Basis of Evolution*. New Haven: Yale University Press, 1940.

[5] Goldschmidt, R. "Evolution as Viewed by One Geneticist." *American Scientist* 40 (1952): 84-135.

[6] Goudge, T. A. *The Ascent of Life,* Toronto: University of Toronto Press, 1961.

[7] Hull, D. L. *Philosophy of Biological Science.* Englewood Cliffs: Prentice-Hall, 1974.

[8] Lack, D. *Darwin's Finches.* Cambridge: Cambridge University Press, 1947.

[9] ———. *The Natural Regulation of Animal Numbers.* Oxford: Oxford University Press, 1954.

[10] ———. *Population Studies of Birds.* Oxford: Oxford University Press, 1966.

[11] Lewontin, R. "Evolution and the Theory of Games." *Journal of Theoretical Biology* 1 (1962): 382-403. Reprinted in M. Grene and E. Mendelsohn (eds). *Topics in the Philosophy of Biology.* Dordrecht: Reidel, 1976, pp. 286-311.

[12] ———. "The Bases of Conflict in Biological Explanation." *Journal of the History of Biology* 2 (1969): 35-45.

[13] ———. *The Genetic Basis of Evolutionary Change.* New York: Columbia University Press, 1974.

[14] Manser, A. R. "The Concept of Evolution." *Philosophy* 40 (1965): 18-34.

[15] Mayr, E. "The Emergence of Evolutionary Novelties." In S. Tax (ed.), *Evolution After Darwin.* Chicago: University of Chicago Press, 1960, pp. 349-380.

[16] ———. *Animal Species and Evolution.* Cambridge, Mass.: Belknap, 1963.

[17] Monod, J. "On the Molecular Theory of Evolution." In R. Harré (ed.), *Problems of Scientific Revolution: Progress and Obstacles to Progress in the Sciences.* Oxford: Oxford University Press, 1975.

[18] Popper, K. *The Poverty of Historicism.* London: Routledge and Kegan Paul, 1957, corrected version, 1963.

[19] ———. *Objective Knowledge: An Evolutionary Approach.* Oxford: Oxford University Press, 1972; corrected version, 1975.

[20] ———. "Darwinism as a Metaphysical Research Programme." In P. A. Schilpp (ed.), *The Philosophy of Karl Popper.* Vol. I. La Salle, Ill.: Open Court, 1974, pp. 133-143.

[21] ———. "The Rationality of Scientific Revolutions." In R. Harré (ed.), *Problems of Scientific Revolution: Progress and Obstacles to Progress in the Sciences.* Oxford: Oxford University Press, 1975.

[22] Ruse, M. "Confirmation and Falsification of Theories of Evolution." *Scientia* 104 (1969): 329-357.

[23] ———. "Are There Laws in Biology?" *Australasian Journal of Philosophy* 48 (1970): pp. 234-246.

[24] ———. "Is the Theory of Evolution Different?" *Scientia* 106 (1972): 765-783. 1069-1093.

[25] ———. *The Philosophy of Biology.* London: Hutchinson, 1973.

[26] ———. "Charles Darwin's Theory of Evolution: An Analysis." *Journal of the History of Biology* 8 (1975): 219-241.

[27] ———. "The Relationship Between Science and Religion in Britain, 1830-1870." *Church History* 46 (1975): 506–522.

[28] ———. "William Whewell and the Argument from Design." *Monist* 60 (1977): 227-257.

[29] Sheppard, P. M. *Natural Selection and Heredity.* Fourth Edition. London: Hutchinson, 1975.

[30] Simpson, G. G. *The Major Features of Evolution.* New York: Columbia University

Press, 1953.

[31] ——. "Historical Science." In *The Fabric of Geology*. Standford: Freeman, Cooper, 1963, pp. 24-48.

[32] Stanley, S. M. "An Explanation for Cope's Rule." *Evolution* 27 (1973): 1-26.

[33] Stebbins, G. L. *Variation and Evolution in Plants*. New York: Columbia University Press, 1950.

[34] Waddington, C. H. *The Strategy of the Genes*. London: Allen and Unwin, 1957.

[35] Williams, G. C. *Adaptation and Natural Selection*. Princeton: Princeton University Press, 1966.

[36] Williams, M. B. "Falsifiable Predictions of Evolutionary Theory." *Philosophy of Science* 40 (1973): 518-537.

[37] Wilson, E. O. *Sociobiology*. Cambridge, Mass.: Belknap, 1975.

[38] Wynne-Edwards, V. C. *Animal Dispersion in Relation to Social Behavior*. Edinburgh: Oliver and Boyd, 1962.

12

Is a New and General Theory of Evolution Emerging?

Stephen Jay Gould

I. THE MODERN SYNTHESIS

In one of the last skeptical books written before the Darwinian tide of the modern synthesis asserted its hegemony, Robson and Richards characterized the expanding orthodoxy that they deplored:

> The theory of Natural Selection . . . postulates that the evolutionary process is unitary, and that not only are groups formed by the multiplication of single variants having survival value, but also that such divergences are amplified to produce adaptations (both specializations and organization). It has been customary to admit that certain ancillary processes are operative (isolation, correlation), but the importance of these, as active principles, is subordinate to selection (1936, pp. 370–371).

Darwinism, as a set of ideas, is sufficiently broad and variously defined to include a multitude of truths and sins. Darwin himself disavowed many interpretations made in his name (1880, for example). The version known as the "modern synthesis" or "Neo-Darwinism" (different from what the late 19th century called Neo-Darwinism—see Romanes, 1900) is, I think, fairly characterized in its essentials by Robson and Richards. Its foundation rests upon two major premises: (1) Point mutations (micromutations) are the ultimate source of variability. Evolutionary change is a process of gradual allelic sub-

From *Paleobiology* 6, no. 1 (1980): 119 30. Reprinted by permission of the publisher and the author.

stitution within a population. Events at broader scale, from the origin of new species to long-ranging evolutionary trends, represent the same process, extended in time and effect—large numbers of allelic substitutions incorporated sequentially over long periods of time. In short, gradualism, continuity and evolutionary change by the transformation of populations. (2) Genetic variation is raw material only. Natural selection directs evolutionary change. Rates and directions of change are controlled by selection with little constraint exerted by raw material (slow rates are due to weak selection, not insufficient variation). All genetic change is adaptive (though some phenotypic effects, due to pleiotropy, etc., may not be). In short, selection leading to adaptation.

All these statements, as Robson and Richards also note, are subject to recognized exceptions—and this imposes a great frustration upon anyone who would characterize the modern synthesis in order to criticize it. All the synthesists recognized exceptions and "ancillary processes," but they attempted both to prescribe a low relative frequency for them and to limit their application to domains of little evolutionary importance. Thus, genetic drift certainly occurs— but only in populations so small and so near the brink that their rapid extinction will almost certainly ensue. And phenotypes include many non-adaptive features by allometry and pleiotropy, but all are epiphenomena of primarily adaptive genetic changes and none can have any marked effect upon the organism (for, if inadaptive, they will lead to negative selection and elimination and, if adaptive, will enter the model in their own right). Thus, a synthesist could always deny a charge of rigidity by invoking these official exceptions, even though their circumscription, both in frequency and effect, actually guaranteed the hegemony of the two cardinal principles. This frustrating situation had been noted by critics of an earlier Darwinian orthodoxy, by Romanes writing of Wallace, for example (1900, p. 21):

> [For Wallace,] the law of utility is, to all intents and purposes, universal, with the result that natural selection is virtually the only cause of organic evolution. I say "to all intents and purposes," or "virtually," because Mr. Wallace does not expressly maintain the abstract impossibility of laws and causes other than those of utility and natural selection; indeed, at the end of his treatise, he quotes with approval Darwin's judgment, that "natural selection has been the most important, but not the exclusive means of modification." Nevertheless, as he nowhere recognizes any other law or cause of adaptive evolution, he practically concludes that, on inductive or empirical grounds, there *is* no such other law or cause to be entertained.

Lest anyone think that Robson and Richards, as doubters, had characterized the opposition unfairly, or that their two principles represent too simplistic or unsubtle a view of the synthetic theory, I cite the characterization of one of the architects of the theory himself (Mayr 1963, p. 586—the first statement of his chapter on species and transspecific evolution):

The proponents of the synthetic theory maintain that all evolution is due to the accumulation of small genetic changes, guided by natural selection, and that transspecific evolution is nothing but an extrapolation and magnification of the events that take place within populations and species.

The early classics of the modern synthesis—particularly Dobzhansky's first edition (1937) and Simpson's first book (1944)—were quite expansive generous and pluralistic. But the synthesis hardened throughout the late 40' and 50's, and later editions of the same classics (Dobzhansky 1951; Simpsor 1953) are more rigid in their insistence upon micromutation, gradua transformation and adaptation guided by selection (see Gould 1980 for an analysis of changes between Simpson's two books). When Watson and Crick then determined the structure of DNA, and when the triplet code was cracked a few years later, everything seemed to fall even further into place. Chromosomes are long strings of triplets coding, in sequence, for the proteins that build organisms. Most point mutations are simple base substitutions. A physics and chemistry had been added, and it squared well with the prevailing orthodoxy.

I well remember how the synthetic theory beguiled me with its unifying power when I was a graduate student in the mid-1960s. Since then I have been watching it slowly unravel as a universal description of evolution. The molecular assault came first, followed quickly by renewed attention to unorthodox theories of speciation and by challenges at the level of macro-evolution itself. I have been reluctant to admit it - since beguiling is often forever but if Mayr's characterization of the synthetic theory is accurate, then that theory, as a general proposition, is effectively dead, despite its persistence as textbook orthodoxy.

II. REDUCTION AND HIERARCHY

The modern synthetic theory embodies a strong faith in reductionism. It advocates a smooth extrapolation across all levels and scales—from the base substitution to the origin of higher taxa. The most sophisticated of leading introductory textbooks in biology still proclaims:

> [Can] more extensive evolutionary change, macroevolution, be explained as an outcome of these microevolutionary shifts. Did birds really arise from reptiles by an accumulation of gene substitutions of the kind illustrated by the raspberry eye-color gene.
>
> The answer is that it is entirely plausible, and no one has come up with a better explanation. . . . The fossil record suggests that macroevolution is indeed gradual, paced at a rate that leads to the conclusion that it is based upon hundreds or thousands of gene substitutions no different in kind from the ones examined in our case histories (Wilson et al. 1973, pp. 793-794).

The general alternative to such reductionism is a concept of hierarchy—a world constructed not as a smooth and seamless continuum, permitting simple extrapolation from the lowest level to the highest, but as a series of ascending levels, each bound to the one below it in some ways and independent in others. Discontinuities and seams characterize the transitions; "emergent" features, not implicit in the operation of processes at lower levels, may control events at higher levels. The basic processes—mutation, selection, etc.—may enter into explanations at all scales (and in that sense we may still hope for a general theory of evolution), but they work in different ways on the characteristic material of divers levels (see Bateson 1978 and Koestler 1978, for all its other inadequacies, for good discussions of hierarchy and its anti-reductionistic implications; Eldredge and Cracraft 1980).

The molecular level, which once seemed through its central dogma and triplet code to provide an excellent "atomic" basis for smooth extrapolation, now demands hierarchical interpretation itself. The triplet code is only machine language (I thank E. Yates for this appropriate metaphor). The program resides at a higher level of control and regulation—and we know virtually nothing about it. With its inserted sequences and jumping genes, the genome contains sets of scissors and pots of glue to snip and unite bits and pieces from various sources. Thirty to seventy percent of the mammalian genome consists of repetitive sequences, some repeated hundreds or thousands of times. What are they for (if anything)? What role do they play in the regulation of development? Molecular biologists are groping to understand this higher control upon primary products of the triplet code. In that understanding, we will probably obtain a basis for styles of evolutionary change radically different from the sequential allelic substitutions, each of minute effect, that the modern synthesis so strongly advocated. The uncovering of hierarchy on the molecular level will probably exclude smooth continuity across other levels. (We may find, for example, that structural gene substitutions control most small-scale, adaptive variation within local populations, while disruption of regulation lies behind most key innovations in macroevolution.)

The modern synthesis drew most of its direct conclusions from studies of local populations and their immediate adaptations. It then extrapolated the postulated mechanism of these adaptations—gradual, allelic substitution—to encompass all larger-scale events. The synthesis is now breaking down on both sides of this argument. Many evolutionists now doubt exclusive control by selection upon genetic change within local populations. Moreover, even if local populations alter as the synthesis maintains, we now doubt that the same style of change controls events at the two major higher levels: speciation and patterns of macroevolution.

III. A NOTE ON LOCAL POPULATIONS AND NEUTRALITY

At the level of populations, the synthesis has broken on the issue of amounts of genetic variation. Selection, though it eliminates variation in both its classical modes (directional and, especially, stabilizing) can also act to preserve variation through such phenomena as overdominance, frequency dependence, and response to small-scale fluctuation of spatial and temporal environments. Nonetheless, the copiousness of genetic variation, as revealed first in the electrophoretic techniques that resolve only some of it (Lewontin and Hubby 1966; Lewontin 1974), cannot be encompassed by our models of selective control (of course, the models, rather than nature, may be wrong). This fact has forced many evolutionists, once stout synthesists themselves, to embrace the idea that alleles often drift to high frequency or fixation, and that many common variants are therefore neutral or just slightly deleterious. This admission lends support to a previous interpretation of the approximately even ticking of the molecular clock (Wilson 1977)—that it reflects the neutral status of most changes in structural genes rather than a grand averaging of various types of selection over time.

None of this evidence, of course, negates the role of conventional selection and adaptation in molding parts of the phenotype with obvious importance for survival and reproduction. Still, it rather damps Mayr's enthusiastic claim for "all evolution . . . guided by natural selection." The question, as with so many issues in the complex sciences of natural history, becomes one of relative frequency. Are the Darwinian substitutions merely a surface skin on a sea of variation invisible to selection, or are the neutral substitutions merely a thin bottom layer underlying a Darwinian ocean above? Or where in between?

In short, the specter of stochasticity has intruded upon explanations of evolutionary *change*. This represents a fundamental challenge to Darwinism, which holds, as its very basis, that random factors enter only in the production of raw material, and that the deterministic process of selection produces change and direction (see Nei 1975).

IV. THE LEVEL OF SPECIATION AND THE GOLDSCHMIDT BREAK

Ever since Darwin called his book *The Origin of Species*, evolutionists have regarded the formation of reproductively isolated units by speciation as a fundamental process of large-scale change. Yet speciation occurs at too high a level to be observed directly in nature or produced by experiment in most cases. Therefore, theories of speciation have been based on analogy, extrapolation and inference. Darwin himself focused on artificial selection and geographic variation. He regarded subspecies as incipient species and viewed their gradual, accumulating divergence as the primary mode of origin for new taxa. The modern synthesis continued this tradition of extrapolation from local populations and used the accepted model for adaptive geographic variation—

gradual allelic substitution directed by natural selection—as a paradigm for the origin of species. Mayr's (1942, 1963) model of allopatric speciation did challenge Darwin's implied notion of sympatric continuity. It emphasized the crucial role of isolation from gene flow and did promote the importance of small founding populations and relatively rapid rates of change. Thus, the small peripheral isolate, rather than the large local population in persistent contact with other conspecifics, became the incipient species. Nonetheless, despite this welcome departure from the purest form of Darwinian gradualism, the allopatric theory held firmly to the two major principles that permit smooth extrapolation from the *Biston betularia* model of adaptive, allelic substitution: (i) The accumulating changes that lead to speciation are adaptive. Reproductive isolation is a consequence of sufficient accumulation. (ii) Although aided by founder effects and even (possibly) by drift, although dependent upon isolation from gene flow, although proceeding more rapidly than local differentiation within large populations, successful speciation is still a cumulative and sequential process powered by selection through large numbers of generations. It is, if you will, Darwinism a little faster.

I have no doubt that many species originate in this way; but it now appears that many, perhaps most, do not. The new models stand at variance with the synthetic proposition that speciation is an extension of microevolution within local populations. Some of the new models call upon genetic variations of a different kind, and they regard reproductive isolation as potentially primary and non-adaptive rather than secondary and adaptive. Insofar as these new models be valid in theory and numerically important in application, speciation is not a simple "conversion" to larger effect of processes occurring at the lower level of adaptive modeling within local populations. It represents a discontinuity in our hierarchy of explanations, as the much maligned Richard Goldschmidt argued explicitly in 1940.

There are many ways to synthesize the swirling set of apparently disparate challenges that have rocked the allopatric orthodoxy and established an alternative set of models for speciation. The following reconstruction is neither historically sequential nor the only logical pathway of linkage, but it does summarize the challenges—on population structure, place of origin, genetic style, rate, and relation to adaptation—in some reasonable order.

1. Under the allopatric orthodoxy, species are viewed as integrated units which, if not actually panmictic, are at least sufficiently homogenized by gene flow to be treated as entities. This belief in effective homogenization within central populations underlies the allopatric theory with its emphasis on *peripheral* isolation as a precondition for speciation. But many evolutionists now believe that gene flow is often too weak to overcome selection and other intrinsic processes within local demes (Ehrlich and Raven 1969). Thus, the model of a large, homogenized central population preventing local differentiation and requiring allopatric "flight" of isolated demes for speciation may not be generally valid. Perhaps most local demes have the required independence for potential speciation.

2. The primary terms of reference for theories of speciation—allopatry and sympatry—lose their meaning if we accept the first statement. Objections to sympatric speciation centered upon the homogenizing force of gene flow. But if demes may be independent in all geographic domains of a species, then sympatry loses its meaning and allopatry its necessity. Independent demes within the central range (sympatric by location) function, in their freedom from gene flow, like the peripheral isolates of allopatric theory. In other words, the terms make no sense outside a theory of population structure that contrasts central panmixia with marginal isolation. They should be abandoned.

3. In this context "sympatric" speciation loses its status as an extremely improbable event. If demes are largely independent, new species may originate anywhere within the geographic range of an ancestral form. Moreover, many evolutionists now doubt that parapatric distributions (far more common than previously thought) must represent cases of secondary contact. White (1978, p. 342) believes that many, if not most, are primary and that speciation can also occur between populations continually in contact if gene flow can be overcome either by strong selection or by the sheer rapidity of potential fixation for major chromosomal variants (see White, 1978, p. 17 on clinal speciation).

4. Most "sympatric" models of speciation are based upon rates and styles of genetic change inconsistent with the reliance placed by the modern synthesis on slow, or at least sequential change.

The most exciting entry among punctuational models for speciation in ecological time is the emphasis, now coming from several quarters, on chromosomal alterations as isolating mechanisms (White 1978; Bush 1975; Carson 1975, 1978; Wilson et al. 1975; Bush et al. 1977)—sometimes called the theory of chromosomal speciation. In certain population structures, particularly in very small and circumscribed groups with high degrees of inbreeding, major chromosomal changes can rise to fixation in less than a handful of generations (mating of heterozygous F_1 sibs to produce F_2 homozygotes for a start).

Allan Wilson, Guy Bush and their colleagues (Wilson et al. 1975; Bush et al. 1977) find a strong correlation between rates of karyotypic and anatomical change, but no relation between amounts of substitution in structural genes and any conventional assessment of phenotypic modification, either in speed or extent. They suggest that speciation may be more a matter of gene regulation and rearrangement than of changes in structural genes that adapt local populations in minor ways to fluctuating environments (the *Biston betularia* model).

Carson (1975, 1978) has also stressed the importance of small demes, chromosomal change, and extremely rapid speciation in his founder-flush theory with its emphasis on extreme bottlenecking during crashes of the flush-crash cycle (see Powell 1978 for experimental support). Explicitly contrasting this view with extrapolationist models based on sequential substitution of structural genes, he writes (1975, p. 88):

Most theories of speciation are wedded to gradualism, using the mode of origin of intraspecific adaptations as a model . . . I would nevertheless like to propose . . . that speciational events may be set in motion and important genetic saltations towards species formation accomplished by a series of catastrophic, stochastic genetic events . . . initiated when an unusual forced reorganization of the epistatic supergenes of the closed variability system occurs . . . I propose that this cycle of disorganization and reorganization be viewed as the essence of the speciation process.

5. Another consequence of such essentially saltational origin is even more disturbing to conventional views than the rapidity of the process itself, as Carson has forcefully stated. The control of evolution by selection leading to adaptation lies at the heart of the modern synthesis. Thus, reproductive isolation, the definition of speciation, is attained as a by-product of adaptation—that is, a population diverges by sequential adaptation and eventually becomes sufficiently different from its ancestor to foreclose interbreeding. (Selection for reproductive isolation may also be direct when two imperfectly-separate forms come into contact.) But in saltational, chromosomal speciation, reproductive isolation comes first and cannot be considered as an adaptation at all. It is a stochastic event that establishes a species by the technical definition of reproductive isolation. To be sure, the later success of this species in competition may depend upon its subsequent acquisition of adaptations; but the origin itself may be non-adaptive. We can, in fact, reverse the conventional view and argue that speciation, by forming new entities stochastically, provides raw material for selection.

These challenges can be summarized in the claim that a discontinuity in explanation exists between allelic substitutions in local populations (sequential, slow and adaptive) and the origin of new species (often discontinuous and non-adaptive). During the heyday of the modern synthesis, Richard Goldschmidt was castigated for his defense of punctuational speciation. I was told as a graduate student that the great geneticist had gone astray because he had been a lab man with no feel for nature, a person who hadn't studied the adaptation of local populations and couldn't appreciate its potential power, by extrapolation, to form new species. But I discovered, in writing *Ontogeny and Phylogeny,* that Goldschmidt had spent a good part of his career studying geographic variations, largely in the coloration of lepidopteran larva (where he developed the concept of rate genes to explain minor changes in pattern). I then turned to his major book (1940) and found that his defense of saltational speciation is not based on ignorance of geographic variation, but on an explicit study of it; half the book is devoted to this subject. Goldschmidt concludes that geographic variation is ubiquitous, adaptive, and essential for the persistence of established species. But it is simply not the stuff of speciation; it is a different process. Speciation, Goldschmidt argues, occurs at different rates and uses different kinds of genetic variation. We do not now accept all his arguments about the nature of variation, but his explicit anti-extrapolationist statement is the epitome and foundation of emerging views on speciation discussed in this section. There is a discontinuity in cause and explanation between adaptation in local populations and speciation; they represent two distinct, though interacting, levels of evolution. We might refer to this discontinuity as the *Goldschmidt break*, for he wrote:

> The characters of subspecies are of a gradient type, the species limit is characterized by a gap, an unbridged difference in many characters. This gap cannot be bridged by theoretically continuing the subspecific gradient or cline beyond its actually existing limits. The subspecies do not merge into the species either actually or ideally . . . Microevolution by accumulation of micromutations—we may also say neo-Darwinian evolution—is a process which leads to diversification strictly within the species, usually, if not exclusively, for the sake of adaptation of the species to specific conditions within the area which it is able to occupy. . . Subspecies are actually, therefore, neither incipient species nor models for the origin of species. They are more or less diversified blind alleys within the species. The decisive step in evolution, the first step towards macroevolution, the step from one species to another, requires another evolutionary method than that of sheer accumulation of micromutations (1940, p. 183).

V. MACROEVOLUTION AND THE WRIGHT BREAK

The extrapolationist model of macroevolution views trends and major transitions as an extension of allelic substitution within populations—the march of frequency distributions through time. Gradual change becomes the normal state of species. The discontinuities of the fossil record are all attributed to its notorious imperfection; the remarkable stasis exhibited by most species during millions of years is ignored (as no data), or relegated to descriptive sections of taxonomic monographs. But gradualism is not the only important implication of the extrapolationist model. Two additional consequences have channeled our concept of macroevolution, both rather rigidly and with unfortunate effect. First, the trends and transitions of macroevolution are envisaged as events in the phyletic mode—populations transforming themselves steadily through time. Splitting and branching are acknowledged to be sure, lest life be terminated by its prevalent extinctions. But splitting becomes a device for the generation of diversity upon designs attained through "progressive" processes of transformation. Splitting, or cladogenesis, becomes subordinate in importance to transformation, or anagenesis (see Ayala 1976, p. 141; but see also Mayr 1963, p. 621 for a rather lonely voice in the defense of copious speciation as an input to "progressive" evolution). Secondly, the adaptationism that prevails in interpreting change in local populations gains greater confidence in extrapolation. For if allelic substitutions in ecological time have an adaptive basis, then surely a unidirectional trend that persists for millions of years within a single lineage cannot bear any other interpretation.

This extrapolationist model of adaptive, phyletic gradualism has been vigorously challenged by several paleobiologists—and again with a claim for discontinuity in explanation at different levels. The general challenge embodies three loosely united themes:

1. Evolutionary trends as a higher level process: Eldredge and I have argued (1972, and Gould and Eldredge 1977) that imperfections of the record cannot explain all discontinuity (and certainly cannot encompass stasis). We regard stasis and discontinuity as an expression of how evolution works when translated into geological time. Gradual change is not the normal state of a species. Large, successful central populations undergo minor adaptive modifications of fluctuating effect through time (Goldschmidt's "diversified blind alleys"), but they will rarely transform *in toto* to something fundamentally new. Speciation, the basis of macroevolution, is a process of branching. And this branching, under any current model of speciation—conventional allopatry to chromosomal saltation—is so rapid in geological translation (thousands of years at most compared with millions for the duration of most fossil species) that its results should generally lie on a bedding plane, not through the thick sedimentary sequence of a long hillslope. (The expectation of gradualism emerges as a kind of double illusion. It represents, first of all, an incorrect translation of conventional allopatry. Allopatric speciation seems so slow and gradual in ecological time that most paleontologists never recognized it as a challenge to the style of gradualism—steady change over millions of years— promulgated by custom as a model for the history of life. But it now appears that "slow" allopatry itself may be less important than a host of alternatives that yield new species rapidly even in ecological time.) Thus, our model of "punctuated equilibria" holds that evolution is concentrated in events of speciation and that successful speciation is an infrequent event punctuating the stasis of large populations that do not alter in fundamental ways during the millions of years that they endure.

But if species originate in geological instants and then do not alter in major ways, then evolutionary trends cannot represent a simple extrapolation of allelic substitution within a population. Trends must be the product of differential success among species (Eldredge and Gould 1972; Stanley 1975). In other words, species themselves must be inputs, and trends the result of their differential origin and survival. Speciation interposes itself as an irreducible level between change in local populations and trends in geological time. Macroevolution is, as Stanley argues (1975, p. 648), decoupled from microevolution.

Sewall Wright recognized the hierarchical implications of viewing species as irreducible inputs to macroevolution when he argued (1967, p. 121) that the relationship between change in local populations and evolutionary trends can only be analogical. Just as mutation is random with respect to the direction of change within a population, so too might speciation be random with respect to the direction of a macroevolutionary trend. A higher form of selection, acting directly upon species through differential rates of extinction, may then be the analog of natural selection working within populations through differential mortality of individuals.

Evolutionary trends therefore represent a third level superposed upon speciation and change within demes. Intrademic events cannot encompass speciation because rates, genetic styles, and relation to adaptation differ for the two processes. Likewise, since trends "use" species as their raw material, they represent a process at a higher level than speciation itself. They reflect a sorting out of speciation events. With apologies for the pun, the hierarchical rupture between speciation and macroevolutionary trends might be called the Wright break.[1]

As a final point about the extrapolation of methods for the study of events within populations, the cladogenetic basis of macroevolution virtually precludes any direct application of the primary apparatus for microevolutionary theory: classical population genetics. I believe that essentially all macroevolution is cladogenesis and its concatenated effects. What we call "anagenesis," and often attempt to delineate as a separate phyletic process leading to "progress," is just accumulated cladogenesis filtered through the directing force of species selection (Stanley 1975)—Wright's higher level analogy to natural selection. Carson (1978, p. 925) made the point forcefully, again recognizing Sewall Wright as its long and chief defender:

Investigation of cladistic events as opposed to phyletic (anagenetic) ones requires a different perspective from that normally assumed in classical population genetics. The statistical and mathematical comfort of the Hardy-Weinberg equilibrium in large populations has to be abandoned in favor of the vague realization that nearly everywhere in nature we are faced with data suggesting the partial or indeed complete sundering of gene pools. If we are to deal realistically with cladogenesis we must seek to delineate each genetic and environmental factor which may promote isolation. The most important devices are clearly those which operate at the very lowest population level: sib from sib, family from family, deme from deme. Formal population genetics just cannot deal with such things, as Wright pointed out long ago.

Eldredge (1979) has traced many conceptual errors and prejudicial blockages to our tendency for conceiving of evolution as the transformation of *characters* within phyletic lineages, rather than as the origin of new *taxa* by cladogenesis (the transformational versus the taxic view in his terms). I believe that, in ways deeper than we realize, our preference for transformational thinking represents a cultural tie to the controlling Western themes of progress and ranking by intrinsic merit—an attitude that can be traced in evolutionary thought to Lamarck's distinction between the march up life's ladder promoted by the *pouvoir de la vie* and the tangential departures imposed by *l'influence des circonstances,* with the first process essential and the second deflective. Nonetheless, macroevolution is fundamentally about the origin of taxa by splitting.

2. The saltational initiation of major transitions: The absence of fossil evidence for intermediary stages between major transitions in organic design,

indeed our inability, even in our imagination, to construct functional intermediates in many cases, has been a persistent and nagging problem for gradualistic accounts of evolution. St. George Mivart (1871), Darwin's most cogent critic, referred to it as the dilemma of "the incipient stages of useful structures"—of what possible benefit to a reptile is two percent of a wing? The dilemma has two potential solutions. The first, preferred by Darwinians because it preserves both gradualism and adaptation, is the principle of preadaptation: the intermediary stages functioned in another way but were, by good fortune in retrospect, preadapted to a new role they could play only after greater elaboration. Thus, if feathers first functioned "for" insulation and later "for" the trapping of insect prey (Ostrom 1979), a proto-wing might be built without any reference to flight.

I do not doubt the supreme importance of preadaptation, but the other alternative, treated with caution, reluctance, disdain or even fear by the modern synthesis, now deserves a rehearing in the light of renewed interest in development: perhaps, in many cases, the intermediates never existed. I do not refer to the saltational origin of entire new designs, complete in all their complex and integrated features—a fantasy that would be truly anti-Darwinian in denying any creativity to selection and relegating it to the role of eliminating old models. Instead, I envisage a potential saltational origin for the essential features of key adaptations. Why may we not imagine that gill arch bones of an ancestral agnathan moved forward in one step to surround the mouth and form proto-jaws? Such a change would scarcely establish the *Bauplan* of the gnathostomes. So much more must be altered in the reconstruction of agnathan design—the building of a true shoulder girdle with bony, paired appendages, to say the least. But the discontinuous origin of a proto-jaw might set up new regimes of development and selection that would quickly lead to other, coordinated modifications. Yet Darwin, conflating gradualism with natural selection as he did so often, wrongly proclaimed that any such discontinuity, even for organs (much less taxa) would destroy his theory:

> If it could be demonstrated that any complex organ existed, which could not possibly have been formed by numerous, successive, slight modifications my theory would absolutely break down (1859, p. 189).

During the past 30 years, such proposals have generally been treated as a fantasy signifying surrender—an invocation of hopeful monsters rather than a square facing of a difficult issue. But our renewed interest in development, the only discipline of biology that might unify molecular and evolutionary approaches into a coherent science, suggests that such ideas are neither fantastic, utterly contrary to genetic principles, nor untestable.

Goldschmidt conflated two proposals as causes for hopeful monsters—"systemic mutations" involving the entire genome (a spinoff from his fallacious belief that the entire genome acted as an integrated unit), and small mutations with large impact upon adult phenotypes because they work upon early stages of ontogeny and lead to cascading effects throughout embryology. We reject his first proposal, but the second, eminently plausible theme might unite a Darwinian insistence upon continuity of genetic change with a macroevolutionary suspicion of phenetic discontinuity. It is, after all, a major focus in the study of heterochrony (effects, often profound, of small changes in developmental rate upon adult phenotypes); it is also implied in the emphasis now being placed upon regulatory genes in the genesis of macroevolutionary change (King and Wilson 1975)—for regulation is fundamentally about timing in the complex orchestration of development. Moreover, although we cannot readily build "hopeful monsters," the subject of major change through alteration of developmental rate can be treated, perhaps more than analogically, both by experiment and comparative biology. The study of spontaneous anomalies of development (teratology) and experimental perturbations of embryogenic rates explores the tendencies and boundaries of developmental systems and allows us to specify potential pathways of macroevolutionary change (see, for example, the stunning experiment of Hampé 1959, on recreation of reptilian patterns in birds, after 200 million years of their phenotypic absence, by experimental manipulations that amount to alterations in rate of development for the fibula). At the very least, these approaches work with real information and seem so much more fruitful than the construction of adaptive stories or the invention of hypothetical intermediates.

3. The importance of non-adaptation: The emphasis on natural selection as the only directing force of any importance in evolution led inevitably to an analysis of all attributes of organisms as adaptations. Indeed, the tendency has infected our language, for, without thinking about what it implies, we use "adaptation" as our favored, *descriptive* term for designating any recognizable bit of changed morphology in evolution. I believe that this "adaptationist program" has had decidedly unfortunate effects in biology (Gould and Lewontin, 1979). It has led to a reliance on speculative storytelling in preference to the analysis of form and its constraints; and, if wrong, in any case, it is virtually impossible to dislodge because the failure of one story leads to invention of another rather than abandonment of the enterprise.

Yet, as I argued earlier, the hegemony of adaptation has been broken at the two lower levels of our evolutionary hierarchy: variation within populations, and speciation. Most populations may contain too much variation for selection to maintain; moreover, if the neutralists are even part right, much allelic substitution occurs without controlling influence from selection, and with no direct relationship to adaptation. If species often form as a result of major chromosomal alterations, then their origin—the establishment of reproductive isolation—may require no reference to adaptation. Similarly, at this third level of macroevolution, both arguments previously cited against the conventional extrapolationist view require that we abandon strict adaptationism.

i) If trends are produced by the unidirectional transformation of populations (orthoselection), then they can scarcely receive other than a conventional adaptive explanation. After all, if adaptation lies behind single allelic substitutions in the *Biston betularia* model for change in local populations, what else but even stronger, more persistent selection and adaptive orientation can render a trend that persists for millions of years? But if trends represent a higher-level process of differential origin and mortality among species, then a suite of potentially non-adaptive explanations must be considered. Trends, for example, may occur because some kinds of species tend to speciate more often than others. This tendency may reside in the character of environments or in attributes of behavior and population structure bearing no relationship to morphologies that spread through lineages as a result of higher speciation rates among some of their members. Or trends may arise from the greater longevity of certain kinds of species. Again, this greater persistence may have little to do with the morphologies that come to prevail as a result. I suspect that many morphological trends in paleontology—a bugbear of the profession because we have been unable to explain them in ordinary adaptive terms—are non-adaptive sequelae of differential species success based upon environments and population structures.

ii) If transitions represent the continuous and gradual transformation of populations, then they must be regulated by adaptation throughout (even though adaptive orientation may alter according to the principle of preadaptation). But if discontinuity arises through shifts in development, then directions of potential change may be limited and strongly constrained by the inherited program and developmental mechanics of an organism. Adaptation may determine whether or not a hopeful monster survives, but primary constraint upon its genesis and direction resides with inherited ontogeny, not with selective modeling.

VI. QUO VADIS?

My crystal ball is clouded both by the dust of these growing controversies and by the mists of ignorance emanating from molecular biology, where even the basis of regulation in eukaryotes remains shrouded in mystery. I think I can see what is breaking down in evolutionary theory—the strict construction of the modern synthesis with its belief in pervasive adaptation, gradualism, and extrapolation by smooth continuity from causes of change in local populations to major trends and transitions in the history of life. I do not know what will take its place as a unified theory, but I would venture to predict some themes and outlines.

The new theory will be rooted in a hierarchical view of nature. It will not embody the depressing notion that levels are fundamentally distinct and necessarily opposed to each other in their identification of causes (as the older paleontologists held in maintaining that macroevolution could not, in principle, be referred to the same causes that regulate microevolution—e.g., Osborn 1922). It will possess a common body of causes and constraints, but will recognize that they work in characteristically different ways upon the material of different levels—intrademic change, speciation, and patterns of macroevolution.

As its second major departure from current orthodoxy, the new theory will restore to biology a concept of organism. In an exceedingly curious and unconscious bit of irony, strict selectionism (which was not, please remember, Darwin's own view) debased what had been a mainstay of biology—the organism as an integrated entity exerting constraint over its history. St. George Mivart expressed the subtle point well in borrowing a metaphor from Galton. I shall call it Galton's polyhedron. Mivart writes (1871, pp. 228–229):

> This conception of such internal and latent capabilities is somewhat like that of Mr. Galton . . . according to which the organic world consists of entities, each of which is, as it were, a spheroid with many facets on its surface, upon one of which it reposes in stable equilibrium. When by the accumulated action of incident forces this equilibrium is disturbed, the spheroid is supposed to turn over until it settles on an adjacent facet once more in stable equilibrium. The internal tendency of an organism to certain considerable and definite changes would correspond to the facets on the surface of the spheroid.

Under strict selectionism the organism is a sphere. It exerts little constraint upon the character of its potential change; it can roll along all paths. Genetic variation is copious, small in its increments, and available in all directions— the essence of the term "random" as used to guarantee that variation serves as raw material only and that selection controls the direction of evolution.

By invoking Galton's polyhedron, I recommend no return to the antiquated and anti-Darwinian view that mysterious "internal" factors provide direction inherently, and that selection only eliminates the unfit (orthogenesis, various forms of vitalism and finalism). Instead, the facets are constraints exerted by the developmental integration of organisms themselves. Change cannot occur in all directions, or with any increment; the organism is not a metaphorical sphere. When the polyhedron tumbles, selection may usually be the propelling force. But if adjacent facets are few in number and wide in spacing, then we cannot identify selection as the only, or even the primary control upon evolution. For selection is channeled by the form of the polyhedron it pushes, and these constraints may exert a more powerful influence upon evolutionary directions than the external push itself. This is the legitimate sense of a much maligned claim that "internal factors" are important in evolution. They channel and constrain Darwinian forces; they do not stand in opposition to them. Most of the other changes in evolutionary viewpoint that I have advocated throughout this paper fall out of Galton's metaphor: punctuational change at all levels (the flip from facet to facet, since homeostatic systems change by abrupt shifting to new equilibria); essential non-adaptation, even in major parts of the phenotype (change in an integrated organism often has effects that reverberate throughout the system); channeling of direction by constraints of history and developmental architecture. Organisms are not billiard balls, struck in deterministic fashion by the cue of natural selection, and rolling to optimal positions on life's table. They influence their own destiny in interesting, complex, and comprehensible ways. We must put this concept of organism back into evolutionary biology.

NOTE

1. I had the honor—not a word I use frequently, but inescapable in this case—of spending a long evening with Dr. Wright last year. I discovered that his quip about macroevolution, just paraphrased, was no throwaway statement but an embodiment of his deep commitment to a hierarchical view of evolutionary causation. (The failure of many evolutionists to think hierarchically is responsible for the most frequent misinterpretation of Wright's views. He never believed that genetic drift—the Sewall Wright effect as it once was called—is an important agent of evolutionary *change*. He regards it as input to the directional process of interdemic selection for evolution within species. Drift can push a deme off an adaptive peak; selection can then draw it to another peak.)

REFERENCES

Ayala, F. J. 1976. "Molecular Genetics and Evolution." Pp. 1-20. In: Ayala, F. J., ed. *Molecular Evolution*. Sinauer Associates; Sunderland, Mass.

Bateson, G. 1978. *Mind and Nature*. E. P. Dutton; New York.

Bush, G. L. 1975. "Modes of Animal Speciation." *Annual Review of Ecological Systems* 339-364.

Bush, G. L., S. M. Case, A. C. Wilson, and J. L. Patton. 1977. "Rapid Speciation and Chromosomal Evolution in Mammals," *Proceedings of the National Academy of Science*. 74:3942-3946.

Carson, H. L. 1975. "The Genetics of Speciation at the Diploid Level," *American Naturalist 109*:83-92.

———. 1978. "Chromosomes and Species Formation," *Evolution*. 32:925-927.

Darwin, C. 1859. *On the Origin of Species*. 490 pp. John Murray; London.

———. 1880. "Sir Wyville Thomson and Natural Selection, *Nature 23*:32.

Dobzhansky, T. 1937. *Genetics and the Origin of Species*. 364 pp. Columbia University Press; New York.

———. 1951. *Genetics and the Origin of Species*. (3rd ed.) 364 pp. Columbia University Press; New York.

Ehrlich, D. R. and P. H. Raven. 1969. "Differentiation of Populations," *Science. 165:* 1228-1232.

Eldredge, N. 1979. "Alternative Approaches to Evolutionary Theory," *Bulletin of the Carnegie Museum of History*, pp. 7-19.

Eldredge, N. and J. Cracraft, 1980. *Phlagenetic Patterns and the Evolutionary Process*. Columbia University Press; New York.

Eldredge, N. and S. J. Gould. 1972. "Punctuated Equilibria; An Alternative to Phyletic Gradualism. Pp. 82-115. In Schopf, T. J. M., ed. *Models in Paleobiology*. Freeman, Cooper and Co.; San Francisco, Calif.

Goldschmidt, R. 1940. *The Material Basis of Evolution*. 436 pp. Yale University Press;

New Haven, Conn.

Gould, S. J. 1980. "G. G. Simpson. Paleontology and the Modern Synthesis." In: Mayr, E., ed. *Conference on the Making of the Modern Synthesis.* Harvard University Press; Cambridge, Mass.

Gould, S. J. and R. C. Lewontin, 1979. "The Spandrels of San Marco and the Panglossian Paradigm: A Critique of the Adaptationist Program." *Proceedings of the Royal Society.* London. *205:* 581-598.

Hampe, A. 1959. "Contribution à l' étude du dèvelopement et de la regulation des déficiences et des excédents dans la patte de l'embryon de poulet." *Arch. Anat. Microsc. Morphol. Exp. 48:*345-478.

King, M. C. and A. C. Wilson. 1975. "Evolution at Two Levels in Humans and Champanziees," *Science. 188:*107-116.

Koestler, A. 1978. *Janus: A Summing Up.* Random House; New York.

Lewontin, R. C. 1974. *The Genetic Basis of Evolutionary Change.* 346 pp. Columbia University Press; New York.

Lewontin, R. C. and J. L. Hubby. 1966. "A Molecular Approach to the Study of Genic Heterozygosity in Natural Populations. Amount of Variation and Degree of Heterozygosity in Natural Populations of *Drosophila pseudoobscura.*" *Genetics. 54:*595-60.

Mayr, E. 1942. *Systematics and the Origin of Species.* 33-4 pp. Columbia University Press; New York.

———. *Animal Species and Evolution.* 797 pp. Belknap Press of Harvard University Press; Cambridge, Mass.

Mivart, St. G. 1871. *On the Genesis of Species.* 296 pp. MacMillan; London.

Nei, M. 1975. *Molecular Population Genetics and Evolution.* American Elsevier; New York.

Osborn, H. F. 1922. "Orthogenesis as Observed from Paleontological Evidence Beginning in the Yyear 1889," *American Naturalist. 56:*134-143.

Ostrom. J. H. 1979. "Bird Flight: How Did It Begin. *American Scientist. 67:*46-56.

Powell, J. R. 1978. "The Founder-Flush Speciation Theory: An Experimental Approach. *Evolution. 32:*465-474.

Robson, G. C. and O. W. Richards, 1936. *The Variation of Animals in Nature.* Longmans, Green, and Co.; London.

Romanes, G. J. 1900. *Darwin and After Darwin,* vol. 2 *Post Darwinian Questions. Heredity and Utility.* 344 pp. Longmans, Green, and Co.; London.

Simpson, G. G. 1944. *Tempo and Mode in Evolution.* 237 pp. Columbia University Press; New York.

———. 1953. *The Major Features of Evolution.* 434 pp. Columbia University Press; New York.

Stanley, S. M. 1975. "A Theory of Evolution Above the Species Level," *Proceedings of the National Academy of Science. 72:*646-650.

White, M. J. D. 1978. *Modes of Speciation.* 455 pp. W. H. Freeman; San Francisco, Calif.

Wilson, A. C., G. L. Bush, S. M. Case, and M. C. King. 1975. "Social Structuring of Mammalian Populations and Rates of Chromosomal Evolution, *Proceedings of the National Academy of Science. 72:*5061-5065.

Wilson, A. C. S. S. Carlson, and T. J. White. 1977, "Bio Chemical Evolution." *Annual Review of Biochemistry. 46:*573-639.

Wilson, E. O. et al. 1973. *Life on Earth.* Sinauer Associates; Sunderland, Mass.

Wright, S. 1967. "Comments on the Preliminary Working Papers of Eden and Waddington." In: Moorehead, P. S. and M. M. Kaplan, eds. *Mathematical Challenges to the Neo-Darwinian Theory of Evolution.* Witsar Inst. Symp. *5:*117-120.

13

Did Darwin Get It Right?

John Maynard Smith

I think I can see what is breaking down in evolutionary theory—the strict construction of the modern synthesis with its belief in pervasive adaptation, gradualism and extrapolation by smooth continuity from causes of change in local populations to major trends and transitions in the history of life.

A new and general evolutionary theory will embody this notion of hierachy and stress a variety of themes either ignored or explicitly rejected by the modern synthesis.

These quotations come from a recent paper in *Palaeobiology* by Stephen Jay Gould. What is the new theory? Is it indeed likely to replace the currently orthodox "neo-Darwinian" view? Proponents of the new view make a minimum and a maximum claim. The minimum claim is an empirical one concerning the nature of the fossil record. It is that species, once they come into existence, persist with little or no change, often for millions of years ("stasis"), and that evolutionary change is concentrated into relatively brief periods ("punctuation"), these punctuational changes occurring at the moment when a single species splits into two. The maximal claim is a deduction from this, together with arguments drawn from the study of development: it is that evolutionary change, when it does occur, is not caused by natural selection operating on the genetic differences between members of populations, as Darwin argued and as most contemporary evolutionists would agree, but by some other process. I will discuss these claims in turn; as will be apparent, it would be possible to accept the first without being driven to accept the second.

From *London Review of Books* 3, no. 11 (1981): 10 11. Reprinted by permission of the publisher and the author.

The claim of stasis and punctuation will ultimately be settled by a study of the fossil record. I am not a palaeontologist, and it might therefore be wiser if I were to say merely that some palaeontologists assert that it is true, and others are vehemently denying it. There is something, however, that an outsider can say. It is that the matter can be settled only by a statistical analysis of measurements of fossil populations from different levels in the rocks, and not by an analysis of the lengths of time for which particular named species or genera persist in the fossil record. The trouble with the latter method is that one does not know whether one is studying the rates of evolution of real organisms, or merely the habits of the taxonomists who gave the names to the fossils. Suppose that in some lineage evolutionary change took place at a more or less steady rate, to such an extent that the earliest and latest forms are sufficiently different to warrant their being placed in different species. If there is at some point a gap in the record, because suitable deposits were not being laid down or have since been eroded, then there will be a gap in the sequence of forms, and taxonomists will give fossils before the gap one name and after it another. It follows that an analysis of named forms tells us little: measurements of populations, on the other hand, would reveal whether change was or was not occurring before and after the gap.

My reason for making this rather obvious point is that the only extended presentation of the punctuationist view—Stanley's book, *Macroevolution*—rests almost entirely on an analysis of the durations of named species and genera. When he does present population measurements, they tend to support the view that changes are gradual rather than sudden. I think that at least some of the changes he presents as examples of sudden change will turn out on analysis to point the other way. I was unable to find any evidence in the book which supported, let alone established, the punctuationist view.

Of course, that is not to say that the punctuationist view is not correct. One study, based on a proper statistical analysis, which does support the minimal claim, but not the maximal one, is Williamson's study of the freshwater molluscs (snails and bivalves) of the Lake Turkana region of Africa over the last five million years. Of the 21 species studied, most showed no substantial evolutionary change during the whole period: "stasis" was a reality. The remaining six species were more interesting. They also showed little change for most of the period. There was, however, a time when the water table fell and the lake was isolated from the rest of the rift valley. When this occurred, these six species changed rather rapidly. Through a depth of deposit of about one meter, corresponding roughly to 50,000 years, successive populations show changes of shape great enough to justify placing the later forms in different species. Later, when the lake was again connected to the rest of the rift valley, these new forms disappear suddenly, and are replaced by the original forms, which presumably re-entered the lake from outside, where they had persisted unchanged.

This is a clear example of stasis and punctuation. However, it offers no support for the view that changes, when they do occur, are not the result of selection acting within populations. Williamson does have intermediate populations, so we know that the change did not depend on the occurrence of a "hopeful monster" (see below), or on the existence of an isolated population small enough to permit random changes to outweigh natural selection. The example is also interesting in showing how we may be misled if we study the fossil record only in one place. Suppose that, when the water table rose again, the new form had replaced the original one in the rest of the rift valley, instead of the other way round. Then, if we had examined the fossil record anywhere else but in Lake Turkana, we would have concluded, wrongly, that an effectively instantaneous evolutionary change had occurred.

Williamson's study suggests an easy resolution of the debate. Both sides are right, and the disagreement is purely semantic. A change taking 50,000 years is sudden to a palaeontologist but gradual to a population geneticist. My own guess is that there is not much more to the argument than that. However, the debate shows no signs of going away.

One question that arises is how far the new ideas are actually new. Much less so, I think, than their proponents would have us believe. They speak and write as if the orthodox view is that evolution occurs at a rate which is not only "gradual" but uniform. Yet George Gaylord Simpson, one of the main architects of the "modern synthesis" now under attack, wrote a book, *Tempo and Mode in Evolution,* devoted to emphasizing the great variability of evolutionary rates. It has never been part of the modern synthesis that evolutionary rates are uniform.

Yet there is a real point at issue. If it turns out to be the case that all, or most, evolutionary change is concentrated into brief periods, and associated with the splitting of lineages, that would require some serious rethinking. Oddly enough, it is not so much the sudden changes which would raise difficulties, but the intervening stasis. Why should a species remain unchanged for millions of years? The explanation favored by most punctuationists is that there are "developmental constraints" which must be overcome before a species can change. The suggestion is that the members of a given species share a developmental pathway which can be modified so as to produce some kinds of change in adult structure rather easily, and other kinds of change only with great difficulty, or not at all. I do not doubt that this is true: indeed, in my book *The Theory of Evolution,* published in 1958 and intended as a popular account of the modern synthesis, I spent some time emphasizing that "the pattern of development of a given species is such that there are only a limited number of ways in which it can be altered without causing complete breakdown." Neo-Darwinists have never supposed that genetic mutation is equally likely to produce changes in adult structure in any direction: all that is assumed is that mutations do not, as a general rule, adapt organisms to withstand the agents which caused them. What is at issue, then, is not whether there are developmental constraints, because clearly there are, but whether such constraints can account for stasis in evolution.

I find it hard to accept such an explanation for stasis, for two reasons. The first is that artificial selection can and does produce dramatic morphological change: one has only to look at the breeds of dogs to appreciate that. The second is that species are not uniform in space. Most species with a wide geographical range show differences between regions. Often these differences are so great that one does not know whether the extreme forms would behave as a single species if they met. Occasionally we know that they would not. This requires that a ring of forms should arise, with the terminal links overlapping. The Herring Gull and Lesser Black-Backed Gull afford a familiar example. In Britain and Scandinavia they behave as distinct species, without hybridizing, but they are linked by a series of forms encircling the Arctic.

Stasis in time is, therefore, a puzzle, since it seems not to occur in space. The simplest explanation is that species remain constant in time if their environments remain constant. It is also worth remembering that the hard parts of marine invertebrates, on which most arguments for stasis are based, tell us relatively little about the animals within. There are on our beaches two species of periwinkle whose shells are indistinguishable, but which do not interbreed and of which one lays eggs and the other bears live young.

The question of stasis and punctuation will be settled by a statistical analysis of the fossil record. But what of the wider issues? Is mutation plus natural selection within populations sufficient to explain evolution on a large scale, or must new mechanisms be proposed?

It is helpful to start by asking why Darwin himself was a believer in gradual change. The reason lies, I believe, in the nature of the problem he was trying to solve. For Darwin, the outstanding characteristic of living organisms which called for an explanation was the detailed way in which they are adapted to their forms of life. He knew that "sports"—structural novelties of large extent—did arise from time to time, but felt that fine adaptation could not be explained by large changes of this kind: it would be like trying to perform a surgical operation with a mechanically-controlled scalpel which could only be moved a foot at a time. Gruber has suggested that Darwin's equating of gradual with natural and of sudden with supernatural was a permanent feature of this thinking, which predated his evolutionary views and his loss of religious faith. It may have originated with Archbishop Sumner's argument (on which Darwin made notes when a student at Cambridge) that Christ must have been a divine rather than a human teacher because of the suddenness with which his teachings were accepted. Darwin seems to have retained the conviction that sudden changes are supernatural long after he had rejected Sumner's application of the idea.

Whatever the source of Darwin's conviction, I think he was correct both in his emphasis on detailed adaptation as the phenomenon to be explained, and in his conviction that to achieve such adaptation requires large numbers of selective events. It does not, however, follow that all the steps had to be small. I have always had a soft spot for "hopeful monsters": new types arising by genetic mutation, strikingly different in some respects from their parents, and taking a first step in the direction of some new adaptation, which could then be perfected by further smaller changes. We know that mutations of large effect occur: our only problem is whether they are ever incorporated during evolution, or are always eliminated by selection. I see no *a priori* reason why such large steps should not occasionally happen in evolution. What genetic evidence we have points the other way, however. On the relatively few occasions when related species differing in some morphological feature have been analyzed genetically, it has turned out, as Darwin would have expected had he known of the possibility, that the difference is caused by a number of genes, each of small effect.

As I see it, a hopeful monster would still stand or fall by the test of natural selection. There is nothing here to call for radical rethinking. Perhaps the greatest weakness of the punctuationists is their failure to suggest a plausible alternative mechanism. The nearest they have come is the hypothesis of "species selection." The idea is that when a new species arises, it differs from its ancestral species in ways which are random relative to any long-term evolutionary trends. Species will differ, however, in their likelihood of going extinct, and of splitting again to form new species. Thus selection will operate between species, favoring those characteristics which make extinction unlikely and splitting likely. In "species selection," as compared to classical individual selection, the species replaces the individual organism, extinction replaces death, the splitting of species into two replaces birth, and mutation is replaced by punctuational changes at the time of splitting.

Some such process must take place. I have argued elsewhere that it may have been a relevant force in maintaining sexual reproduction in higher animals. It is, however, a weak force compared to typical Darwinian between-individual selection, basically because the origin and extinction of species are rare events compared to the birth and death of individuals. Some critics of Darwinism have argued that the perfection of adaptation is too great to be accounted for by the selection of random mutations. I think, on quantitative grounds, that they are mistaken. If, however, they were to use the same argument to refute species selection as the major cause of evolutionary trends, they might well be right. For punctuationists, one way out of the difficulty would be to argue that adaptation is in fact less precise than biologists have supposed. Gould has recently tried this road. As it happens, I think he is right to complain of some of the more fanciful adaptive explanations that have been offered, but I also think that he will find that the residue of genuine adaptive fit between structure and function is orders of magnitude too great to be explained by species selection.

One other extension of the punctuationist argument is worth discussing. As explained above, stasis has been explained by developmental constraints. This amounts to saying that the developmental processes are such that only certain kinds of animal are possible and viable. The extension is to apply the same idea to explain the existence of the major patterns of organization, or "bauplans," observable in the natural world. The existence of such bau-plans is not at issue. For example, all vertebrates, whether swimming, flying, creeping or burrowing, have the same basic pattern of an internal jointed backbone with a hollow nerve cord above it and segmented body muscles on either side of it, and the vast majority have two pairs of fins, or of legs which are derived from fins (although a few have lost one or both pairs of appendages). Why should this be so?

Darwin's opinion is worth quoting. In *The Origin of Species,* he wrote:

> It is generally acknowledged that all organic beings have been formed on two laws—Unity of Type, and the Conditions of Existence. By unity of type is meant that fundamental agreement in structure which we see in organic beings of the same class, and which is quite independent of their habits of life. On my theory, unity of type is explained by unity of descent. The expression of conditions of existence, so often insisted on by the illustrious Cuvier, is fully embraced by the principle of natural selection. For natural selection acts by either now adapting the varying parts of each being to its organic and inorganic conditions of life; or by having adapted them during the long-past periods of time. . . Hence, in fact, the law of Conditions of Existence is the higher law; as it includes, through the inheritance of former adaptations, that of Unity of Type.

That is, we have two pairs of limbs because our remote ancestors had two pairs of fins, and they had two pairs of fins because that is an efficient number for a swimming animal to have.

I fully share Darwin's opinion. The basic vertebrate pattern arose in the first place as an adaptation for sinusoidal swimming. Early fish have two pairs of fins for the same reason that most early aeroplanes had wings and tailplane: two pairs of fins is the smallest number that can produce an upward or downward force through any point in the body. In the same vein, insects (which are descended from animals with many legs) have six legs because that is the smallest number which permits an insect to take half its legs off the ground and not fall over.

The alternative view would be that there are (as yet unknown) laws of form or development which permit only certain kinds of organisms to exist—for example, organisms with internal skeletons, dorsal nerve cords and four legs, or with external skeletons, ventral nerve cords and six legs—and which forbid all others, in the same way that the laws of physics permit only elliptical planetary orbits, or the laws of chemistry permit only certain compounds. This view is a manifestation of the "physics envy" which still infects some biologists. I believe it to be mistaken. In some cases it is demonstrably false. For example, some of the earliest vertebrates had more than two pairs of fins (just as some early aeroplanes had a noseplane as well as a tailplane). Hence there is no general law forbidding such organisms.

What I have said about bauplans does not rule out the possibility that there may be a limited number of kinds of unit developmental process which occur, and which are linked together in various ways to produce adult structures. The discovery of such processes would be of profound importance for biology, and would no doubt influence our views about evolution.

One last word needs to be said about bauplans. They may, as Darwin thought, have arisen in the first place as adaptations to particular ways of life, but, once having arisen, they have proved to be far more conservative in evolution than the way of life which gave them birth. Apparently it has been easier for organisms to adapt to new ways of life by modifying existing structures than by scrapping them and starting afresh. It is for this reason that comparative anatomy is a good guide to relationship.

Punctuationist views will, I believe, prove to be a ripple rather than a revolution. Their most positive achievement may be to persuade more people to study populations of fossils with adequate statistical methods. In the meanwhile, those who would like to believe that Darwin is dead, whether because they are creationists, or because they dislike the apparently Thatcherite conclusions which have been drawn from his theory, or find the mathematics of population genetics too hard for them, would be well advised to be cautious: the reports of his death have been exaggerated.

14

Universal Darwinism

Richard Dawkins

It is widely believed on statistical grounds that life has arisen many times all around the universe (Asimov, 1979; Billingham, 1981). However varied in detail alien forms of life may be, there will probably be certain principles that are fundamental to all life, everywhere. I suggest that prominent among these will be the principles of Darwinism. Darwin's theory of evolution by natural selection is more than a local theory to account for the existence and form of life on Earth. It is probably the only theory that *can* adequately account for the phenomena that we associate with life.

My concern is not with the details of other planets. I shall not speculate about alien biochemistries based on silicon chains, or alien neurophysiologies based on silicon chips. The universal perspective is my way of dramatizing the importance of Darwinism for our own biology here on Earth, and my examples will be mostly taken from Earthly biology. I do, however, also think that "exobiologists" speculating about extraterrestrial life should make more use of evolutionary reasoning. Their writings have been rich in speculation about how extraterrestrial life might work, but poor in discussion about how it might *evolve*. This essay should, therefore, be seen firstly as an argument for the general importance of Darwin's theory of natural selection; secondly as a preliminary contribution to a new discipline of "evolutionary exobiology."

The "growth of biological thought" (Mayr, 1982) is largely the story of Darwinism's triumph over alternative explanations of existence. The chief weapon of this triumph is usually portrayed as *evidence*. The thing that is said to be wrong with Lamarck's theory is that its assumptions are factually

From *Evolution from Molecules to Men*, ed. D. S. Bendall (Cambridge: Cambridge University Press, 1983), pp. 403–25. Reprinted by permission of the publisher and the author.

wrong. In Mayr's words: "Accepting his premises. Lamarck's theory was as legitimate a theory of adaptation as that Darwin. Unfortunately, these premises turned out to be invalid." But I think we can say something stronger: *even accepting his premises,* Lamarck's theory is *not* as legitimate a theory of adaptation as that of Darwin because, unlike Darwin's, it is *in principle* incapable of doing the job we ask of it—explaining the evolution of organized, adaptive complexity. I believe this is so for all theories that have ever been suggested for the mechanism of evolution except Darwinian natural selection, in which case Darwinism rests on a securer pedestal than that provided by facts alone.

Now, I have made reference to theories of evolution "doing the job we ask of them." Everything turns on the question of what that job is. The answer may be different for different people. Some biologists, for instance, get excited about "the species problem," while I have never mustered much enthusiasm for it as a "mystery of mysteries." For some, the main thing that any theory of evolution has to explain is the diversity of life—cladogenesis. Others may require of their theory an explanation of the observed changes in the molecular constitution of the genome. I would not presume to try to convert any of these people to my point of view. All I can do is to make my point of view clear, so that the rest of my argument is clear.

I agree with Maynard Smith (1969) that "The main task of any theory of evolution is to explain adaptive complexity, i.e. to explain the same set of facts which Paley used as evidence of a Creator." I suppose people like me might be labeled neo-Paleyists, or perhaps "transformed Paleyists." We concur with Paley that adaptive complexity demands a very special kind of explanation: either a Designer as Paley taught, or something such as natural selection that does the job of a designer. Indeed, adaptive complexity is probably the best diagnostic of the presence of life itself.

ADAPTIVE COMPLEXITY AS A DIAGNOSTIC CHARACTER OF LIFE

If you find something, anywhere in the universe, whose structure is complex and gives the strong appearance of having been designed for a purpose, then that something either is alive, or was once alive, or is an artifact created by something alive. It is fair to include fossils and artifacts since their discovery on any planet would certainly be taken as evidence for life there.

Complexity is a statistical concept (Pringle, 1951). A complex thing is a statistically improbable thing, something with a very low *a priori* likelihood of coming into being. The number of possible ways of arranging the 10^{27} atoms of a human body is obviously inconceivably large. Of these possible ways, only very few would be recognized as a human body. But this is not, by itself, the point. Any existing configuration of atoms is, *a posteriori*, unique, as "improbable," with hindsight, as any other. The point is that, of all possible ways of arranging those 10^{27} atoms, only a tiny minority would constitute anything remotely resembling a machine that worked to keep itself in being, and to reproduce its kind. Living things are not just statistically improbable in the trivial sense of hindsight; their statistical improbability is limited by the *a priori* constraints of design. They are *adaptively* complex.

The term "adaptationist" has been coined as a pejorative name for one who assumes "without further proof that all aspects of the morphology, physiology and behavior of organisms are adaptive optimal solutions to problems" (Lewontin, 1979). I have responded to this elsewhere (Dawkins, 1982*a*, Chapter 3). Here, I shall be an adaptationist in the much weaker sense that I shall only be *concerned* with those aspects of the morphology, physiology and behavior of organisms that are undisputedly adaptive solutions to problems. In the same way a zoologist may specialize on vertebrates without denying the existence of invertebrates. I shall be preoccupied with undisputed adaptations because I have defined them as my working diagnostic characteristic of all life, anywhere in the universe, in the same way as the vertebrate zoologist might be preoccupied with backbones because backbones are the diagnostic character of all vertebrates. From time to time I shall need an example of an undisputed adaptation, and the time-honored eye will serve the purpose as well as ever (Paley, 1828; Darwin, 1859; any fundamentalist tract). "As far as the examination of the instrument goes, there is precisely the same proof that the eye was made for vision, as there is that the telescope was made for assisting it. They are made upon the same principles; both being adjusted to the laws by which the transmission and refraction of rays of light are regulated" (Paley 1828, V. 1, p. 17).

If a similiar instrument were found upon another planet, some special explanation would be called for. Either there is a God, or, if we are going to explain the universe in terms of blind physical forces, those blind physical forces are going to have to be deployed in a very peculiar way. The same is not true of non-living objects, such as the moon or the solar system (see below). Paley's instincts here were right.

> My opinion of Astronomy has always been, that it is *not* the best medium through which to prove the agency of an intelligent Creator . . . The very simplicity of [the heavenly bodies'] appearance is against them . . . Now we deduce design from relation, aptitude, and correspondence of *parts*. Some degree therefore of *complexity* is necessary to render a subject fit for this species of argument. But the heavenly bodies do not, except perhaps in the instance of Saturn's ring, present themselves to our observation as compounded of parts at all (1828, Vol. 2, pp. 146–7).

A transparent pebble, polished by the sea, might act as a lens, focusing a real image. The fact that it is an efficient optical device is not particularly interesting because, unlike an eye or a telescope, it is too simple. We do not feel the need to invoke anything remotely resembling the concept of design. The eye and the telescope have many parts, all coadapted and working together to achieve the same functional end. The polished pebble has far fewer coadapted features: the coincidence of transparency, high refractive index and mechanical forces that polish the surface in a curved shape. The odds against such a threefold coincidence are not particularly great. No special explanation is called for.

Compare how a statistician decides what P value to accept as evidence for an effect in an experiment. It is a matter of judgment and dispute, almost of taste, exactly when a coincidence becomes too great to stomach. But, no matter whether you are cautious statistician or a daring statistician, there are some complex adaptations whose "P value," whose coincidence rating, is so impressive that nobody would hesitate to diagnose life (or an artifact designed by a living thing). My definition of living complexity is, in effect, "that complexity which is too great to have come about through a single coincidence." For the purposes of this paper, the problem that any theory of evolution has to solve is how living adaptive complexity comes about.

In the book referred to above, Mayr (1982) helpfully lists what he sees as the six clearly distinct theories of evolution that have ever been proposed in the history of biology. I shall use this list to provide me with my main headings in this paper. For each of the six, instead of asking what the evidence is, for or against, I shall ask whether the theory is *in principle* capable of doing the job of explaining the existence of adaptive complexity. I shall take the six theories in order, and will conclude that only Theory 6, Darwinian selection, matches up to the task.

Theory 1. *Built-in capacity for, or drive toward, increasing perfection*

To the modern mind this is not really a theory at all, and I shall not bother to discuss it. It is obviously mystical, and does not explain anything that it does not assume to start with.

Theory 2. *Use and disuse plus inheritance of acquired characters*

It is convenient to discuss this in two parts.

USE AND DISUSE

It is an observed fact that on this planet living bodies sometimes become better adapted as a result of use. Muscles that are exercised tend to grow bigger. Necks that reach eagerly towards the treetops may lengthen in all their parts. Conceivably, if on some planet such acquired improvements could be incorporated into the hereditary information, adaptive evolution could result. This is the theory often associated with Lamarck, although there was more to what Lamarck said. Crick (1982, p. 59) says of the idea: "As far as I know, no one has given *general* theoretical reasons why such a mechanism must be less efficient than natural selection. . . ." In this section and the next I shall give two general theoretical objections to Lamarckism of the sort which,

I suspect, Crick was calling for. I have discussed both before (Dawkins, 1982b), so will be brief here. First the shortcomings of the principle of use and disuse.

The problem is the crudity and imprecision of the adaptation that the principle of use and disuse is capable of providing. Consider the evolutionary improvements that must have occurred during the evolution of an organ such as an eye, and ask which of them could conceivably have come about through use and disuse. Does "use" increase the transparency of a lens? No, photons do not wash it clean as they pour through it. The lens and other optical parts must have reduced, over evolutionary time, their spherical and chromatic aberration; could this come about through increased use? Surely not. Exercise might have strengthened the muscles of the iris, but it could not have built up the fine feedback control system which controls those muscles. The mere bombardment of a retina with colored light cannot call color-sensitive cones into existence, nor connect up their outputs so as to provide color vision.

Darwinian types of theory, of course, have no trouble in explaining all these improvements. Any improvement in visual accuracy could significantly affect survival. Any tiny reduction in spherical aberration may save a fast flying bird from fatally misjudging the position of an obstacle. Any minute improvement in an eye's resolution of acute colored detail may crucially improve its detection of camouflaged prey. The genetic basis of any improvement, however slight, will come to predominate in the gene pool. The relationship between selection and adaptation is a direct and close-coupled one. The Lamarckian theory, on the other hand, relies on a much cruder coupling: the rule that the more an animal uses a certain bit of itself, the bigger that bit ought to be. The rule occasionally might have some validity but not generally, and, as a sculptor of adaptation it is a blunt hatchet in comparison to the fine chisels of natural selection. This point is universal. It does not depend on detailed facts about life on this particular planet. The same goes for my misgivings about the inheritance of acquired characters.

INHERITANCE OF ACQUIRED CHARACTERS

The problem here is that acquired characters are not always improvements. There is no reason why they should be, and indeed the vast majority of them are injuries. This is not just a fact about life on earth. It has a universal rationale. If you have a complex and reasonably well-adapted system, the number of things you can do to it that will make it perform less well is vastly greater than the number of things you can do to it that will improve it (Fisher, 1958). Lamarckian evolution will move in adaptive directions only if some mechanism—selection—exists for distinguishing those acquired characters that are improvements from those that are not. Only the improvements should be imprinted into the germ line.

Although he was not taking about Lamarckism, Lorenz (1966) emphasized a related point for the case of learned behavior, which is perhaps the most important kind of acquired adaptation. An animal learns to be a better animal during its own lifetime. It learns to eat sweet foods, say, thereby increasing its survival chances. But there is nothing inherently nutritious about a sweet taste. Something, presumably natural selection, has to have built into the nervous system the arbitrary rule: "treat sweet taste as reward," and this works because saccharine does not occur in nature whereas sugar does.

Similarly, most animals learn to avoid situations that have, in the past, led to pain. The stimuli that animals treat as painful tend, in nature, to be associated with injury and increased chance of death. But again the connection must ultimately be built into the nervous system by natural selection, for it is not an obvious, necessary connection (M. Dawkins, 1980). It is easy to imagine artificially selecting a breed of animals that enjoyed being injured, and felt pain whenever their physiological welfare was being improved. If learning is adaptive *improvement,* there has to be, in Lorenz's phrase, an innate teaching mechanism, or "innate schoolmarm." The principle holds even where the reinforcers are "secondary," learned by association with primary reinforcers (P. P. G. Bateson, 1983).

It holds, too, for morphological characters. Feet that are subjected to wear and tear grow tougher and more thick-skinned. The thickening of the skin is an acquired adaptation, but it is not obvious why the change went in this direction. In man-made machines, parts that are subjected to wear get thinner not thicker, for obvious reasons. Why does the skin on the feet do the opposite? Because, fundamentally, natural selection has worked in the past to ensure an adaptive rather than a maladaptive response to wear and tear.

The relevance of this for would-be Lamarckian evolution is that there has to be a deep Darwinian underpinning even if there is a Lamarckian surface structure: a Darwinian choice of which potentially acquirable characters shall in fact be acquired and inherited. As I have argued before (Dawkins, 1982*a,* pp. 164-77), this is true of a recent, highly publicized immunological theory of Lamarckian adaptation (Steele, 1979). Lamarckian mechanisms cannot be fundamentally responsible for adaptive evolution. Even if acquired characters are inherited on some planet, evolution there will still rely on a Darwinian guide for its adaptive direction.

Theory 3. Direct induction by the environment

Adaptation, as we have seen, is a fit between organism and environment. The set of conceivable organisms is wider than the actual set. And there is a set of conceivable environments wider than the actual set. These two subsets match each other to some extent, and the matching is adaptation. We can re-express the point by saying that information from the environment is present in the organism. In a few cases this is vividly literal—a frog carries a picture of its environment around on its back. Such information is usually carried by an animal in the less literal sense that a trained observer, dissecting a new animal, can reconstruct many details of its natural environment.

Now, how could the information get from the environment into the animal? Lorenz (1966) argues that there are two ways, natural selection and reinforcement learning, but that these are both *selective* processes in the broad sense (Pringle, 1951). There is, in theory, an alternative method for the environment to imprint its information on the organism, and that is by direct "instruction" (Danchin, 1979). Some theories of how the immune system works are "instructive": antibody molecules are thought to be shaped directly by molding themselves around antigen molecules. The currently favored theory is, by contrast, selective (Burnet, 1969). I take "instruction" to be synonymous with the "direct induction by the environment" of Mayr's Theory 3. It is not always clearly distinct from Theory 2.

Instruction is the process whereby information flows directly from its environment into an animal. A case could be made for treating imitation learning, latent learning and imprinting (Thorpe, 1963) as instructive, but for clarity it is safer to use a hypothetical example. Think of an animal on some planet, deriving camouflage from its tiger-like stripes. It lives in long dry "grass," and its stripes closely match the typical thickness and spacing of local grass blades. On our own planet such adaptation would come about through the selection of random genetic variation, but on the imaginary planet it comes about through direct instruction. The animals go brown except where their skin is shaded from the "sun" by blades of grass. Their stripes are therefore adapted with great precision, not just to any old habitat, but to the precise habitat in which they have sunbathed, and it is this same habitat in which they are going to have to survive. Local populations are automatically camouflaged against local grasses. Information about the habitat, in this case about the spacing patterns of the grass blades, has flowed into the animals, and is embodied in the spacing pattern of their skin pigment.

Instructive adaptation demands the inheritance of acquired characters if it is to give rise to permanent or progressive evolutionary change. "Instruction" received in one generation must be "remembered" in the genetic (or equivalent) information. This process is in principle cumulative and progressive. However, if the genetic store is not to become overloaded by the accumulations of generations, some mechanism must exist for discarding unwanted "instructions," and retaining desirable ones. I suspect that this must lead us, once again, to the need for some kind of selective process.

Imagine, for instance, a form of mammal-like life in which a stout "umbilical nerve" enabled a mother to "dump" the entire contents of her memory in the brain of her fetus. The technology is available even to our nervous systems: the corpus callosum can shunt large quantities of information from right hemisphere to left. An umbilical nerve could make the experience and wisdom of each generation automatically available to the next, and this might seem very desirable. But without a selective filter, it would take few generations for the load of information to become unmanageably large. Once again we come up against the need for a selective underpinning. I will leave this now, and make one more point about instructive adaptation (which applies equally to all Lamarckian types of theory).

The point is that there is a logical link-up between the two major theories of adaptive evolution—selection and instruction—and the two major theories of embryonic development—epigenesis and preformationism. Instructive evolution can work only if embryology is preformationistic. If embryology is epigenetic, as it is on our planet, instructive evolution cannot work. I have expounded the argument before (Dawkins, 1982a, pp. 174–6), so I will abbreviate it here.

If acquired characters are to be inherited, embryonic processes must be reversible: phenotypic change has to be read back into the genes (or equivalent). If embryology is preformationistic—the genes are a true blueprint—then it may indeed be reversible. You can translate a house back into its blueprint. But if embryonic development is epigenetic: if, as on this planet, the genetic information is more like a recipe for a cake (Bateson, 1976) than a blueprint for a house, it is irreversible. There is no one-to-one mapping between bits of genome and bits of phenotype, any more than there is mapping between crumbs of cake and words of recipe. The recipe is not a blueprint that can be reconstructed from the cake. The transformation of recipe into cake cannot be put into reverse, and nor can the process of making a body. Therefore acquired adaptations cannot be read back into the "genes," on any planet where embryology is epigenetic.

This is not to say that there could not, on some planet, be a form of life whose embryology was preformationistic. That is a separate question. How likely is it? The form of life would have to be very different from ours, so much so that it is hard to visualize how it might work. As for reversible embryology itself, it is even harder to visualize. Some mechanism would have to scan the detailed form of the adult body, carefully noting down, for instance, the exact location of brown pigment in a sun-striped skin, perhaps turning it into a linear stream of code numbers, as in a television camera. Embryonic development would read the scan out again, like a television receiver. I have an intuitive hunch that there is an objection in principle to this kind of embryology, but I cannot at present formulate it clearly. All I am saying here is that, if planets are divided into those where embryology is preformationistic and those, like Earth, where embryology is epigenetic, Darwinian evolution could be supported on both kinds of planet, but Lamarckian evolution, even if there were no other reasons for doubting its existence, could be supported only on the preformationistic planets—if there are any.

The close theoretical link that I have demonstrated between Lamarckian evolution and preformationistic embryology gives rise to a mildly entertaining irony. Those with ideological reasons for hankering after a neo-Lamarckian view of evolution are often especially militant partisans of epigenetic, "interactionist," ideas of development, possibly—and here is the irony—for the very same ideological reasons (Koestler, 1967; Ho & Saunders, 1982).

Theory 4. Saltationism

The great virtue of the idea of evolution is that it explains, in terms of blind physical forces, the existence of undisputed adaptations whose statistical improbability is enormous, without recourse to the supernatural or the mystical. Since we *define* an undisputed adaptation as an adaptation that is too complex to have come about by chance, how is it possible for a theory to invoke only blind physical forces in explanation? The answer—Darwin's answer— is astonishingly simple when we consider how self-evident Paley's Divine Watchmaker must have seemed to his contemporaries. The key is that the coadapted parts do not have to be assembled *all at once*. They can be put together in small stages. But they really do have to be *small* stages. Otherwise we are back again with the problem we started with: the creation by chance of complexity that is too great to have been created by chance!

Take the eye again, as an example of an organ that contains a large number of independent coadapted parts, say N. The *a priori* probability of any one of these N features coming into existence by chance is low, but not incredibly low. It is comparable to the chance of a crystal pebble being washed by the sea so that it acts as a lens. Any one adaptation on its own could, plausibly, have come into existence through blind physical forces. If each of the N coadapted features confers some slight advantage on its own, then the whole many-parted organ can be put together over a long period of time. This is particularly plausible for the eye—ironically in view of that organ's niche of honor in the creationist pantheon. The eye is, *par excellence,* a case where a fraction of an organ is better than no organ at all; an eye without a lens or even a pupil, for instance, could still detect the looming shadow of a predator.

To repeat, the key to the Darwinian explanation of adaptive complexity is the replacement of instantaneous, coincidental, multi-dimensional luck, by gradual, inch by inch, smeared-out luck. Luck is involved, to be sure. But a theory that bunches the luck up into major steps is more incredible than a theory that spreads the luck out in small stages. This leads to the following general principle of universal biology. Wherever in the universe adaptive complexity shall be found, it will have come into being gradually through a series of small alterations, never through large and sudden increments in adaptive complexity. We must reject Mayr's 4th theory, saltationism, as a candidate for explanation of the evolution of complexity.

It is almost impossible to dispute this rejection. It is implicit in the definition of adaptive complexity that the only alternative to gradualistic evolution is supernatural magic. This is not to say that the argument in favor of gradualism is a worthless tautology, an unfalsifiable dogma of the sort that creationists and philosophers are so fond of jumping about on. It is not *logically* impossible for a full-fashioned eye to spring *de novo* from virgin bare skin. It is just that the possibility is statistically negligible.

Now it has recently been widely and repeatedly publicized that some modern evolutionists reject "gradualism," and espouse what Turner (1982) has called theories of evolution by jerks. Since these are reasonable people without mystical leanings, they must be gradualists in the sense in which I am here using the term: the "gradualism" that they oppose must be defined differently. There are actually two confusions of language here, and I intend to clear them up in turn. The first is the common confusion between "punctuated equilibrium" (Eldredge & Gould, 1972) and true saltationism. The second is a confusion between two theoretically distinct kinds of saltation.

Punctuated equilibrium is not macromutation, not saltation at all in the traditional sense of the term. It is, however, necessary to discuss it here, because it is popularly regarded as a theory of saltation, and its partisans quote, with approval, Huxley's criticism of Darwin for upholding the principle of *Natura non facit saltum* (Gould, 1980). The punctuationist theory is portrayed as radical and revolutionary and at variance with the "gradualistic" assumptions of both Darwin and the neo-Darwinian synthesis (e.g. Lewin, 1980). Punctuated equilibrium, however, was originally conceived as what the orthodox neo-Darwinian synthetic theory should truly predict, on a palaeontological timescale, if we take its embedded ideas of allopatric speciation seriously (Eldredge & Gould, 1972). It derives its "jerks" by taking the "stately unfolding" of the neo-Darwinian synthesis, and *inserting* long periods of stasis separating brief bursts of gradual, albeit rapid, evolution.

The plausibility of such "rapid gradualism" is dramatized by a thought experiment of Stebbins (1982). He imagines a species of mouse, evolving larger body size at such an imperceptibly slow rate that the differences between the means of successive generations would be utterly swamped by sampling error. Yet even at this slow rate Stebbin's mouse lineage would attain the body size of a large elephant in about 60,000 years, a time-span so short that it would be regarded as instantaneous by palaeontologists. Evolutionary change too *slow* to be detected by microevolutionists can nevertheless be too *fast* to be detected by microevolutionists. What a palaeontologist sees as a "saltation" can in fact be a smooth and gradual change so slow as to be undetectable to the microevolutionist. This kind of palaeontological "saltation" has nothing to do with the one-generation macromutations that, I suspect, Huxley and Darwin had in mind when they debated *Natura non facit saltum*. Confusion has arisen here, possibly because some individual champions of punctuated equilibrium have also, incidentally, championed macromutation (Gould, 1982). Other "punctuationists" have either confused their theory with macromutationism, or have explicitly invoked macromutation as one of the mechanisms of punctuation (e.g. Stanley, 1981).

Turning to macromutation, or true saltation itself, the second confusion that I want to clear up is between the two kinds of macromutation that we might conceive of. I could name them, unmemorably, saltation (1) and saltation (2), but instead I shall pursue an earlier fancy for airliners as metaphors, and label them "Boeing 747" and "Stretched DC-8" saltation. 747 saltation is the inconceivable kind. It gets its name from Sir Fred Hoyle's much quoted metaphor for his own cosmic misunderstanding of Darwinism (Hoyle & Wickramasinghe, 1981). Hoyle compared Darwinian selection to a tornado, blowing through a junkyard and assembling a Boeing 747 (what he overlooked, of course, was the point about luck being "smeared-out" in small steps— see above). Stretched DC-8 saltation is quite different. It is not in principle hard to believe in at all. It refers to large and sudden changes in *magnitude* of some biological measure, without an accompanying large increase in adaptive information. It is named after an airliner that was made by elongating the fuselage of an existing design, not adding significant new complexity. The change from DC-8 is a big change in magnitude—a saltation not a gradualistic series of tiny changes. But, unlike the change from junk-heap to 747, it is not a big increase in information content or complexity, and that is the point I am emphasizing by the analogy.

An example of DC-8 saltation would be the following. Suppose the giraffe's neck shot out in one spectacular mutational step. Two parents had necks of standard antelope length. They had a freak child with a neck of modern giraffe length, and all giraffes are descended from this freak. This is unlikely to be true on Earth, but something like it may happen elsewhere in the universe. There is no objection to it in principle, in the sense that there is a profound objection to the (747) idea that a complex organ like an eye could arise from bare skin by a single mutation. The crucial difference is one of complexity.

I am assuming that the change from short antelope's neck to long giraffe's neck is *not* an increase in complexity. To be sure, both necks are exceedingly complex structures. You couldn't go from *no*-neck to either kind of neck in one step: that would be 747 saltation. But once the complex organization of the antelope's neck already exists, the step to giraffe's neck is just an elongation: various things have to grow faster at some stage in embryonic development; existing complexity is preserved. In practice, of course, such a drastic change in magnitude would be highly likely to have deleterious repercussions which would render the macromutant unlikely to survive. The existing antelope heart probably could not pump the blood up to the newly elevated giraffe head. Such practical objections to evolution by "DC-8 saltation" can only help my case in favor of gradualism, but I still want to make a separate, and more universal, case against 747 saltation.

It may be argued that the distinction between 747 and DC-8 saltation is impossible to draw in practice. After all, DC-8 saltations, such as the proposed macromutational elongation of the giraffe's neck, may appear very complex: myotomes, vertebrae, nerves, blood vessels, all have to elongate together. Why does this not make it a 747 saltation, and therefore rule it out? But although this type of "coadaptation" has indeed often been thought of as a problem for any evolutionary theory, not just macromutational ones (see Ridley, 1982, for a history), it is so only if we take an impoverished view of developmental mechanisms. We know that single mutations can orchestrate changes in growth rates of many diverse parts of organs, and, when we think about developmental processes, it is not in the least surprising that this should be so. When a single mutation causes a *Drosophila* to grow a leg where an antenna ought to be, the leg grows in all its formidable complexity. But this is not mysterious or surprising, not a 747 saltation, because the organization of a leg is already present in the body before the mutation. Wherever, as in embryogenesis, we have a hierarchically branching tree of causal relationships, a small alteration at a senior node of the tree can have large and complex ramified effects on the tips of the twigs. But although the change may be large in magnitude, there can be no large and sudden increments in adaptive information. If you think you have found a particular example of a large and sudden increment in adaptively complex information in practice, you can be certain the adaptive information was already there, even if it is an atavistic "throwback" to an earlier ancestor.

There is not, then, any objection in principle to theories of evolution by jerks, even the theory of hopeful monsters (Goldschmidt, 1940), provided that it is DC-8 saltation, not 747 saltation that is meant. Gould (1982) would clearly agree: "I regard forms of macromutation which include the sudden origin of new species with all their multifarious adaptations intact *ab initio*, as illegitimate." No educational biologist actually believes in 747 saltation, but not all have been sufficiently explicit about the distinction between DC-8 and 747 saltation. An unfortunate consequence is that creationists and their journalistic fellow-travelers have been able to exploit saltationist-sounding statements of respected biologists. The biologist's intended meaning may have been what I am calling DC-8 saltation, or even non-saltatory punctuation; but the creationist *assumes* saltation in the sense that I have dubbed 747, and 747 saltation would, indeed, be a blessed miracle.

I also wonder whether an injustice is not being done to Darwin, owing to this same failure to come to grips with the distinction between DC-8 and 747 saltation. It is frequently alleged that Darwin was wedded to gradualism, and therefore that, if some form of evolution by jerks is proved, Darwin will have been shown wrong. This is undoubtedly the reason for the ballyhoo and publicity that has attended the theory of punctuated equilibrium. But was Darwin really opposed to all jerks? Or was he, as I suspect, strongly opposed only to 747 saltation?

As we have already seen, punctuated equilibrium has nothing to do with saltation, but anyway I think it is not at all clear that, as is often alleged, Darwin would have been discomfited by punctuationist interpretations of the fossil record. The following passage, from later editions of the *Origin,* sounds like something from a current issue of *Paleobiology:* "the periods during which species have been undergoing modification, though very long as measured by years, have probably been short in comparison with the periods during which these same species remained without undergoing any change."

Gould (1982) shrugs this off as somehow anomalous and away from the mainstream of Darwin's thought. As he correctly says: "You cannot do history by selective quotation and search for qualifying footnotes. General tenor and historical impact are the proper criteria. Did his contemporaries or descendants ever read Darwin as a saltationist?" Certainly nobody ever accused Darwin of being a saltationist. But to most people saltation means macromutation, and, as Gould himself stresses, "Punctuated equilibrium is not a theory of macromutation." More importantly, I believe we can reach a better understanding of Darwin's general gradualistic bias if we invoke the distinction between 747 and DC-8 saltation.

Perhaps part of the problem is that Darwin himself did not have the distinction. In some anti-saltation passages it seems to be DC-8 saltation that he has in mind. But on those occasions he does not seem to feel very strongly about it: "About sudden jumps," he wrote in a letter in 1860, "I have no objection to them—they would aid me in some cases. All I can say is, that I went into the subject and found no evidence to make me believe in jumps [as a source of new species] and a good deal pointing in the other direction" (quoted in Gillespie, 1979). This does not sound like a man fervently opposed, in principle, to sudden jumps. And of course there is no reason why he *should* have been fervently opposed, if he only had DC-8 saltations in mind.

But at other times he really is pretty fervent, and on those occasions, I suggest, he is thinking of 747 saltation: ". . . it is impossible to imagine so many co-adaptations being formed all by a chance blow" (quoted in Ridley, 1982). As the historian Neal Gillespie puts it: "For Darwin, monstrous births, a doctrine favored by Chambers, Owen, Argyll, Mivart, and others, from clear theological as well as scientific motives, as an explanation of how new species or even higher taxa, had developed, was no better than a miracle: 'it leaves the case of the co-adaptation of organic beings to each other and to their physical conditions of life, untouched and unexplained.' It was 'no explanation' at all, of no more scientific value than creation 'from the dust of the earth'" (Gillespie, 1979, p. 118).

As Ridley (1982) says of the "religious tradition of idealist thinkers [who] were committed to the explanation of complex adaptive contrivances by intelligent design," "The greatest concession they could make to Darwin was that the Designer operated by tinkering with the generation of diversity, designing the variation." Darwin's response was: "If I were convinced that I required such additions to the theory of natural selection, I would reject it as rubbish . . . I would give nothing for the theory of Natural selection, if it requires miraculous additions at any one stage of descent."

Darwin's hostility to monstrous saltation, then, makes sense if we assume that he was thinking in terms of 747 saltation—the sudden invention of new adaptive complexity. It is highly likely that that is what he was thinking of, because that is exactly what many of his opponents had in mind. Saltationists such as the Duke of Argyll (though presumably not Huxley) wanted to believe in 747 saltation, precisely because it *did* demand supernatural intervention. Darwin did not believe in it, for exactly the same reason. To quote Gillespie again (p. 120): ". . . for Darwin, designed evolution, whether manifested in saltation, monstrous births, or manipulated variations, was but a disguised form of special creation."

I think this approach provides us with the only sensible reading of Darwin's well known remark that "If it could be demonstrated that any complex organ existed, which could not possibly have been formed by numerous, successive, slight modifications, my theory would absolutely break down." That is not a plea for gradualism, as a modern palaeobiologist uses the term. Darwin's theory is falsifiable, but he was much too wise to make his theory *that* easy to falsify! Why on earth *should* Darwin have committed himself to such an arbitrarily restrictive version of evolution, a version that positively invites falsification? I think it is clear that he didn't. His use of the term "complex" seems to me to be clinching. Gould (1982) describes this passage from Darwin as "clearly invalid." So it is invalid if the alternative to slight modifications is seen as DC-8 saltation. But if the alternative is seen as 747 saltation, Darwin's remark is valid and very wise. Notwithstanding those whom Miller (1982) has unkindly called Darwin's more foolish critics, his theory is indeed falsifiable, and in the passage quoted he puts his finger on one way in which it might be falsified.

There are two kinds of imaginable saltation, then, DC-8 saltation and 747 saltation. DC-8 saltation is perfectly possible, undoubtedly happens in the laboratory and the farmyard, and may have made important contributions to evolution. 747 saltation is statistically ruled out unless there is supernatural intervention. In Darwin's own time, proponents and opponents of saltation often had 747 saltation in mind, because they believed in—or were arguing against—divine intervention. Darwin was hostile to (747) saltation, because he correctly saw natural selection as an *alternative* to the miraculous as an explanation for adaptive complexity. Nowadays saltation either means punctuation (which isn't saltation at all) or DC-8 saltation, neither of which Darwin would have had strong objections to in principle, merely doubts about the facts. In the modern context, therefore, I do not think Darwin should be labeled a strong gradualist. In the modern context, I suspect that he would be rather open-minded.

It is in the anti-747 sense that Darwin was a passionate gradualist, and it is in the same sense that we must all be gradualists, not just with respect to life on earth, but with respect to life all over the universe. Gradualism in this sense is essentially synonymous with evolution. The sense in which we may be non-gradualists is a much less radical, although still quite interesting, sense. The theory of evolution by jerks has been hailed on television and elsewhere as radical and revolutionary, a paradigm shift. There is, indeed, an interpretation of it which is revolutionary, but that interpretation (the 747 macromutation version) is certainly wrong, and is apparently not held by its original proponents. The sense in which the theory might be right is not particularly revolutionary. In this field you may choose your jerks so as to be revolutionary, *or* so as to be correct, but not both.

Theory 5. Random evolution

Various members of this family of theories have been in vogue at various times. The "mutationists" of the early part of this century—De Vries, W. Bateson and their colleagues—believed that selection served only to weed out deleterious freaks, and that the real driving force in evolution was mutation pressure. Unless you believe mutations are directed by some mysterious life force, it is sufficiently obvious that you can be a mutationist only if you forget about adaptive complexity—forget, in other words, most of the consequences of evolution that are of any interest! For historians there remains the baffling enigma of how such distinguished biologists as De Vries, W. Bateson and T. H. Morgan could rest satisfied with such a crassly inadequate theory. It is not enough to say that De Vries's view was blinkered by his working only on the evening primrose. He only had to look at the adaptive complexity in his own body to see that "mutationism" was not just a wrong theory: it was an obvious non-starter.

These post-Darwinian mutationists were also saltationists and anti-gradualists, and Mayr treats them under that heading, but the aspect of their view that I am criticizing here is more fundamental. It appears that they actually thought that mutation, on its own without selection, was sufficient to explain evolution. This *could* not be so on any non-mystical view of mutation, whether gradualist or saltationist. If mutation is undirected, it is clearly unable to explain the adaptive directions of evolution. If mutation is directed in adaptive ways we are entitled to ask how this comes about. At least Lamarck's principle of use and disuse makes a valiant attempt at explaining how variation might be directed. The "mutationists" didn't even seem to see that there was a problem, possibly because they under-rated the importance of adaptation—and they were not the last to do so. The irony with which we must now read W. Bateson's dismissal of Darwin is almost painful: "the transformation of masses of populations by imperceptible steps guided by selection is, as most of us now see, so inapplicable to the fact that we can only marvel . . . at the want of penetration displayed by the advocates of such a proposition . . ." (1913, quoted in Mayr, 1982).

Nowadays some population geneticists describe themselves as supporters of "non-Darwinian evolution." They believe that a substantial number of the gene replacements that occur in evolution are non-adaptive substitutions of alleles whose effects are indifferent relative to one another (Kimura, 1968). This may well be true, if not in Israel (Nevo, 1983) maybe somewhere in the Universe. But it obviously has nothing whatever to contribute to solving the problem of the evolution of adaptive complexity. Modern advocates of neutralism admit that their theory cannot account for adaptation, but that doesn't seem to stop them from regarding the theory as interesting. Different people are interested in different things.

The phrase "random genetic drift" is often associated with the name of Sewall Wright, but Wright's conception of the relationship between random drift and adaptation is altogether subtler than the others I have mentioned (Wright, 1980). Wright does not belong in Mayr's fifth category, for he clearly sees selection as the driving force of adaptive evolution. Random drift may make it easier for selection to do its job by assisting the escape from local optima (Dawkins, 1982a, p. 40), but it is still selection that is determining the rise of adaptive complexity.

Recently palaeontologists have come up with fascinating results when they perform computer simulations of "random phylogenies" (e.g. Raup, 1977). These random walks through evolutionary time produce trends that look uncannily like real ones, and it is disquietingly easy, and tempting, to read into the random phylogenies apparently adaptive trends which, however, are not there. But this does not mean that we can admit random drift as an explanation of real adaptive trends. What it might mean is that some of us have been too facile and gullible in what we think are adaptive trends. This does not alter the fact that there are some trends that really *are* adaptive— even if we don't always identify them correctly in practice—and those real adaptive trends can't be produced by random drift. They must be produced by some non-random force, presumably selection.

So, finally, we arrive at the sixth of Mayr's theories of evolution.

Theory 6. Direction (order) imposed on random variation by natural selection

Darwinism—the non-random selection of randomly varying replicating entities by reason of their "phenotypic" effects—is the only force I know that can, in principle, guide evolution in the direction of adaptive complexity. It works on this planet. It doesn't suffer from any of the drawbacks that beset the other five classes of theory, and there is no reason to doubt its efficacy throughout the universe.

The ingredients in a general recipe for Darwinian evolution are replicating entities of some kind, exerting phenotypic "power" of some kind over their replication success. I have referred to these necessary entities as "active germline replicators" or "optimons" (Dawkins, 1982a, chapter 5). It is important to keep their replication conceptually separate from their phenotypic effects, even though, on some planets, there may be a blurring in practice. Phenotypic adaptations can be seen as tools of replicator propagation.

Gould (1983) disparages the replicator's-eye view of evolution as preoccupied with "book-keeping." The metaphor is a superficially happy one: it is easy to see the genetic changes that accompany evolution as book-keeping entries, mere accountant's records of the really interesting phenotypic events going on in the outside world. Deeper consideration, however, shows that the truth is almost the exact opposite. It is central and essential to Darwinian (as opposed to Lamarckian) evolution that there shall be causal arrows flowing from genotype to phenotype, but not in the reverse direction. Changes in gene frequencies are not passive book-keeping records of phenotypic changes: it is precisely because (and to the extent that) they actively *cause* phenotypic changes that evolution of the phenotype can occur. Serious errors flow, both from a failure to understand the importance of this one-way flow (Dawkins, 1982*a,* chapter 6), and from an over-interpretation of it as inflexible and undeviating "genetic determinism" (Dawkins, 1982*a,* chapter 2).

The universal perspective leads me to emphasize a distinction between what may be called "one-off selection" and "cumulative selection." Order in the non-living world may result from processes that can be portrayed as a rudimentary kind of selection. The pebbles on a seashore become sorted by the waves, so that larger pebbles come to lie in layers separate from smaller ones. We can regard this as an example of the selection of a stable configuration out of initially more random disorder. The same can be said of the "harmonious" orbital patterns of planets around stars, and electrons around nuclei, of the shapes of crystals, bubbles and droplets, even, perhaps, of the dimensionality of the universe in which we find ourselves (Atkins, 1981). But this is all one-off selection. It does not give rise to progressive evolution because there is no replication, no succession of generations. Complex adaptation requires many generations of cumulative selection, each generation's change building upon what has gone before. In one-off selection, a stable state develops and is then maintained. It does not multiply, does not have offspring.

In life the selection that goes on *in any one generation* is one-off selection, analogous to the sorting of pebbles on a beach. The peculiar feature of life is that successive generations of such selection build up, progressively and cumulatively, structures that are eventually complex enough to foster the strong illusion of design. One-off selection is a commonplace of physics and cannot give rise to adaptive complexity. Cumulative selection is the hallmark of biology and is, I believe, the force underlying all adaptive complexity.

OTHER TOPICS FOR A FUTURE SCIENCE
OF UNIVERSAL DARWINISM

Active germ-line replicators together with their phenotypic consequences, then, constitute the general recipe for life, but the form of the system may vary greatly from planet to planet, both with respect to the replicating entities themselves, and with respect to the "phenotypic" means by which they ensure their survival. Indeed, the very distinction between "genotype" and "phenotype" may be blurred (L. Orgel, personal communication). The replicating entities do not have to be DNA or RNA. They do not have to be organic molecules at all. Even on this planet it is possible that DNA itself is a late usurper of the role, taking over from some earlier, inorganic crystalline replicator (Cairns-Smith, 1982). It is also arguable that today selection operates on several levels, for instance, the levels of the gene and the species or lineage, and perhaps some unit of cultural transmission (Lewontin, 1970).

A full science of Universal Darwinism might consider aspects of replicators transcending their detailed nature and the time-scale over which they are copied.

For instance, the extent to which they are "particulate" as opposed to "blending" probably has a more important bearing on evolution than their detailed molecular or physical nature. Similarly, a universe-wide classification of replicators might make more reference to their dimensionality and coding principles than to their size and structure. DNA is a digitally coded one-dimensional array. A "genetic" code in the form of a two-dimensional matrix is conceivable. Even a three-dimensional code is imaginable, although students of Universal Darwinism will probably worry about how such a code could be "read." (DNA is, of course, a molecule whose 3-dimensional structure determines how it is replicated and transcribed, but that doesn't make it a 3-dimensional code. DNA's meaning depends upon the 1-dimensional sequential arrangement of its symbols, not upon their 3-dimensional position relative to one another in the cell.) There might also be theoretical problems with analogue, as opposed to digital codes, similar to the theoretical problems that would be raised by a purely analogue nervous system (Rushton, 1961).

As for the phenotypic levers of power by which replicators influence their survival, we are so used to their being bound up into discrete organisms or "vehicles" that we forget the possibility of a more diffuse extra-corporeal or "extended" phenotype. Even on this Earth a large amount of interesting adaptation can be interpreted as part of the extended phenotype (Dawkins, 1982a, Chapters 11, 12 and 13). There is, however, a general theoretical case that can be made in favor of the discrete organismal body, with its own recurrent life cycle, as a necessity in any process of evolution of advanced adaptive complexity (Dawkins, 1982a, Chapter 14), and this topic might have a place in a full account of Universal Darwinism.

Another candidate for full discussion might be what I shall call divergence, and convergence or recombination of replicator lineages. In the case of Earth-bound DNA, "convergence" is provided by sex and related processes. Here the DNA "converges" within the species after having very recently "diverged." But suggestions are now being made that a different kind of convergence can occur among lineages that originally diverged an exceedingly long time ago. For instance there is evidence of gene transfer between fish and bacteria (Jacob, 1983). The replicating lineages on other planets may permit very varied kinds of recombination, on very different time-scales. On Earth the rivers of phylogeny are almost entirely divergent: if main tributaries ever recontact each other after branching apart it is only through the tiniest of trickling cross-streamlets, as in the fish/bacteria case. There is, of course, a richly anastomosing delta of divergence and convergence due to sexual recombination *within* the species, but only within the species. There may be planets on which the "genetic" system permits much more cross-talk at all levels of the branching hierarchy, one huge fertile delta.

I have not thought enough about the fantasies of the previous paragraphs to evaluate their plausibility. My general point is that there is one limiting constraint upon all speculations about life in the universe. If a life-form displays adaptive complexity, it must possess an evolutionary mechanism capable of generating adaptive complexity. However diverse evolutionary mechanisms may be, if there is no other generalization that can be made about life all around the Universe, I am betting it will always be recognizable as Darwinian life. The Darwinian Law (Eigen, 1983) may be as universal as the great laws of physics.

NOTE

As usual I have benefitted from discussions with many people, including especially Mark Ridley, who also criticized the manuscript, and Alan Grafen. Dr. F. J. Ayala called attention to an important error in the original spoken version of the paper.

REFERENCES

Asimov, I. (1979). *Extraterrestrial Civilizations.* Pan; London.

Atkins, P. W. (1981). *The Creation.* W. H. Freeman; Oxford.

Bateson, P. P. G. (1976). "Specificity and the Origins of Behavior," *Advances in the Study of Behavior, 6,* 1–20.

———. (1983) "Rules for Changing the Rules." In *Evolution from Molecules to Men,* ed. D. S. Bendall, pp. 483–501. Cambridge University Press; Cambridge.

Billingham, J. (1981). *Life in the Universe.* MIT Press; Cambridge, Mass.

Burnet, F. M. (1969). *Cellular immunology.* Melbourne University Press: Melbourne.

Cairns-Smith, A. G. (1982). *Genetic Takeover.* Cambridge University Press; Cambridge.

Crick, F. H. C. (1982). *Life Itself.* Macdonald; London.

Danchin A. (1979). "Themes de la biologie: theories instructives et theories selectives," *Revue des Questions Scientifiques, 150,* 151–64.

Darwin, C. R. (1859). *The Origin of Species.* 1st edition, reprinted (1968). Penguin; London.

Dawkins, M. (1980). *Animal Suffering: The Science of Animal Welfare.* Chapman & Hall; London.

Dawkins, R. (1982*a*). *The Extended Phenotype.* (W. H. Freeman: Oxford.

———. (1982*b*). "The Necessity of Darwinism," *New Scientist, 94,* 130–32, reprinted in *Darwin Up to Date,* ed. J. Cherfas. New Scientist; London.

Eigen, M. (1983). "Self-replication and Molecular Evolution." In *Evolution from Molecules to Men,* ed. D. S. Bendall, pp. 105–30. Cambridge University Press; Cambridge.

Eldredge, N. & Gould, S. J. (1972). "Punctuated Equilibria: An Alternative to Phyletic Gradualism." In *Models in Paleobiology,* ed. T. J. M. Schopf. Freeman Cooper; San Francisco.

Fisher, R. A. (1958). *The Genetical Theory of Natural Selection.* Dover; New York.

Gillespie, N. C. (1979). *Charles Darwin and the Problem of Creation.* University of Chicago Press; Chicago.

Goldschmidt, R. (1940). *The Material Basis of Evolution.* Yale University Press; New Haven.

Gould, S. J. (1980). *The Panda's Thumb.* W. W. Norton; New York.

———. (1982). "The Meaning of Punctuated Equilibrium and Its Role in Validating a Hierarchical Approach to Macroevolution." In *Perspectives on Evolution*, ed. R. Milkman, pp. 83–104. Sinauer; Sunderland, Mass.

Gould, S. J. (1983). "Irrelevance, Submission and Partnership: The Changing Role of Palaeontology in Darwin's Three Centennials and a Modest Proposal for Macroevolution." In *Evolution from Molecules to Men*, ed. D. S. Bendall, pp. 387–402. Cambridge University Press; Cambridge.

Ho, M-W. & Saunders, P. T. (1982). "Adaptation and Natural Selection: Mechanism and Teleology." In *Towards a Liberatory Biology* (Dialectics of Biology Group, general editor S. Rose), 85–102.

Hoyle, F. & Wickramasinghe, N. C. (1981). *Evolution from Space*. J. M. Dent; London.

Jacob, F. (1983). "Molecular Tinkering in Evolution." In *Evolution from Molecules to Men*, ed. D. S. Bendall, pp. 131–44. Cambridge University Press; Cambridge.

Kimura, M. (1967). "Evolutionary Rate at the Molecular Level," *Nature, 217,* 624–6.

Koestler, A. (1967). *The Ghost in the Machine*. Hutchinson; London.

Lewin, R. (1980). "Evolutionary Theory Under Fire," *Science, 210,* 883–7.

Lewontin, R. C. (1970). "The Units of Selection," *Annual Review of Ecology and Systematics,* **1,** 1–18.

———. (1979). "Sociobiology as an Adaptationist Program," *Behavioral Science, 24,* 5–14.

Lorenz, K. (1966). *Evolution and Modification of Behavior*. Methuen; London.

Maynard Smith, J. (1969). "The Status of Neo-Darwinism." In *Towards a Theoretical Biology*, ed. C. H. Waddington. University Press; Edinburgh.

Mayr, E. (1982). *The Growth of Biological Thought*. Harvard University Press; Cambridge, Mass.

Miller, J. (1982). *Darwin for Beginners*. Writers and Readers; London.

Nevo, E. (1983). "Population Genetics and Ecology: The Interface." In *Evolution from Molecules to Men*, ed. D. S. Bendall, pp. 287–44. Cambridge University Press; Cambridge.

Paley, W. (1828). *Natural Theology*. 2nd edition. J. Vincent; Oxford.

Pringle, J. W. S. (1951). "On the Parallel Between Learning and Evolution," *Behaviour, 3,* 90–110.

Raup, D. M. (1977). "Stochastic Models in Evolutionary Palaeontology." In *Patterns of Evolution*, ed. A. Hallam. Elsevier; Amsterdam.

Ridley, M. (1982). "Coadaptation and the Inadequacy of Natural Selection," *British Journal for the History of Science, 15,* 45–68.

Rushton, W. A. H. (1961). "Peripheral Coding in the Nervous System." In *Sensory Communication*, ed. W. A. Rosenblith. MIT Press; Cambridge, Mass.

Stanley, S. M. (1981). *The New Evolutionary Timetable*. Basic Books; New York.

Stebbins, G. L. (1982). *Darwin to DNA, Molecules to Humanity*. W. H. Freeman; San Francisco.

Steele, E. J. (1979). *Somatic Selection and Adaptive Evolution*. Williams and Wallace; Toronto.

Thorpe, W. H. (1963). *Learning and Instinct in Animals*, 2nd edition. Methuen; London.

Turner, J. R. G. (1982). Review of R. J. Berry, *Neo-Darwinism. New Scientist, 94,* 160–2.

Wright, S. (1980). "Genic and Organismic Selection," *Evolution, 34,* 825–43.

Part Three

The Creationist Challenge

Introduction

What is creationism and where does it come from? The second of these questions is answered in the superb essay by Ronald L. Numbers, "The Creationists." There is nothing more I need add to the content, but do let me draw out one implication. However you may decide about the status of creationism—science, religion, or whatever—do not pretend to yourself that this, and this alone, represents accepted Christianity (or Judaism). Perhaps God is a creationist and Genesis is literally true. The fact of the matter is that many Christians and Jews think otherwise. Nor is such dissent confined to liberal theologians from trendy New England divinity schools. No less a conservative on doctrinal matters than John Paul II has said:

> The Bible itself speaks to us of the origin of the universe and its make-up, not in order to provide us with a scientific treatise but in order to state the correct relationships of man with God and with the universe. Sacred Scripture wishes simply to declare that the world was created by God, and in order to reach this truth it expresses itself in the terms of the cosmology in use at the time of the writer. The Sacred Book likewise wishes to tell men that the world was not created as the seat of the gods, as was taught by other cosmogonies and cosmologies, but was rather created for the service of man and the glory of God. Any other teaching about the origin and make-up of the universe is alien to the intentions of the Bible, which does not wish to teach how the heavens were made but how one goes to heaven. (Address to the Pontifical Academy of Science, October 1981)

To present the ideas of the creationists, I offer an extract from one of my own books, *Darwinism Defended*. This is simply a *precis* of the major text, *Scientific Creationism*. You should get a good overview of the position and I sincerely trust that it is not distorted. Complementing this, I also present an article by the leading creationist, Duane T. Gish. I shall make no comment here about the content of Gish's discussion, except to point out that he at once explicitly moves the creation/evolution clash to the level of status and methodology. Is creationism scientific? Is evolutionism scientific? In some respects Gish seems to think that neither are scientific.

With the stage now set, we move on to the Arkansas case, an account of which I have given already in the "Prologue: A Philosopher's Day in Court." The issue at stake was whether creationism (or creation science) is genuine science, with, it must be admitted, the background question whether evolutionism (particularly Darwinism) is likewise genuine science. First, I give the bill that was enacted into law in Arkansas in 1981. Note that Section 4 contains a clear definition of creation science, together with a definition of so-called "Evolution-Science." Then, I move on to my own contribution to the case, where I, as a historian and philosopher of science, argued strongly that whereas Darwinian evolutionary theory is scientific, creation science is not. It is religion.

As I mentioned in the Prologue, I prepared three position papers for the American Civil Liberties Union lawyers. These covered my views on the science/religion relationship, together with items discussed in my contributions to Parts One and Two of this collection. From these papers was prepared my script for the trial, known informally as my "Questions and Answers," which I include here. This was put together by the lawyers, and was the basis for my direct testimony at the trial. As I remember, we stuck to it almost line by line.

Finally, we have Judge Overton's ruling. The pertinent section is 4 (C). I might add that none of the other witnesses on either side spoke very much at all to philosophical issues. I might add also—with a little surprise, having reread the material following an interim of some years—how important the notion of falsifiability proved to be. (Apart from the fact that it was dwelt on at great length in cross-examination.) I believed, and still believe, that it is an important mark of the scientific, but forces other than I gave falsifiability major significance in Arkansas. In particular, for all that Popper has criticized in Darwinism and would (no doubt) criticize even more in creationism, both Darwinians and creationists had already staked their case on falsifiability. Ayala (also an Arkansas witness), for instance, tied himself to Popper's criterion (see the reference in my critique of Popper), and we see Gish do just the same. For these historical reasons alone, falsifiability was practically bound to come to the fore, and it did.

15

The Creationists

Ronald L. Numbers

Scarcely twenty years after the publication of Charles Darwin's *Origin of Species* in 1859 special creationists could name only two working naturalists in North America, John William Dawson (1820–99) of Montreal and Arnold Guyot (1806–84) of Princeton, who had not succumbed to some theory of organic evolution (Pfeifer 1974, 203; Gray 1963, 202–3). The situation in Great Britain looked equally bleak for creationists, and on both sides of the Atlantic liberal churchmen were beginning to follow their scientific colleagues into the evolutionist camp.[1] By the closing years of the nineteenth century evolution was infiltrating even the ranks of the evangelicals, and, in the opinion of many observers, belief in special creation seemed destined to go the way of the dinosaur. However, contrary to the hopes of liberals and the fears of conservatives, creationism did not become extinct. The majority of late-nineteenth-century Americans remained true to a traditional reading of Genesis, and as late as 1982 a public-opinion poll revealed that 44 percent of Americans, nearly a fourth of whom were college graduates, continued to believe that "God created man pretty much in his present form at one time within the last 10,000 years" ("Poll" 1982, 22).[2]

Such surveys failed, however, to disclose the great diversity of opinion among those professing to be creationists. Risking oversimplification, we can divide creationists into two main camps: "strict creationists," who interpret the days of Genesis literally, and "progressive creationists," who construe the Mosaic days to be immense periods of time. Yet, even within these camps

From *God and Nature,* edited by David C. Lindberg and Ronald L. Numbers (Berkeley, Calif.: University of California Press, 1986). Originally published as "Creationism in 20th Century America," *Science* 218 (November 5, 1982): 538–544. Copyright © 1982 by the American Association for the Advancement of Science. Reprinted by permission of the publisher and the author.

substantial differences exist. Among strict creationists, for example, some believe that God created all terrestrial life—past and present—less than ten thousand years ago, while others postulate one or more creations prior to the seven days of Genesis. Similarly, some progressive creationists believe in numerous creative acts, while others limit God's intervention to the creation of life and perhaps the human soul. Since this last species of creationism is practically indistinguishable from theistic evolutionism, this essay focuses on the strict creationists and the more conservative of the progressive creationists, particularly on the small number who claimed scientific expertise. Drawing on their writings, it traces the ideological development of creationism from the crusade to outlaw the teaching of evolution in the 1920s to the current battle for equal time. During this period the leading apologists for special creation shifted from an openly biblical defense of their views to one based largely on science. At the same time they grew less tolerant of notions of an old earth and symbolic days of creation, common among creationists early in the century, and more doctrinaire in their insistence on a recent creation in six literal days and on a universal flood.

THE LOYAL MAJORITY

The general acceptance of organic evolution by the intellectual elite of the late Victorian era has often obscured the fact that the majority of Americans remained loyal to the doctrine of special creation (Dillenberger & Welch 1954, 227). In addition to the masses who said nothing, there were many people who vocally rejected kinship with the apes and other, more reflective, persons who concurred with the Princeton theologian Charles Hodge (1797–1878) that Darwinism was atheism. Among the most intransigent foes of organic evolution were the premillennialists, whose predictions of Christ's imminent return depended on a literal reading of the Scriptures (Whalen 1972, 219-29; Numbers 1975, 18-23). Because of their conviction that one error in the Bible invalidated the entire book, they had little patience with scientists who, as described by the evangelist Dwight L. Moody (1837-99), "Dug up old carcasses . . . to make them testify against God" (McLoughlin 1959, 213).

Such an attitude did not, however, prevent many biblical literalists from agreeing with geologists that the earth was far older than six thousand years. They did so by identifying two separate creations in the first chapter of Genesis: the first, "in the beginning," perhaps millions of years ago, and the second, in six actual days, approximately four thousand years before the birth of Christ. According to the so-called gap theory, most fossils were relics of the first creation, destroyed by God prior to the Adamic restoration (Numbers 1977, 89-90; Ramm 1954, 195-98). In 1909 the *Scofield Reference Bible,* the most authoritative biblical guide in fundamentalist circles, sanctioned this view.[3]

Scientists like Guyot and Dawson, the last of the reputable nineteenth-century creationists, went still further to accommodate science by interpreting the days of Genesis as ages and by correlating them with successive epochs in the natural history of the world (O'Brien 1971; Numbers 1977, 91-100). Although they believed in special creative acts, especially of the first humans, they tended to minimize the number of supernatural interventions and to maximize the operation of natural law. During the late nineteenth century the theory of progressive creation circulated widely in the colleges and seminaries of America.[4]

The early Darwinian debate focused largely on the implications of evolution for natural theology (Moore 1979); and so long as these discussions remained confined to scholarly circles, those who objected to evolution on biblical grounds saw little reason to participate. However, when the debate spilled over into the public arena during the 1880s and 1890s, creationists grew alarmed. "When these vague speculations, scattered to the four winds by the million-tongued press, are caught up by ignorant and untrained men," declared one premillennialist in 1889, "it is time for earnest Christian men to call a halt" (Hastings 1889).

The questionable scientific status of Darwinism undoubtedly encouraged such critics to speak up ("Evolutionism in the Pulpit" 1910-15; Bowler 1983). Although the overwhelming majority of scientists after 1880 accepted a long earth history and some form of organized evolution, many in the late nineteenth century were expressing serious reservations about the ability of Darwin's particular theory of natural selection to account for the origin of species. Their published criticisms of Darwinism led creationists mistakenly to conclude that scientists were in the midst of discarding evolution. The appearance of books with such titles as *The Collapse of Evolution* and *At the Death Bed of Darwinism* bolstered this belief and convinced anti-evolutionists that liberal Christians had capitulated to evolution too quickly. In view of this turn of events it seemed likely that those who had "abandoned the stronghold of faith out of sheer fright will soon be found scurrying back to the old and impregnable citadel, when they learn that 'the enemy is is full retreat'" (Young 1909, 41).

For the time being, however, those conservative Christians who would soon call themselves fundamentalists perceived a greater threat to orthodox faith than evolution—higher criticism, which treated the Bible more as a historical document than as God's inspired Word. Their relative apathy toward evolution is evident in *The Fundamentals,* a mass-produced series of twelve booklets published between 1910 and 1915 to revitalize and reform Christianity around the world. Although one contributor identified evolution as the principal cause of disbelief in the Scriptures and another traced the roots of higher criticism to Darwin, the collection as a whole lacked the strident anti-evolutionism that would characterize the fundamentalist movement of the 1920s (Mauro 1910-15; Reeve 1910-15).

This is particularly true of the writings of George Frederick Wright (1838-1921), a Congregational minister and amateur geologist of international repute (Wright 1916). At first glance his selection to represent the fundamentalist point of view seems anomalous. As a prominent Christian Darwinist in the 1870s he had argued that the intended purpose of Genesis was to protest polytheism, not teach science (Wright 1898). By the 1890s, however, he had come to espouse the progressive creationism of Guyot and Dawson, partly, it seems, in reaction to the claims of higher critics regarding the accuracy of the Pentateuch (Wright 1902). Because of his standing as a scientific authority and his conservative view of the Scriptures, the editors of *The Fundamentals* selected him to address the question of the relationship between evolution and the Christian faith.

In an essay misleadingly titled "The Passing of Evolution" Wright attempted to steer a middle course between the theistic evolution of his early days and the traditional views of some special creationists. On the one hand, he argued that the Bible itself taught evolution, "an orderly progress from lower to higher forms of matter and life." On the other hand, he limited evolution to the origin of species, pointing out that even Darwin had postulated the supernatural creation of several forms of plants and animals, endowed by the Creator with a "marvelous capacity for variation." Furthermore, he argued that, despite the physical similarity between human beings and the higher animals, the former "came into existence as the Bible represents, by the special creation of a single pair, from whom all the varieties of the race have sprung" (Wright 1910-15).[5]

Although Wright represented the left wing of fundamentalism, his moderate views on evolution contributed to the conciliatory tone that prevailed during the years leading up to World War I. Fundamentalists may not have liked evolution, but few, if any, at this time saw the necessity or desirability of launching a crusade to eradicate it from the schools and churches in America.

THE ANTI-EVOLUTION CRUSADE

Early in 1922 William Jennings Bryan (1860-1925), Presbyterian layman and thrice-defeated Democratic candidate for the president of the United States, heard of an effort in Kentucky to ban the teaching of evolution in public schools. "The movement will sweep the country," he predicted hopefully, "and we will drive Darwinism from our schools" (Levine 1965, 277). His prophecy proved overly optimistic, but before the end of the decade more than twenty state legislatures did debate anti-evolution laws, and four—Oklahoma, Tennessee, Mississippi, and Arkansas—banned the teaching of evolution in public schools (Shipley 1927; 1930). At times the controversy became so tumultuous that it looked to some as though "America might go mad" (Nelson 1964, 319). Many persons shared responsibility for these events, but none more than Bryan. His entry into the fray had a catalytic effect (Szasz 1982, 107-16) and gave anti-evolutionists what they needed most: "a spokesman with a national reputation, immense prestige, and a loyal following" (Levine 1965, 272).

The development of Bryan's own attitude toward evolution closely paralleled that of the fundamentalist movement. Since early in the century he had occasionally alluded to the silliness of believing in monkey ancestors and to the ethical dangers of thinking that might makes right, but until the outbreak of World War I he saw little reason to quarrel with those who disagreed. The war, however, exposed the darkest side of human nature and shattered his illusions about the future of Christian society. Obviously something had gone awry, and Bryan soon traced the source of the trouble to the paralyzing influence of Darwinism on the human conscience. By substituting the law of the jungle for the teaching of Christ, it threatened the principles he valued most: democracy and Christianity. Two books in particular confirmed his suspicion. The first, Vernon Kellogg's *Headquarters Nights* in 1917, recounted firsthand conversations with German officers that revealed the role Darwin's biology had played in persuading the Germans to declare war. The second, Benjamin Kidd's *Science of Power* in 1918, purported to demonstrate the historical and philosophical links between Darwinism and German militarism (Levine 1965, 216-65).

About the time that Bryan discovered the Darwinian origins of the war, he also became aware, to his great distress, of unsettling effects the theory of evolution was having on America's own young people. From frequent visits to college campuses and from talks with parents, pastors, and Sunday school teachers, he heard about an epidemic of unbelief that was sweeping the country. Upon investigating the cause, his wife reported, "he became convinced that the teaching of Evolution as a fact instead of a theory caused the students to lose faith in the Bible, first, in the story of creation, and later in other doctrines, which underlie the Christian religion" (Williams 1936, 448). Again Bryan found confirming evidence in a recently published book, *Belief in God and Immortality,* authored in 1916 by the Bryn Mawr psychologist James H. Leuba, who demonstrated statistically that college attendance endangered traditional religious beliefs (Levine 1965, 266-67).

Armed with this information about the cause of the world's and the nation's moral decay, Bryan launched a nationwide crusade against the offending doctrine. In one of his most popular and influential lectures, "The Menace of Darwinism," he summed up his case against evolution, arguing that it was both un-Christian and unscientific. Darwinism, he declared, was nothing but "guesses strung together," and poor guesses at that. Borrowing from a turn-of-the-century tract, he illustrated how the evolutionist explained the origin of the eye:

> The evolutionist guesses that there was a time when eyes were unknown—that is a necessary part of the hypothesis. . . . a piece of pigment, or, as some say, a freckle appeared upon the skin of an animal that had no eyes. This piece of pigment or freckle converged the rays of the sun upon that spot and when the little animal felt the heat on that spot it turned the spot to the sun to get more heat. The increased heat irritated the skin—so the evolutionists guess, and a nerve came there and out of the nerve came the eye! (Bryan 1922, 94, 97-98).[6]

"Can you beat it?" he asked incredulously—and that it happened not once but twice? As for himself, he would take one verse in Genesis over all that Darwin wrote.

Throughout his political career Bryan had placed his faith in the common people, and he resented the attempt of a few thousand scientists "to establish an oligarchy over the forty million American Christians," to dictate what should be taught in the schools (Coletta 1969, 230). To a democrat like Bryan it seemed preposterous that this "scientific soviet" (Levine 1965, 289) would not only demand to teach its insidious philosophy but impudently insist that society pay its salaries. Confident that nine-tenths of the Christian citizens agreed with him, he decided to appeal directly to them, as he had done so successfully in fighting the liquor interests.[7] "Commit your case to the people," he advised creationists. "Forget, if need be, the highbrows both in the political and college world, and carry this cause to the people. They are the final and efficiently corrective power" ("Progress" 1929, 13).

Who were the people who joined Bryan's crusade? As recent studies have shown, they came from all walks of life and from every region of the country. They lived in New York, Chicago, and Los Angeles as well as in small towns and in the country. Few possessed advanced degrees, but many were not without education. Nevertheless, Bryan undeniably found his staunchest supporters and won his greatest victories in the conservative and still largely rural South, described hyperbolically by one fundamentalist journal as "the last stronghold of orthodoxy on the North American continent," a region where the "masses of the people in all denominations 'believe the Bible from lid to lid'" ("Fighting Evolution" 1925, 5).[8]

The strength of Bryan's following within the churches is perhaps more difficult to determine, because not all fundamentalists were creationists and many creationists refused to participate in the crusade against evolution. However, a 1929 survey of the theological beliefs of seven hundred Protestant ministers provides some valuable clues (Betts 1929, 26, 44). The question "Do you believe that the creation of the world occurred in the manner and time recorded in Genesis?" elicited the following positive responses:

Lutheran	89%
Baptist	63%
Evangelical	62%
Presbyterian	35%
Methodist	24%
Congregational	12%
Episcopalian	11%
Other	60%

Unfortunately, these statistics tell us nothing about the various ways respondents may have interpreted the phrase "in the manner and time recorded in Genesis," nor do they reveal anything about the level of political involvement in the campaign against evolution. Lutherans, for example, despite their overwhelming rejection of evolution, generally preferred education to legislation and tended to view legal action against evolution as "a dangerous mingling of church and state" (Rudnick 1966, 88-90; Szasz 1969, 279). Similarly, premillennialists, who saw the spread of evolution as one more sign of the world's impending end, sometimes lacked incentive to correct the evils around them (Sandeen 1971, 266-68).[9]

Baptists and Presbyterians, who dominated the fundamentalist movement, participated actively in the campaign against evolution. The Southern Baptist Convention, spiritual home of some of the most outspoken foes of evolution, lent encouragement to the creationist crusaders by voting unanimously in 1926 that "this Convention accepts Genesis as teaching that man was the special creation of God, and rejects every theory, evolution or other, which teaches that man originated in, or came by way of, a lower animal ancestry" (Clark 1952, 154; Thompson 1975-76). The Presbyterian Church contributed Bryan and other leaders to the creationist cause but, as the above survey indicates, also harbored many evolutionists. In 1923 the General Assembly turned back an attempt by Bryan and his fundamentalist cohorts to cut off funds to any church school found teaching human evolution, approving instead a compromise measure that condemned only materialistic evolution (Loetscher 1954, 111). The other major Protestant bodies paid relatively little attention to the debate over evolution; and Catholics, though divided on the question of evolution, seldom favored restrictive legislation (Morrison 1953).[10]

Leadership of the anti-evolution movement came not from the organized churches of America but from individuals like Bryan and interdenominational organizations such as the World's Christian Fundamentals Association, a predominantly premillennialist body founded in 1919 by William Bell Riley (1861-1947), pastor of the First Baptist Church in Minneapolis.[11] Riley became active as an anti-evolutionist after discovering, to his apparent surprise, that evolutionists were teaching their views at the University of Minnesota. The early twentieth century witnessed an unprecedented expansion of public education; enrollment in public high schools nearly doubled between 1920 and 1930 (Bailey 1964, 72-73). Fundamentalists like Riley and Bryan wanted to make sure that students attending these institutions would not lose their faith. Thus they resolved to drive every evolutionist from the public school payroll. Those who lost their jobs as a result deserved little sympathy, for, as one rabble-rousing creationist put it, the German soldiers who killed Belgian and French children with poisoned candy were angels compared with the teachers and textbook writers who corrupted the souls of children and thereby sentenced them to eternal death (Martin 1923, 164-65).

The creationists, we should remember, did not always act without provocation. In many instances their opponents displayed equal intolerance and insensitivity. In fact, one contemporary observer blamed the creation-evolution controversy in part on the "intellectual flapperism" of irresponsible and poorly informed teachers who delighted in shocking naive students with unsupportable statements about evolution. It was understandable, wrote an Englishman, that American parents would resent sending their sons and daughters to public institutions that exposed them to "a multiple assault upon traditional faiths" (Beale 1936, 249-51).

CREATIONIST SCIENCE AND SCIENTISTS

In 1922 Riley outlined the reasons why fundamentalists opposed the teaching of evolution. "The first and most important reason for its elimination," he explained, "is the unquestioned fact that evolution is not a science; it is a hypothesis only, a speculation" ([Riley] 1922, 5). Bryan often made the same point, defining true science as "classified knowledge . . . the explanation of facts" (Bryan 1922, 94). Although creationists had far more compelling reasons for rejecting evolution than its alleged unscientific status, their insistence on this point was not merely an obscurantist ploy. Rather it stemmed from their commitment to a once-respected tradition, associated with the English philosopher Sir Francis Bacon (1561-1626), that emphasized the factual, nontheoretical nature of science (Marsden 1977, 214-15). By identifying with the Baconian tradition, creationists could label evolution as false science, could claim equality with scientific authorities in comprehending facts, and could deny the charge of being anti-science. "It is not 'science' that orthodox Christians oppose," a fundamentalist editor insisted defensively. "No! no! a thousand times, No! They are opposed only to the theory of evolution, which has not yet been proved, and therefore is not to be called by the sacred name of *science*" (K[eyser] 1925, 413).

Because of their conviction that evolution was unscientific, creationists assured themselves that the world's best scientists agreed with them. They received an important boost at the beginning of their campaign from an address by the distinguished British biologist William Bateson (1861-1926) in 1921, in which he declared that scientists had *not* discovered "the actual mode and process of evolution" (Bateson 1922).[12] Although he warned creationists against misinterpreting his statement as a rejection of evolution, they paid no more attention to that caveat than they did to the numerous pro-evolution resolutions passed by scientific societies (Shipley 1927, 384).

Unfortunately for the creationists, they could claim few legitimate scientists of their own: a couple of self-made men of science, one or two physicians, and a handful of teachers who, as one evolutionist described them, were "trying to hold down, not a chair, but a whole settee, of 'Natural Science' in some little institution."[13] Of this group the most influential were Harry Rimmer (1890-1952) and George McCready Price (1870-1963).

Rimmer, Presbyterian minister and self-styled "research scientist," obtained his limited exposure to science during a term or two at San Francisco's Hahnemann Medical College, a small homeopathic institution that required no more than a high school diploma for admission. As a medical student he picked up a vocabulary of "double-jointed, twelve cylinder, knee-action words" that later served to impress the uninitiated (Rimmer 1945, 14). After his brief stint in medical school he attended Whittier College and the Bible Institute of Los Angeles for a year each before entering full-time evangelistic work. About 1919 he settled in Los Angeles, where he set up a small laboratory at the rear of his house to conduct experiments in embryology and related sciences. Within a year or two he established the Research Science Bureau "to prove through findings in biology, paleontology, and anthropology that science and the literal Bible were not contradictory." The bureau staff—that is, Rimmer— apparently used income from the sale of memberships to finance anthropological field trips in the western United States, but Rimmer's dream of visiting Africa to prove the dissimilarity of gorillas and humans failed to materialize. By the late 1920s the bureau lay dormant, and Rimmer signed on with Riley's World's Christian Fundamentals Associations as a field secretary.[14]

Besides engaging in research, Rimmer delivered thousands of lectures, primarily to student groups, on the scientific accuracy of the Bible. Posing as a scientist, he attacked Darwinism and poked fun at the credulity of evolutionists. To attract attention, he repeatedly offered one hundred dollars to anyone who could discover a scientific error in the Scriptures; not surprisingly, the offer never cost him a dollar ("World Religious Digest" 1939, 215). He also, by his own reckoning, never lost a public debate. Following one encounter with an evolutionist in Philadelphia, he wrote home gleefully that "the debate was a simple walkover, a massacre—murder pure and simple. The eminent professor was simply scared stiff to advance any of the common arguments of the evolutionists, and he fizzled like a wet firecracker" (Edmondson 1969, 329-30, 333-34).

Price, a Seventh-Day Adventist geologist, was less skilled at debating than Rimmer but more influential scientifically. As a young man Price attended an Adventist college in Michigan for two years and later completed a teacher training course at the provincial normal school in his native New Brunswick. The turn of the century found him serving as principal of a small high school in an isolated part of eastern Canada, where one of his few companions was a local physician. During their many conversations, the doctor almost converted his fundamentalist friend to evolution, but each time Price wavered, he was saved by prayer and by reading the works of the Seventh-Day Adventist prophetess Ellen G. White (1827-1915), who claimed divine inspiration for her view that Noah's flood accounted for the fossil record on which evolutionists based their theory. As a result of these experiences, Price vowed to devote his life to promoting creationism of the strictest kind.[15]

By 1906 he was working as a handyman at an Adventist sanitarium in southern California. That year he published a slim volume entitled *Illogical Geology: The Weakest Point in the Evolution Theory,* in which he brashly offered one thousand dollars "to any one who will, in the face of the facts here presented, show me how to prove that one kind of fossil is older than another." (Like Rimmer, he never had to pay.) According to Price's argument,

Darwinism rested "logically and historically on the succession of life idea as taught by geology" and "if this succession of life is not an actual scientific fact, then Darwinism . . . is a most gigantic hoax" (Price 1906, 9).[16]

Although a few fundamentalists praised Price's polemic, David Starr Jordan (1851-1931), president of Stanford University and an authority on fossil fishes, warned him that he should not expect "any geologist to take [his work] seriously." Jordan conceded that the unknown author had written "a very clever book" but described it as "a sort of lawyer's plea, based on scattering mistakes, omissions and exceptions against general truths that anybody familiar with the facts in a general way cannot possibly dispute. It would be just as easy and just as plausible and just as convincing if one should take the facts of European history and attempt to show that all the various events were simultaneous."[17] As Jordan recognized, Price lacked any formal training or field experience in geology. He was, however, a voracious reader of geological literature, an armchair scientist who self-consciously minimized the importance of field experience.

During the next fifteen years Price occupied scientific settees in several Seventh-Day Adventist schools and authored six more books attacking evolution, particularly its geological foundation. Although not unknown outside his own church before the early 1920s, he did not attract national attention until then. Shortly after Bryan declared war on evolution, Price published in 1923 *The New Geology,* the most systematic and comprehensive of his many books. Uninhibited by false modesty, he presented his "great *law of conformable stratigraphic sequences* . . . by all odds the most important law ever formulated with reference to the order in which the strata occur." This law stated that "*any kind of fossiliferous beds whatever, 'young' or 'old,' may be found occurring conformably on any other fossiliferous beds, 'older' or 'younger'*" (Price 1923, 637-38).[18] To Price, so-called deceptive conformities (where strata seem to be missing) and thrust faults (where the strata are apparently in the wrong order) proved that there was no natural order to the fossil-bearing rocks, all of which he attributed to the Genesis flood.

A Yale geologist reviewing the book for *Science* accused Price of "harboring a geological nightmare" (Schuchert 1924). Despite such criticism from the scientific establishment—and the fact that his theory contradicted both the day-age and gap interpretations of Genesis—Price's reputation among fundamentalists rose dramatically. Rimmer, for example, hailed *The New Geology* as "a masterpiece of REAL science [that] explodes in a convincing manner some of the ancient fallacies of science 'falsely so called'" (Rimmer 1925, 28). By the mid-1920s Price's byline was appearing with increasing frequency in a broad spectrum of conservative religious periodicals, and the editor of *Science* could accurately describe him as "the principal scientific authority of the Fundamentalists" *(Science* 1926).

THE SCOPES TRIAL AND BEYOND

In the spring of 1925 John Thomas Scopes, a high school teacher in the small town of Dayton, Tennessee, confessed to having violated the state's recently passed law banning the teaching of human evolution in public schools. His subsequent trial focused international attention on the anti-evolution crusade and brought William Jennings Bryan to Dayton to assist the prosecution. In anticipation of arguing the scientific merits of evolution, Bryan sought out the best scientific minds in the creationist camp to serve as expert witnesses. The response to his inquiries could only have disappointed the aging crusader. Price, then teaching in England, sent his regrets—along with advice for Bryan to stay away from scientific topics (Numbers 1979, 24). Howard A. Kelly a prominent Johns Hopkins physician who had contributed to *The Fundamentals,* confessed that, except for Adam and Eve, he believed in evolution. Louis T. More, a physicist who had just written a book in 1925 on *The Dogma of Evolution,* replied that he accepted evolution as a working hypothesis. Alfred W. McCann, author in 1922 of *God—or Gorilla,* took the opportunity to chide Bryan for supporting prohibition in the past and for now trying "to bottle-up the tendencies of men to think for themselves."[19]

At the trial itself things scarcely went better. When Bryan could name only Price and the deceased Wright as scientists for whom he had respect, the caustic Clarence Darrow (1857-1938), attorney for the defense, scoffed· "You mentioned Price because he is the only human being in the world so far as you know that signs his name as a geologist that believes like you do. . . . every scientist in this country knows [he] is a mountebank and a pretender and not a geologist at all." Eventually Bryan conceded that the world was indeed far more than six thousand years old and that the six days of creation had probably been longer than twenty-four hours each—concessions that may have harmonized with the progressive creationism of Wright but hardly with the strict creationism of Price (Numbers 1979, 24; Levine 1965, 349).

Though one could scarcely have guessed it from some of his public pronouncements, Bryan had long been a progressive creationist. In fact, his beliefs regarding evolution diverged considerably from those of his more conservative supporters. Shortly before his trial he had confided to Dr. Kelly that he, too, had no objection to "evolution before man but for the fact that a concession as to the truth of evolution up to man furnishes our opponents with an argument which they are quick to use, namely, if evolution accounts for all the species up to man, does it not raise a presumption in behalf of evolution to include man?" Until biologists could actually demonstrate the evolution of one species into another, he thought it best to keep them on the defensive.[20]

Bryan's admission at Dayton spotlighted a serious and long-standing problem among anti-evolutionists: their failure to agree on a theory of creation. Even the most visible leaders could not reach a consensus. Riley, for example, followed Guyot and Dawson (and Bryan) in viewing the days of Genesis as ages, believing that the testimony of geology necessitated this interpretation.

Rimmer favored the gap theory, which involved two separate creations, in part because his scientific mind could not fathom how, given Riley's scheme, plants created on the third day could have survived thousands of years without sunshine, until the sun appeared on the fourth. According to the testimony of acquaintances, he also believed that the Bible taught a local rather than a universal flood (Culver 1955, 7). Price, who cared not a whit about the opinion of geologists, insisted on nothing less than a recent creation in six literal days and a worldwide deluge. He regarded the day-age theory as "the devil's counterfeit" and the gap theory as only slightly more acceptable (Price 1902, 125-27; 1954, 39). Rimmer and Riley, who preferred to minimize the differences among creationists, attempted the logically impossible, if ecumenically desirable, task of incorporating Price's "new geology" into their own schemes (Riley & Rimmer n.d.; Riley 1930, 45).

Although the court in Dayton found Scopes guilty as charged, creationists had little cause for rejoicing. The press had not treated them kindly, and the taxing ordeal no doubt contributed to Bryan's death a few days after the end of the trial. Nevertheless, the anti-evolutionists continued their crusade, winning victories in Mississippi in 1926 and in Arkansas two years later (Shipley 1930, 330-32). By the end of the decade, however, their legislative campaign had lost its steam. The presidential election of 1928, pitting a Protestant against a Catholic, offered fundamentalists a new cause, and the onset of the depression in 1929 further diverted their attention (Szasz 1981, 117-25).

Contrary to appearances, the creationists were simply changing tactics, not giving up. Instead of lobbying state legislatures, they shifted their attack to local communities, where they engaged in what one critic described as "the emasculation of textbooks, the 'purging' of libraries, and above all the continued hounding of teachers" (Shipley 1930, 330). Their new approach attracted less attention but paid off handsomely, as school boards, textbook publishers, and teachers in both urban and rural areas, North and South, bowed to their pressure. Darwinism virtually disappeared from high school texts, and for years many American teachers feared being identified as evolutionists (Beale 1936, 228-37; Gatewood 1969, 39; Grabiner & Miller 1974; Laba & Gross 1950).

CREATIONISM UNDERGROUND

During the heady days of the 1920s, when their activities made front-page headlines, creationists dreamed of converting the world; a decade later, forgotten and rejected by the establishment, they turned their energies inward and began creating an institutional base of their own. Deprived of the popular press and frustrated by their inability to publish their views in organs controlled by orthodox scientists, they determined to organize their own societies and edit their own journals (Carpenter 1980).[21] Their early efforts, however, encountered two problems: the absence of a critical mass of scientifically trained creationists and lack of internal agreement.

In 1935 Price, along with Dudley Joseph Whitney, a farm journalist, and L. Allen Higley, a Wheaton College science professor, formed a Religion and Science Association to create "a united front against the theory of evolution." Among those invited to participate in the association's first—and only—convention were representatives of the three major creationist parties, including Price himself, Rimmer, and one of Dawson's sons, who, like his father, advocated the day-age theory.[22] But as soon as the Price faction discovered that its associates had no intention of agreeing on a short earth history, it bolted the organization, leaving it a shambles.[23]

Shortly thereafter, in 1938, Price and some Seventh-Day Adventist friends in the Los Angeles area, several of them physicians associated with the College of Medical Evangelists (now part of Loma Linda University), organized their own Deluge Geology Society and, between 1941 and 1945, published a *Bulletin of Deluge Geology and Related Science.* As described by Price, the group consisted of "a very eminent set of men. . . . In no other part of this round globe could anything like the number of scientifically educated believers in Creation and opponents of evolution be assembled, as here in Southern California"(Numbers 1979, 26). Perhaps the society's most notable achievement was its sponsorship in the early 1940s of a hush-hush project to study giant fossil footprints, believed to be human, discovered in rocks far older than the theory of evolution would allow. This find, the society announced excitedly, thus demolished that theory "at a single stroke" and promised to *"astound the scientific world!"* Yet despite such activity and the group's religious homogeneity, it, too, soon foundered   on "the same rock," complained a disappointed member, that wrecked the Religion and Science Association, that is *"pre-Genesis time for the earth."*[24]

By this time creationists were also beginning to face a new problem: the presence within their own ranks of young university-trained scientists who wanted to bring evangelical Christianity more into line wih mainstream science. The encounter between the two generations often proved traumatic, as is illustrated by the case of Harold W. Clark (b. 1891). A former student of Price's, he had gone on to earn a master's degree in biology from the University of California and taken a position at a small Adventist college in northern California. By 1940 his training and field experience had convinced him that Price's *New Geology* was "entirely out of date and inadequate" as a text, especially in its rejection of the geological column. When Price learned of this, he angrily accused his former disciple of suffering from "the modern mental disease of university-itis" and of currying the favor of "tobacco-smoking, Sabbath-breaking, God-defying" evolutionists. Despite Clark's protests that he still believed in a literal six-day creation and universal flood, Price kept up his attack for the better part of a decade, at one point addressing a vitriolic pamphlet, *Theories of Satanic Origin,* to his erstwhile friend and fellow creationist (Numbers 1979, 25).

The inroads of secular scientific training also became apparent in the American Scientific Affiliation (ASA), created by evangelical scientists in 1941.[25]

Although the society took no official stand on creation, strict creationists found the atmosphere congenial during the early years of the society. In the late 1940s, however, some of the more progressive members, led by J. Laurence Kulp, a young geochemist on the faculty of Columbia University, began criticizing Price and his followers for their allegedly unscientific effort to squeeze earth history into less than ten thousand years. Kulp, a Wheaton alumnus and member of the Plymouth Brethren, had acquired a doctorate in physical chemistry from Princeton University and gone on to complete all the requirements, except a dissertation, for a Ph.D. in geology. Although initially suspicious of the conclusions of geology regarding the history and antiquity of the earth, he had come to accept them. As one of the first evangelicals professionally trained in geology, he felt a responsibility to warn his colleagues in the ASA about Price's work, which, he believed, had "infiltrated the greater portion of fundamental Christianty in America primarily due to the absence of trained Christian geologists." In what was apparently the first systematic critique of the "new geology" Kulp concluded that the "major propositions of the theory are contradicted by established physical and chemical laws" (Kulp 1950).[26] Conservatives within the ASA not unreasonably suspected that Kulp's exposure to "the orthodox geological viewpoint" had severely undermined his faith in a literal interpretation of the Bible ("Comment" 1940, 2).

Before long it became evident that a growing number of ASA members, like Kulp, were drifting from strict to progressive creationism and sometimes on to theistic evolutionism. The transition for many involved immense personal stress, as revealed in the autobiographical testimony of another Wheaton alumnus, J. Frank Cassel:

> First to be overcome was the onus of dealing with a "verboten" term and in a "non-existent" area. Then, as each made an honest and objective consideration of the data, he was struck with the validity and undeniability of datum after datum. As he strove to incorporate each of these facts into his biblico-scientific frame of reference, he found that—while the frame became more complete and satisfying—he began to question first the feasibility and then the desirability of an effort to refute the total evolutionary concept, and finally he became impressed by its impossibility on the basis of existing data. This has been a heart-rending, soul-searching experience for the committed Christian as he has seen what he had long considered the *raison d'être* of God's call for his life endeavor fade away, and as he has struggled to release strongly held convictions as to the close limitations of Creationism.

Cassel went on to note that the struggle was "made no easier by the lack of approbation (much less acceptance) of some of his less well-informed colleagues, some of whom seem to question motives or even to imply heresy" (Cassel 1959, 26-27).[27] Strict creationists, who suffered their own agonies, found it difficult not to conclude that their liberal colleagues were simply taking the easy way out. To both parties a split seemed inevitable.

CREATIONISM ABROAD

During the decades immediately following the crusade of the 1920s American anti-evolutionists were buoyed by reports of a creationist revival in Europe, especially in England, where creationism was thought to be all but dead. The Victoria Institute in London, a haven for English creationists in the nineteenth century, had by the 1920s become a stronghold of theistic evolution. When Price visited the institute in 1925 to receive its Langhorne-Orchard Prize for an essay on "Revelation and Evolution," several members protested his attempt to export the fundamentalist controversy to England. Even evangelicals refused to get caught up in the turmoil that engulfed the United States. As historian George Marsden has explained, English evangelicals, always a minority, had developed a stronger tradition of theological toleration than revivalist Americans, who until the twentieth century had never experienced minority status. This, while the displaced Americans fought to recover their lost position, English evangelicals adopted a nonmilitant live-and-let-live philosophy that stressed personal piety (Numbers 1975, 25; Marsden 1977; 1980, 222-26).

The sudden appearance of a small but vocal group of British creationists in the early 1930s caught nearly everyone by surprise. The central figure in this movement was Douglas Dewar (1875-1957), a Cambridge graduate and amateur ornithologist, who had served for decades as a lawyer in the Indian Civil Service. Originally an evolutionist, he had gradually become convinced of the necessity of adopting "a provisional hypothesis of special creation . . . supplemented by a theory of evolution." This allowed him to accept unlimited development within biological families. His published views, unlike those of most American creationists, betrayed little biblical influence (Dewar 1931, 158; Lunn 1947, 1, 154; *Evolution Protest* 1965). His greatest intellectual debt was not to Moses but to a French zoologist, Louis Vialleton (1859-1929), who had attracted considerable attention in the 1920s for suggesting a theory of discontinuous evolution, which anti-evolutionists eagerly—but erroneously—equated with special creation (Paul 1979, 99-100).

Soon after announcing his conversion to creationism in 1931, Dewar submitted a short paper on mammalian fossils to the Zoological Society of London, of which he was a member. The secretary of the society subsequently rejected the piece, noting that a competent referee thought Dewar's evidence "led to no valuable conclusion." Such treatment infuriated Dewar and convinced him that evolution had become "a scientific creed." Those who questioned scientific orthodoxy, he complained, "are deemed unfit to hold scientific offices; their articles are rejected by newspapers or journals; their contributions are refused by scientific societies, and publishers decline to publish their books except at the author's expense. Thus the independents are today pretty effectually muzzled" (Dewar 1932, 142). Because of such experiences Dewar and other British dissidents in 1932 organized the Evolution Protest Movement, which after two decades claimed a membership of two hundred ("EPM" 1972).

HENRY M. MORRIS AND THE REVIVAL OF CREATIONISM

In 1964 one historian predicted that "a renaissance of the [creationist] movement is most unlikely" (Halliburton 1964, 283). And so it seemed. But even as these words were penned, a major revival was under way, led by a Texas engineer, Henry M. Morris (b. 1918). Raised a nominal Southern Baptist, and as such a believer in creation, Morris as a youth had drifted unthinkingly into evolutionism and religious indifference. A thorough study of the Bible following graduation from college convinced him of its absolute truth and prompted him to reevaluate his belief in evolution. After an intense period of soul-searching he concluded that creation had taken place in six literal days, because the Bible clearly said so and "God doesn't lie." Corroborating evidence came from the book of nature. While sitting in his office at Rice Institute, where he was teaching civil engineering, he would study the butterflies and wasps that flew in through the window; being familiar with structural design, he calculated the improbability of such complex creatures developing by chance. Nature as well as the Bible seemed to argue for creation.[28]

For assistance in answering the claims of evolutionists, he found little creationist literature of value apart from the writings of Rimmer and Price. Although he rejected Price's peculiar theology, he took an immediate liking to the Adventist's flood geology and in 1946 incorporated it into a little book, *That You Might Believe,* the first book, so far as he knew, "published since the Scopes trial in which a scientist from a secular university advocated recent special creation and a worldwide flood" (Morris 1978, 10). In the late 1940s he joined the American Scientific Affiliation—just in time to protest Kulp's attack on Price's geology. Yet his words fell largely on deaf ears. In 1953 when he presented some of his own views on the flood to the ASA, one of the few compliments came from a young theologian, John C. Whitcomb, Jr., who belonged to the Grace Brethren. The two subsequently became friends and decided to collaborate on a major defense of the Noachian flood. By the time they finished their project, Morris had earned a Ph.D. in hydraulic engineering from the University of Minnesota and was chairing the civil engineering department at Virginia Polytechnic Institute; Whitcomb was teaching Old Testament studies at Grace Theological Seminary in Indiana.[29]

In 1961 they brought out *The Genesis Flood,* the most impressive contribution to strict creationism since the publication of Price's *New Geology* in 1923. In many respects their book appeared to be simply "a reissue of G. M. Price's views, brought up to date," as one reader described it. Beginning with a testimony to their belief in "the verbal inerrancy of Scripture," Whitcomb and Morris went on to argue for a recent creation of the entire universe, a Fall that triggered the second law of thermodynamics, and a worldwide flood that in one year laid down most of the geological strata. Given this history, they argued, "the last refuge of the case for evolution immediately vanishes away, and the record of the rocks becomes a tremendous witness . . . to the holiness and justice and power of the living God of Creation!" (Whitcomb & Morris 1961, xx, 451).

Despite the book's lack of conceptual novelty, it provoked an intense debate among evangelicals. Progressive creationists denounced it as a travesty on geology that threatened to set back the cause of Christian science a generation, while strict creationists praised it for making biblical catastrophism intellectually respectable. Its appeal, suggested one critic, lay primarily in the fact that, unlike previous creationist works, it "looked *legitimate* as a scientific contribution," accompanied as it was by footnotes and other scholarly appurtenances. In responding to their detractors, Whitcomb and Morris repeatedly refused to be drawn into a scientific debate, arguing that "the real issue is not the correctness of the interpretation of various details of the geological data, but simply what God has revealed in His Word concerning these matters" (Morris & Whitcomb 1964, 60).[30]

Whatever its merits, *The Genesis Flood* unquestionably "brought about a stunning renaissance of flood geology" (Young 1977, 7), symbolized by the establishment in 1963 of the Creation Research Society. Shortly before the publication of his book Morris had sent the manuscript to Walter E. Lammerts (b. 1904), a Missouri-Synod Lutheran with a doctorate in genetics from the University of California. As an undergraduate at Berkeley Lammerts had discovered Price's *New Geology,* and during the early 1940s, while teaching at UCLA, he had worked with Price in the Creation-Deluge Society. After the mid-1940s, however, his interest in creationism had flagged—until awakened by reading the Whitcomb and Morris manuscript. Disgusted by the ASA's flirtation with evolution, he organized in the early 1960s a correspondence network with Morris and eight other strict creationists, dubbed the "team of ten." In 1963 seven of the ten met with a few other like-minded scientists at the home of a team member in Midland, Michigan, to form the Creation Research Society (CRS) (Lammerts 1974).

The society began with a carefully selected eighteen-man "inner-core steering committee," which included the original team of ten. The composition of this committee reflected, albeit imperfectly, the denominational, regional, and professional bases of the creationist revival. There were six Missouri-Synod Lutherans, five Baptists, two Seventh-Day Adventists, and one each from the Reformed Presbyterian Church, the Reformed Christian Church, the Church of the Brethren, and an independent Bible church. (Information about one member is not available.) Eleven lived in the Midwest, three in the South, and two in the Far West. The committee included six biologists but only one geologist, an independent consultant with a master's degree. Seven members taught in church-related colleges, five in state institutions; the others worked for industry or were self-employed.[31]

To avoid the creeping evolutionism that had infected the ASA and to ensure that the society remained loyal to the Price-Morris tradition, the CRS required members to sign a statement of belief accepting the inerrancy of the Bible, the special creation of "all basic types of living things," and a worldwide deluge (*Creation Research* 1964, [13]). It restricted membership to Christians only. (Although creationists liked to stress the scientific evidence for their position, one estimated that "only about five percent of evolutionists-turned-creationists did so on the basis of the overwhelming evidence for creation in the world of nature"; the remaining 95 percent became creationists because they believed in the Bible [Lang 1978, 2]).[32] To legitimate its claim to being a scientific society, the CRS published a quarterly journal and limited full membership to persons possessing a graduate degree in a scientific discipline.

At the end of its first decade the society claimed 450 regular members, plus 1,600 sustaining members, who failed to meet the scientific qualifications. Eschewing politics, the CRS devoted itself almost exclusively to education and research, funded "at very little expense, and . . . with no expenditure of public money" (Lammerts 1974, 63). CRS-related projects included expeditions to search for Noah's ark, studies of fossil human footprints and pollen grains found out of the predicted evolutionary order, experiments on radiation-produced mutations in plants, and theoretical studies in physics demonstrating a recent origin of the earth (Gish 1975). A number of members collaborated in preparing a biology textbook based on creationist principles (Moore & Slusher 1970). In view of the previous history of creation science, it was an auspicious beginning.

While the CRS catered to the needs of scientists, a second, predominantly lay organization carried creationism to the masses. Created in 1964 in the wake of interest generated by *The Genesis Flood,* the Bible-Science Association came to be identified by many with one man: Walter Lang, an ambitious Missouri-Synod pastor who self-consciously prized spiritual insight above scientific expertise. As editor of the widely circulated *Bible-Science Newsletter* he vigorously promoted the Price-Morris line—and occasionally provided a platform for individuals on the fringes of the creationist movement, such as those who questioned the heliocentric theory and who believed that Albert Einstein's theory of relativity "was invented in order to circumvent the evidence that the earth is at rest." Needless to say, the pastor's broad-mindedness greatly embarrassed creationists seeking scientific respectability, who feared that such bizarre behavior would tarnish the entire movement (Lang 1977a, 4-5; 1977b, 2-3; 1978b, 1-3; Wheeler 1976, 101-2).

SCIENTIFIC CREATIONISM

The creationist revival of the 1960s attracted little public attention until late in the decade, when fundamentalists became aroused about the federally funded Biological Sciences Curriculum Study texts (Skoog 1979; "A Critique" 1966, 1), which featured evolution, and the California State Board of Education voted to require public school textbooks to include creation along with evolution. This decision resulted in large part from the efforts of two southern California housewives, Nell Segraves and Jean Sumrall, associates of both the Bible-Science Association and the CRS. In 1961 Segraves learned of the U.S. Supreme Court's ruling in the Madalyn Murray case protecting atheist students from required prayers in public schools. Murray's ability to shield her child from religious exposure suggested to Segraves that creationist parents like herself "were entitled to protect our children from the influence of beliefs that would be offensive to our religious beliefs." It was this line of argument that finally persuaded the Board of Education to grant creationists equal rights (Bates 1975, 58; "Fifteen Years" 1979, 2; Wade 1972; see also Moore 1974; and Nelkin 1982).

Flushed with victory, Segraves and her son Kelly in 1970 joined an effort to organize a Creation-Science Research Center (CSRC), affiliated with Christian Heritage College in San Diego, to prepare creationist literature suitable for adoption in public schools. Associated with them in this enterprise was Morris, who resigned his position at Virginia Polytechnic Institute to help establish a center for creation research. Because of differences in personalities and objectives, the Segraveses in 1972 left the college, taking the CSRC with them; Morris thereupon set up a new research division at the college, the Institute for Creation Research (ICR), which, he announced with obvious relief, would be "controlled and operated by scientists" and would engage in research and education, not political action. During the 1970s Morris added five scientists to his staff and, funded largely by small gifts and royalties from institute publications, turned the ICR into the world's leading center for the propagation of strict creationism (Morris 1972).[33] Meanwhile, the CSRC continued campaigning for the legal recognition of special creation, often citing a direct relationship between the acceptance of evolution and the breakdown of law and order. Its own research, the CSRC announced, proved that evolution fostered "the moral decay of spiritual values which contribute to the destruction of mental health and . . . [the prevalence of] divorce, abortion, and rampant venereal disease" (Segraves 1977, 17; "Fifteen Years" 1979, 2-3).

The 1970s witnessed a major shift in creationist tactics. Instead of trying to outlaw evolution, as they had done in the 1920s, anti-evolutionists now fought to give creation equal time. And instead of appealing to the authority of the Bible, as Morris and Whitcomb had done as recently as 1961, they consciously downplayed the Genesis story in favor of what they called "scientific creationism." Several factors no doubt contributed to this shift. One sociologist has suggested that creationists began stressing the scientific legitimacy of their enterprise because "their theological legitimation of reality was no longer sufficient for maintaining their world and passing on their world view to their children" (Bates 1976, 98). However, there were also practical considerations. In 1968 the U.S. Supreme Court declared the Arkansas anti-evolution law unconstitutional, giving creationists reason to suspect that legislation requiring the teaching of biblical creationism would meet a similar fate. They also feared that requiring the biblical account "would open the door to a wide variety of interpretations of Genesis" and produce demands for the inclusion of non-Christian versions of creation (Morris 1974a, 2; see also Larson 1984).

In view of such potential hazards, Morris recommended that creationists ask public schools to teach "only the scientific aspects of creationism" (Morris

1974a, 2), which in practice meant leaving out all references to the six days of Genesis and Noah's ark and focusing instead on evidence for a recent worldwide catastrophe and on arguments against evolution. Thus the product remained virtually the same; only the packaging changed. The 1974 ICR textbook *Scientific Creationism,* for example, came in two editions: one for public schools, containing no references to the Bible, and another for use in Christian schools that included a chapter on "Creation According to Scripture" (Morris 1974b).

In defending creation as a scientific alternative to evolution, creationists relied less on Francis Bacon and his conception of science and more on two new philosopher-heroes: Karl Popper and Thomas Kuhn. Popper required all scientific theories to be falsifiable; since evolution could not be falsified, reasoned the creationists, it was by definition not science. Kuhn described scientific progress in terms of competing models or paradigms rather than the accumulation of objective knowledge.[34] Thus creationists saw no reason why their flood-geology model should not be allowed to compete on an equal scientific basis with the evolution model. In selling this two-model approach to school boards, creationists were advised: "Sell more SCIENCE. . . . Who can object to teaching more science? What is controversial about that? . . . do not use the word 'creationism.' Speak only of science. Explain that withholding scientific information contradicting evolution amounts to 'censorship' and smacks of getting into the province of religious dogma. . . . Use the 'censorship' label as one who is against censoring science. YOU are for science; anyone else who wants to censor scientific data is an old fogey and too doctrinaire to consider" (Leitch 1980, 2). This tactic proved extremely effective, at least initially. Two state legislatures, in Arkansas and Louisiana, and various school boards adopted the two-model approach, and an informal poll of school board members in 1980 showed that only 25 percent favored teaching nothing but evolution ("Finding" 1980, 52; Segraves 1977, 24). In 1982, however, a federal judge declared the Arkansas law, requiring a "balanced treatment" of creation and evolution, to be unconstitutional ("Creationism in Schools" 1982). Three years later a similar decision was reached regarding the Louisiana law.

Except for the battle to get scientific creationism into public schools, nothing brought more attention to the creationists than their public debates with prominent evolutionists, usually held on college campuses. During the 1970s the ICR staff alone participated in more than a hundred of these contests and, according to their own reckoning, never lost one. Although Morris preferred delivering straight lectures—and likened debates to the bloody confrontations between Christians and lions in ancient Rome—he recognized their value in carrying the creationist message to "more non-Christians and non-creationists than almost any other method" (Morris 1981, iii; 1974d, 2). Fortunately for him, an associate, Duane T. Gish, holder of a doctorate in biochemistry from the University of California, relished such confrontations. If the mild-mannered, professorial Morris was the Darwin of the creationist movement, then the bumptious Gish was its T. H. Huxley. He "hits the floor running" just like a bulldog, observed an admiring colleague; and "I go for the jugular vein," added Gish himself. Such enthusiasm helped draw crowds of up to five thousand.[35]

Early in 1981 the ICR announced the fulfillment of a recurring dream among creationists: a program offering graduate degrees in various creation-oriented sciences ("ICR Schedules" 1981). Besides hoping to fill an anticipated demand for teachers trained in scientific creationism, the ICR wished to provide an academic setting where creationist students would be free from discrimination. Over the years a number of creationists had reportedly been kicked out of secular universities because of their heterodox views, prompting leaders to warn graduate students to keep silent, "because if you don't, in almost 99 percent of the cases you will be asked to leave." To avoid anticipated harassment, several graduate students took to using pseudonyms when writing for creationist publications.[36]

Creationists also feared—with good reason—the possibility of defections while their students studied under evolutionists. Since the late 1950s the Seventh-Day Adventist Church had invested hundreds of thousands of dollars to staff its Geoscience Research Institute with well-trained young scientists, only to discover that in several instances exposure to orthodox science had destroyed belief in strict creationism. To reduce the incidence of apostasy, the church established its own graduate programs at Loma Linda University, where Price had once taught (Numbers 1979, 27-28; Couperus 1980).

TO ALL THE WORLD

It is still too early to assess the full impact of the creationist revival sparked by Whitcomb and Morris, but its influence, especially among evangelical Christians, seems to have been immense. Not least, it has elevated the strict creationism of Price and Morris to a position of apparent orthodoxy. It has also endowed creationism with a measure of scientific respectability unknown since the deaths of Guyot and Dawson. Yet it is impossible to determine how much of the creationists' success stemmed from converting evolutionists as opposed to mobilizing the already converted, and how much it owed to widespread disillusionment with established science. A sociological survey of church members in northern California in 1963 revealed that over a fourth of those polled—30 percent of Protestants and 28 percent of Catholics—were already opposed to evolution when the creationist revival began (Bainbridge & Stark 1980, 20). Broken down by denomination, it showed

Liberal Protestants (Congregationalists, Methodists, Episcopalians, Disciples)	11%
Moderate Protestants (Presbyterians, American Lutherans, American Baptists)	29%
Church of God	57%

Missouri-Synod Lutherans	64%
Southern Baptists	72%
Church of Christ	78%
Nazarenes	80%
Assemblies of God	91%
Seventh-Day Adventists	94%

Thus the creationists launched their crusade having a large reservoir of potential support.

Has belief in creationism increased since the early 1960s? The scanty evidence available suggests that it has. A nationwide Gallup poll in 1982, cited at the beginning of this paper, showed that nearly as many Americans (44 percent) believed in a recent special creation as accepted theistic (38 percent) or nontheistic (9 percent) evolution ("Poll" 1982, 22). These figures, when compared with the roughly 30 percent of northern California church members who opposed evolution in 1963, suggest, in a grossly imprecise way, a substantial gain in the actual number of American creationists. Bits and pieces of additional evidence lend credence to this conclusion. For example, in 1935 only 36 percent of the students at Brigham Young University, a Mormon school, rejected human evolution; in 1973 the percentage had climbed to 81 (Christensen & Cannon 1978). Also, during the 1970s both the Missouri-Synod Lutheran and Seventh-Day Adventist churches, traditional bastions of strict creationism, took strong measures to reverse a trend toward greater toleration of progressive creationism ("Return to Conservatism" 1973, 1; Numbers 1979, 27–28). In at least these instances, strict creationism did seem to be gaining ground.

Unlike the anti-evolution crusade of the 1920s, which remained confined mainly to North America, the revival of the 1960s rapidly spread overseas as American creationists and their books circled the globe. Partly as a result of stimulation from America, including the publication of a British edition of *The Genesis Flood* in 1969, the lethargic Evolution Protest Movement in Great Britain was revitalized; and two new creationist organizations, the Newton Scientific Association and the Biblical Creation Society, sprang into existence (Barker 1979; [Clark] 1972–73; 1977; "British Scientists" 1973; "EPM" 1972).[37] On the Continent the Dutch assumed the lead in promoting creationism, encouraged by the translation of books on flood geology and by visits from ICR scientists (Ouweneel 1978). Similar developments occurred elsewhere in Europe, as well as in Australia, Asia, and South America. By 1980 Morris's books alone had been translated into Chinese, Czech, Dutch, French, German, Japanese, Korean, Portuguese, Russian, and Spanish. Strict creationism had become an international phenomenon.[38]

NOTES

1. Michael Ruse (1979) argues that most British biologists were evolutionists by the mid-

1860s, while David L. Hull, Peter D. Tessner, and Arthur M. Diamond (1978, 721) point out that more than a quarter of British scientists continued to reject the evolution of species as late as 1869. On the acceptance of evolution among religious leaders see e.g., Frank Hugh Foster (1939, 38-58) and Owen Chadwick (1972, 23-24).

2. According to the poll, 9 percent of the respondents favored an evolutionary process in which God played no part, 38 percent believed God directed the evolutionary process, and 9 percent had no opinion.

3. On the influence of the *Scofield Reference Bible* see Ernest R. Sandeen (1971, 222).

4. On the popularity of the Guyot-Dawson view, also associated with the geologist James Dwight Dana, see William North Rice (1904, 101).

5. The Scottish theologian James Orr contributed an equally tolerant essay in *The Fundamentals* (Orr 1910-15).

6. "The Menace of Darwinism" appears in Bryan's book *In His Image* as chapter 4, "The Origin of Man." Bryan apparently borrowed his account of the evolution of the eye from Patterson (1902, 32-33).

7. Bryan gives the estimate of nine-tenths in a letter to W. A. McRae, 5 Apr. 1924 (Bryan Papers, box 29).

8. The best state histories of the anti-evolution crusade are Bailey (1950); Gatewood (1966); and Gray (1970). Szasz (1969, 351) stresses the urban dimension of the crusade.

9. For examples of prominent fundamentalists who stayed aloof from the anti-evolution controversy see Stonehouse (1954, 401-2) and Lewis (1963, 86-88).

10. Furness (1954) includes chapter-by-chapter surveys of seven denominations.

11. On Riley see Riley (1938, 101-2) and Szasz (1980, 89-91). Marsden (1980, 167-70) stresses the interdenominational character of the anti-evolution crusade.

12. The creationists' use of Bateson provoked the evolutionist Henry Fairfield Osborn into repudiating the British scientist (Osborn 1926, 29).

13. Heber D. Curtis to W. J. Bryan, 22 May 1923 (Bryan Papers, box 37). Two physicians, Arthur I. Brown of Vancouver and Howard A. Kelly of Johns Hopkins, achieved prominence in the fundamentalist movement, but Kelly leaned toward theistic evolution.

14. See Edmondson (1969, 276-336); Cole (1931, 264-65); B[oyer] (1939, 6-7); and "Two Great Field Secretaries" (1926, 17).

15. This and the following paragraphs on Price closely follow the account in Numbers (1979 22-24).

16. Price's first anti-evolution book was published four years earlier (Price 1902).

17. David Star Jordan to G. M. Price, 5 May 1911 (Price Papers).

18. The discovery of Price's law was first announced in Price (1913, 119).

19. Howard A. Kelly to W. J. Bryan, 15 June 1925; Louis T. More to W. J. Bryan, 7 July 1925; and Alfred W. McCann to W. J. Bryan, 30 June 1925 (Bryan Papers, box 47).

20. W. J. Bryan to Howard A. Kelly, 22 June 1925 (Bryan Papers, box 47). In a letter to the editor of *The Forum,* Bryan (1923) asserted that he had never taught that the world was made in six literal days. I am indebted to Paul M. Waggoner for bringing this document to my attention.

21. For a typical statement of creationist frustration see Price (1935). The title for this section comes from Morris (1974, 13).

22. See "Announcement of the Religion and Science Association" (Price Papers); "The Religion and Science Association" (1936, 159-60); "Meeting of the Religion and Science Association" (1936, 209); Clark (1977, 168).

23. On the attitude of the Price faction see Harold W. Clark to G. M. Price, 12 Sept. 1937 (Price Papers).

24. Ben F. Allen to the Board of Directors of the Creation-Deluge Society, 12 Aug. 1945 (courtesy of Molleurus Couperus). Regarding the fossil footprints, see the *Newsletters* of the Creation-Deluge Society for 19 Aug. 1944 and 17 Feb. 1945.

25. On the early years of the American Scientific Affiliation see Everest (1951).

26. Kulp (1949, 20) mentions his initial skepticism of geology.

27. For a fuller discussion see Numbers (1984).

28. Interviews with Henry M. Morris, 26 Oct. 1980 and 6 Jan. 1981. See also the autobiographical material in Morris (1984).

29. Interviews with Morris.

30. The statement regarding the appearance of the book comes from Walter Hearn, quoted in Bates (1976, 52). See also Roberts (1964); Van de Fliert (1969); and Lammerts (1964). Among Missouri-Synod Lutherans, John W. Klotz (1955) may have had an even greater influence than Morris and Whitcomb.

31. Names, academic fields, and institutional affiliations are given in Creation Research Society Quarterly (1964 [113]); for additional information I am indebted to Duane T. Gish, John N. Moore, Henry M. Morris, Harold Slusher, and William J. Tinkle.

32. Other creationists have disputed the 5 percent estimate.

33. Information also obtained from the interview with Morris, 6 Jan. 1981.

34. On Popper's influence see, e.g., Roth (1977). In a letter to the editor of New Scientist, Popper (1980) affirmed that the evolution of life on earth was testable and, therefore, scientific. On Kuhn's influence see, e.g., Roth (1975); Brand (1974); and Wheeler (1975, 192–210).

35. The reference to Gish comes from an interview with Harold Slusher and Duane T. Gish, 6 Jan. 1981.

36. Evidence for alleged discrimination and the use of pseudonyms comes from: "Grand Canyon Presents Problems for Long Ages" (1980); interview with Ervil D. Clark, 9 Jan. 1981; interview with Steven A. Austin, 6 Jan. 1981; and interview with Duane T. Gish, 26 Oct. 1980, the source of the quotation.

37. Barker greatly underestimates the size of the E.P.M. in 1966.

38. Notices regarding the spread of creationism overseas appeared frequently in Bible-Science Newsletter and Acts & Facts. On translations see "ICR Books Available in Many Languages" (1980, 2, 7).

REFERENCES

Bailey, Kenneth K. 1950. "The Enactment of Tennessee's Anti-evolution Law," Journal of Southern History 16:472–510.

———. 1964. Southern White Protestantism in the Twentieth Century. Harper & Row; New York.

Bainbridge, William Sims and Rodney Stark. 1980. "Superstitions Old and New," Skeptical Inquirer 4 (Summer).

Barker, Eileen. 1979. "In the Beginning: The Battle for Creationist Science against Evolutionism." In On the Margins of Science: The Social Construction of Rejected Knowledge, ed. Roy Wallis, 197–200. Sociological Review Monograph, no. 27. University of Keele; Keele:

Bates, Vernon Lee. 1976. "Christian Fundamentalism and the Theory of Evolution in Public School Education: A Study of the Creation Science Movement." Ph.D. diss., University of California, Davis.

Bateson, William. 1922. "Evolutionary Faith and Modern Doubts," Science 55:55–61.

Beale, Howard K. 1936. Are American Teachers Free? An Analysis of Restraints upon the Freedom of Teaching in American Schools. Charles Scribner's Sons; New York.

Betts, George Herbert. 1929. The Beliefs of 700 Ministers and Their Meaning for Religious Education. Abingdon Press; New York.

Bowler, Peter J. 1983. *The Eclipse of Darwinism: Anti-Darwinian Evolution Theories in the Decades around 1900.* Johns Hopkins University Press; Baltimore.

B[oyer], F. J. 1939. "Harry Rimmer, D.D." *Christian Faith and Life* 45.

Brand, Leonard R. 1974. "A Philosophic Rationale for a Creation-Flood Model," *Origins* 1:73–83.

"British Scientists Form Creationist Organization." 1973. *Acts & Facts 2* (Nov.–Dec.):3.

Bryan, William Jennings. 1922. *In His Image.* Fleming H. Revell; New York.

———. 1923. Letter to the editor of *The Forum 70*:1852–53.

Bryan Papers. N.d. Library of Congress.

Carpenter, Joel A. 1980. "Fundamentalist Institutions and the Rise of Evangelical Protestantism, 1929–1942," *Church History 49*:62–75.

Cassel, J. Frank. 1959. "The Evolution of Evangelical Thinking on Evolution," *Journal of the American Scientific Affiliation 11* (Dec.):26–27.

Chadwick, Owen. 1972. *The Victorian Church,* Part 2. 2d ed. Adam and Charles Black; London.

Christensen, Harold T. and Kenneth L. Cannon. 1978. "The Fundamental Emphasis at Brigham Young University: 1935–1973," *Journal for the Scientific Study of Religion 17*:53–57.

Clark, Edward Lassiter. 1952. "The Southern Baptist Reaction to the Darwinian Theory of Evolution." Ph.D. diss., Southwestern Baptist Theological Seminary, Fort Worth.

Clark, Harold W. 1977. *The Battle over Genesis.* Review and Herald Publishing Association; Washington.

[Clark, Robert E. D.] 1972–73. "Evolution: Polarization of Views," *Faith and Thought 100*:227–29.

———. 1977. "American and English Creationists," *Faith and Thought 104*:6–8.

Cole, Steward G. 1931. *The History of Fundamentalism.* Richard R. Smith; New York.

Coletta, Paolo E. 1969. *William Jennings Bryan.* Vol. 3, *Political Puritan, 1915–1925.* University of Nebraska Press; Lincoln.

"Comment on the 'Deluge Geology' Paper of J. L. Kulp." 1950. *Journal of the American Scientific Affiliation 2* (June).

Couperus, Molleurus. 1980. "Tensions Between Religion and Science." *Spectrum 10* (Mar.):74–78.

Creation Research Society Quarterly. 1964. 1 (July).

"Creationism in Schools: The Decision in MacLean versus the Arkansas Board of Education." 1982. *Science 215*:934–43.

"A Critique of BSCS Biology Texts." 1966. *Bible-Science Newsletter 4* (15 Mar.).

Culver, Robert D. 1955. "An Evaluation of *The Christian View of Science and Scripture* by Bernard Ramm from the Standpoint of Christian Theology," *Journal of the American Scientific Affiliation 7* (Dec.).

Dewar, Douglas. 1931. *The Difficulties of the Evolution Theory.* Edward Arnold and Co.; London.

———. 1932. "The Limitations of Organic Evolution," *Journal of the Victoria Institute 64.*

Dillenberger, John and Claude Welch. 1954. *Protestant Christianity Interpreted Through Its Development.* Charles Scribner's Sons; New York.

Edmondson, William D. 1969. "Fundamentalist Sects of Los Angeles, 1900-1930." Ph.D. diss., Claremont Graduate School, Claremont.

"EPM—40 Years On; Evolution 114 Years Off." 1972. Supplement to *Creation 1* (May).

Everest, Alton. 1951. "The American Scientific Affiliation—The First Decade," *Journal of the American Scientific Affiliation 3* (Sept.):31–38.

Evolution Protest Movement Pamphlet No. 125. 1965. April.

"Evolutionism in the Pulpit." 1910–15. In *The Fundamentals,* vol. 8, 28–30. Chicago: Testimony.

"Fifteen Years of Creationism." 1979. *Five Minutes with the Bible and Science.* Supplement to *Bible-Science Newsletter 17* (May):2.

"Fighting Evolution at the Fundamentals Convention." 1925. *Christian Fundamentals in School and Church 7* (July-Sept.).

"Finding: Let Kids Decide How We Got Here." 1980. *American School Board Journal 167* (Mar.).

Foster, Frank Hugh. 1939. *The Modern Movement in American Theology: Sketches in the History of American Protestant Thought from the Civil War to the World War.* Fleming H. Revell Co.; New York.

The Fundamentals. 1910–15. 12 vols. Chicago: Testimony.

Furniss, Norman F. 1954. *The Fundamentalist Controversy, 1918-1931.* Yale University Press; New Haven.

Gatewood, William J., Jr. 1966. *Preachers, Pedagogues and Politicians: The Evolution Controvery in North Carolina, 1920-1927.* University of North Carolina Press; Chapel Hill.

———. ed. 1969. *Controversy in the Twenties: Fundamentalism, Modernism, and Evolution.* Vanderbilt University Press; Nashville.

Gish, Duane T. 1975. "A Decade of Creationist Research," *Creation Research Society Quarterly 12* (June):34–36.

Grabiner, Judith V. and Peter D. Miller. 1974. "Effects of the Scopes Trial," *Science 185*:832–37.

"Grand Canyon Presents Problems for Long Ages." 1980. *Five Minutes with the Bible and Science.* Supplement to *Bible-Science Newsletter 18* (June):1–2.

Gray, Asa. 1963. *Darwinism: Essays and Reviews Pertaining to Darwinism.* Ed. A. Hunter Dupree. Harvard University Press; Cambridge, Mass.

Gray, Virginia. 1970. "Anti-Evolution Sentiment and Behavior: The Case of Arkansas," *Journal of American History 57*:352–66.

Halliburton, R., Jr. 1964. "The Adoption of Arkansas' Anti-Evolution Law," *Arkansas Historical Quarterly 23.*

Hastings, H. L. 1889. "Preface." In *The Errors of Evolution: An Examination of the Nebular Theory, Geological Evolution, the Origin of Life, and Darwinism,* by Robert Patterson. 3d ed. Scriptural Tract Repository; Boston.

Hull, David L., Peter D. Tessner, and Arthur M. Diamond. 1978. "Planck's Principle," *Science 202.*

"ICR Books Available in Many Languages." 1980. *Acts & Facts 9* (Feb.).

"ICR Schedules M. S. Programs." 1981. *Acts & Facts 10* (Feb.):1–2.

K[eyser], L. S. 1925. "No War Against Science—Never!" *Bible Champion 31.*

Klotz, John W. 1955. *Genes, Genesis, and Evolution.* Concordia Publishing House; St. Louis.

Kulp, J. Laurence. 1949. "Some Presuppositions of Evolutionary Thinking," *Journal of the American Scientific Affiliation 1* (June).

———. 1950. "Deluge Geology." *Journal of the American Scientific Affiliation 2* (Mar.):1–15.

Laba, Estelle R. and Eugene W. Gross. 1950. "Evolution Slighted in High-School Biology," *Clearing House 24*:396–99.

Lammerts, Walter E. 1964. "Introduction," *Annual*. Creation Research Society.

———. 1974. "The Creationist Movement in the United States: A Personal Account," *Journal of Christian Reconstruction* (Summer):49–63.

Lang, Walter. 1977a. "A Naturalistic Cosmology vs. a Biblical Cosmology." *Bible-Science Newsletter 15* (Jan.-Feb.).

———. 1977b. "Editorial Comments." *Bible-Science Newsletter 15* (Mar.).

———. 1978a. "Editorial Comments." *Bible-Science Newsletter 16* (June).

———. 1978b. "Fifteen Years of Creationism." *Bible-Science Newsletter 16* (Oct.).

Larson, Edward J. 1984. "Public Science vs. Popular Opinion: The Creation-Evolution Legal Controversy." Ph.D. diss., University of Wisconsin, Madison.

Leitch, Russell H. 1980. "Mistakes Creationists Make," *Bible-Science Newsletter 18* (Mar.).

Levine, Lawrence W. 1965. *Defender of the Faith—William Jennings Bryan: The Last Decade, 1915–25*. Oxford University Press; New York.

Lewis, William Bryant, 1963. "The Role of Harold Paul Sloan and his Methodist League for Faith and Life in the Fundamentalist-Modernist Controversy of the Methodist Episcopal Church." Ph.D. diss., Vanderbilt University, Nashville.

Loetscher, Lefferts A. 1954. *The Broadening Church: A Study of Theological Issues in the Presbyterian Church since 1869*. University of Pennsylvania Press; Philadelphia.

Lunn, Arnold, ed. 1947. *Is Evolution Proved? A Debate between Douglas Dewar and H. S. Shelton*. Hollis and Carter; London.

Marsden, George M. 1977. "Fundamentalism as an American Phenomenon: A Comparison with English Evangelicalism," *Church History 46*:215–32.

———. 1980. *Fundamentalism and American Culture: The Shaping of Twentieth Century Evangelicalism, 1870–1925*. Oxford University Press; New York.

Martin, T. T. 1923. *Hell and the High School: Christ or Evolution, Which?* Western Baptist Publishing Co.; Kansas City.

Mauro, Philip. 1910–15. "Modern Philosophy." In *The Fundamentals*, vol. 2, 85–105. Testimony; Chicago.

McLoughlin, William G., Jr. 1959. *Modern Revivalism: Charles Grandison Finney to Billy Graham*. Ronald Press; New York.

"Meeting of the Religion and Science Association." 1936. *Christian Faith and Life* 42.

Moore, James R. 1979. *The Post-Darwinian Controversies: A Study of the Protestant Struggle to Come to Terms with Darwin in Great Britain and America, 1870–1900*. Cambridge University Press; Cambridge.

Moore, John A. 1974. "Creationism in California," *Daedalus 103*:173–89.

Moore, John N. and Harold Schultz Slusher, eds. 1970. *Biology: A Search for Order in Complexity*. Zondervan Publishing House; Grand Rapids, Mich.

Morris, Henry M. 1972. "Director's Column," *Acts & Facts 1* (June–July).

———. 1974a. "Director's Column," *Acts & Facts 3* (Mar.).

———. 1974b. "Director's Column," *Acts & Facts 3* (Sept.).

———, ed. 1974c. *Scientific Creationism*. Gen. ed. Creation-Life Publishers; San Diego.

———. 1974d. *The Troubled Waters of Evolution*. Creation-Life Publishers; San Diego.

———. 1978. *That You Might Believe*. Rev. ed. Creation-Life Publishers; San Diego.

———. 1981a. "Director's Column," *Acts & Facts 1* (June–July).

Morris, Henry M. 1981b. "Two Decades of Creation: Past and Future," *Impact,* Supplement to *Acts & Facts 10* (Jan.).

———. 1984. *History of Modern Creationism.* Master Book Publishers; San Diego.

Morris, Henry M. and John C. Whitcomb, Jr. 1964. "Reply to Reviews in the March 1964 Issue," *Journal of the American Scientific Affiliation 16* (June).

Morrison, John L. 1953. "American Catholics and the Crusade Against Evolution." *Records of American Catholic Historical Society of Philadelphia 64*:59–71.

Nelkin, Dorothy. 1982. *The Creation Controversy: Science or Scriptures in the Schools.* W. W. Norton; New York.

Nelson, Roland T. 1964. "Fundamentalism and the Northern Baptist Convention." Ph.D. diss., University of Chicago, Chicago.

Numbers, Ronald L. 1975. "Science Falsely So-Called: Evolution and the Adventists in the Nineteenth Century," *Journal of the American Scientific Affiliation 27* (Mar.).

———. 1977. *Creation by Natural Law: Laplace's Nebular Hypothesis in American Thought.* University of Washington Press; Seattle.

———. 1979. "Sciences of Satanic Origin: Adventist Attitudes toward Evolutionary Biology and Geology," *Spectrum 9* (Jan.).

———. 1984. "The Dilemma of Evangelical Scientists." In *Evangelism and Modern America,* ed. George M. Marsden, 150–60. William B. Eerdmans; Grand Rapids, Mich.

O'Brien, Charles F. 1971. *Sir William Dawson: A Life in Science and Religion.* American Philosophical Society; Philadelphia.

Orr, James. 1910–15. "Science and Christian Faith." In *The Fundamentals,* vol. 4, 91–104. Testimony; Chicago.

Osborn, Henry Fairfield. 1926. *Evolution and Religion in Education: Polemics of the Fundamentalist Controversy of 1922 to 1926.* Charles Scribner's Sons; New York.

Ouweneel, W. J. 1978. "Creationism in the Netherlands," *Impact.* Supplement to *Acts & Facts 9* (Feb.):i-iv.

Patterson, Alexander. 1902. *The Other Side of Evolution: An Examination of Its Evidences.* Winona Publishing Co.; Chicago.

Patterson, Robert. 1893. *The Errors of Evolution: An Examination of the Nebular Theory, Geological Evolution, the Origin of Life, and Darwinism.* 3d ed. Scriptural Tract Repository; Boston.

Paul, Harry W. 1979. *The Edge of Contingency: French Catholic Reaction to Scientific Change from Darwin to Duhem.* University Presses of Florida; Gainesville, Fla.

Pfeifer, Edward J. 1974. "United States." In *The Comparative Reception of Darwinism,* ed. Thomas F. Glick. University of Texas Press; Austin.

"Poll Finds Americans Split on Creation Idea." 1982. *New York Times,* Aug. 29.

Popper, Karl. 1980. Letter to the editor of *New Scientist* 87 (21 Aug.):611.

Price, George McCready. 1901. *Outlines of Modern Science and Modern Christianity.* Pacific Press; Oakland, Calif.

———. 1906. *Illogical Geology: The Weakest Point in the Evolution Theory.* Modern Heretic Co.; Los Angeles.

———. 1913. *The Fundamentals of Geology and Their Bearings on the Doctrine of a Literal Creation.* Pacific Press; Mountain View, Calif.

———. 1923. *The New Geology.* Pacific Press; Mountain View, Calif.

———. 1935. "Guarding the Sacred Cow," *Christian Faith and Life 41*:124–27.

———. 1954. *The Story of the Fossils.* Pacific Press; Mountain View, Calif.

Price Papers. N.d. Andrews Univ.

"Progress of Anti-Evolution." 1929. *Christian Fundamentalist* 2.

Ramm, Bernard. 1954. *The Christian View of Science and Scripture.* William B. Eerdmans; Grand Rapids, Mich.

Reeve, J. J. 1910–15. "My Personal Experience with the Higher Criticism." In *The Fundamentals,* vol. 3, 98–118. Testimony; Chicago.

"The Religion and Science Association." 1936. *Christian Faith and Life* 42:159–60.

"Return to Conservatism." 1973. *Bible-Science Newsletter 11* (Aug.).

Rice, William North. 1904. *Christian Faith in an Age of Science.* 2d ed. A. C. Armstrong and Son; New York.

Riley, Marie Acomb. 1938. *The Dynamic of a Dream: The Life Story of Dr. William B. Riley.* William B. Eerdmans; Grand Rapids, Mich.

[Riley, William B.] 1922. "The Evolution Controversy," *Christian Fundamentals in School and Church 4* (Apr.-May).

———. 1930. "The Creative Week," *Christian Fundamentalist* 4.

Riley, William B. and Harry Rimmer. N.d. *A Debate Resolved, That the Creative Days in Genesis Were Aeons, Not Solar Days.* Pamphlet.

Rimmer, Harry. 1945. *The Harmony of Science and Scripture.* 11th ed. William B. Eerdmans; Grand Rapids, Mich.

Roberts, Frank H. 1964. "Review of *The Genesis Flood* by Henry M. Morris & John C. Whitcomb, Jr.," *Journal of the American Scientific Affiliation 16* (Mar.):28–29.

Roth, Ariel A. 1975. "The Pervasiveness of the Paradigm," *Origins* 2:55–57.

———. 1977. "Does Evolution Qualify as a Scientific Principle?" *Origins* 4:4–10.

Rudnick, Milton L. 1966. *Fundamentalism and the Missouri Synod: A Historical Study of Their Interaction and Mutual Influence.* Concordia Publishing House; St. Louis.

Ruse, Michael. 1979. *The Darwinian Revolution: Science Red in Tooth and Claw.* University of Chicago Press; Chicago.

Sandeen, Ernest R. 1971. *The Roots of Fundamentalism: British and American Millenarianism, 1800–1930.* University of Chicago Press; Chicago.

Schuchert, Charles. 1924. "Review of *The New Geology* by George McCready Price," *Science* 59:486–87.

Science. 1926. *63*:259.

Segraves, Nell J. 1977. *The Creation Report.* Creation-Science Research Center; San Diego.

Shipley, Maynard. 1927. *The War on Modern Science: A Short History of the Fundamentalist Attacks on Evolution and Modernism.* Alfred A. Knopf; New York.

———. 1930. "Growth of the Anti-Evolution Movement," *Current History 32*.

Skoog, Gerald. 1979. "Topic of Evolution in Secondary School Biology Textbooks: 1900–1977," *Science Education 63*:621–40.

Stonehouse, Ned B. 1954. *J. Gresham Machen: A Biographical Memoir.* William B. Eerdmans; Grand Rapids, Mich.

Szasz, Ferenc Morton. 1969. "Three Fundamentalist Leaders: The Roles of William Bell Riley, John Roach Straton, and William Jennings Bryan in the Fundamentalist-Modernist Controversy." Ph.D. diss., University of Rochester, Rochester.

———. 1982. *The Divided Mind of Protestant America, 1889–1930.* University of Alabama Press: University, Ala.

Thompson, James J., Jr. 1975–76. "Southern Baptists and the Antievolution Controversy of the 1920's," *Mississippi Quarterly 29*:65–81.

"Two Great Field Secretaries—Harry Rimmer and Dr. Arthur I. Brown." 1926. *Christian Fundamentals in School and Church 8* (July–Sept.)

Van de Fliert, J. R. 1969. "Fundamentalism and the Fundamentals of Geology," *Journal of the American Scientific Association 21* (Sept.):69–81.

Wade, Nicholas. 1972. "Creationists and Evolutionists: Confrontation in California," *Science 178*:724–29.

Whalen, Robert D. 1972. "Millenarianism and Millennialism in America, 1790–1880." Ph.D. diss., State University of New York at Stony Brook.

Wheeler, Gerald. 1975. *The Two-Taled Dinosaur: Why Science and Religion Conflict over the Origin of Life*. Southern Publishing Association; Nashville.

———. 1976. "The Third National Creation Science Conference," *Origins 3*.

Whitcomb, John C., Jr., and Henry M. Morris. 1961. *The Genesis Flood: The Biblical Record and Its Scientific Implications*. Presbyterian and Reformed Publishing Co.; Philadelphia.

Whitney, Dudley Joseph. 1928. "'What Theory of Earth History Shall We Adopt?'" *Bible Champion 34*.

Williams, Wayne C. 1936. *William Jennings Bryan*. G. P. Putnam; New York.

"World Religious Digest," 1939. *Christian Faith and Life 45*.

Wright, G. Frederick. 1898. "The First Chapter of Genesis and Modern Science," *Homiletic Review 35*:392–99.

———. 1902. "Introduction." In *The Other Side of Evolution: An Examination of Its Evidences*, by Alexander Patterson, xvii–xix. Winona Publishing Co.; Chicago.

———. 1910–15. "The Passing of Evolution." In *The Fundamentals*, vol. 7, 5–20. Testimony; Chicago.

———. 1916. *Story of My Life and Work*. Bibliotheca Sacra Co.; Oberlin, Ohio.

Young, Davis A. 1977. *Creation and the Flood: An Alternative to Flood Geology and Theistic Evolution*. Baker Book House; Grand Rapids, Mich.

Young, G. L. 1909. "Relation of Evolution and Darwinism to the Question of Origins," *Bible Student and Teacher 11* (July).

16

"Scientific Creationism"

Michael Ruse

Obviously, the present-day Creationists are people to be reckoned with, and "Scientific Creationism" is a doctrine which cannot be ignored. What exactly is being said? . . . Perhaps the closest [the Creationists] have come to producing an "official" statement of their position is in a volume published in 1974: *Scientific Creationism*. This is a work "prepared by the technical staff and consultants of the Institute for Creation Research," under the general editorship of Henry Morris, director of that institute. It comes in two editions, one with Biblical references for its various claims, and the other, the "Public School Edition."

In the body of this second text, care is taken not to let Biblical beliefs openly intrude. Indeed, the explicitly stated intention is to justify all claims "solely on a scientific basis" without reference to any religious beliefs or dogmas (Morris, 1974, p. iv). Hence, since this seems to be the "purest" form of the Creationists' position, or rather that form which they want introduced in the schools, it is on this latter edition that I shall base my exposition here. . . .

The first chapter, *Evolution or Creation?*, presents the two, alternative positions: evolutionism (essentially Darwinism) and Creationism. The authors prefer not to think in terms of "theories," because in fact neither evolution nor Creation can form the basis of a true theory. Creation was a one-of-a-kind phenomenon which occurred in the past, and "thus is inaccessible to the scientific method" (p. 5). Evolution is no true theory for many reasons, a major one being that the central mechanism is a tautology. Natural selection

---

simply states that those which survive in the struggle are the fittest, because the fittest are those which survive! As (the Nobel Prize winner) Peter Medawar has observed: "There are philosophical or methodological objections to evolutionary theory. . . . It is too difficult to imagine or envisage an evolutionary episode which could not be explained by the formulae of neo-Darwinism" (Morris, 1974, p. 7 quoting Medawar, 1967, p. xi).

However, one can properly talk of "model," meaning a conceptual system that tries to correlate data. Although one can always save a model by adding additional claims to diffuse apparently contradictory phenomena, if one model needs less face-saving than another, then it is to be preferred. Viewed in this light, therefore, we have two models. On the one hand, we have the evolutionist position, where it is claimed that organisms evolved naturally, primarily through a process of natural selection. On the other hand, we have the Creationist position, which supposes that in the fairly recent past, the world was created miraculously by God, that animals, plants, and humans were all brought into existence at that time, and that that was it as far as new life is concerned. Additionally, there was some sort of catastrophe or catastrophes, which occurred at some time(s) in the history of the globe.

It is argued that these two models lead to a number of predictions which fall into various categories and which apply at various levels. Some of these predictions are of a fairly general nature. For instance the "Evolution model" implies that natural laws are always in a state of flux and changing; on the contrary, the "Creation model" implies that law is unchanging and never alters. Then we have a set of predictions about the inorganic universe. The evolution model predicts that galaxies are constantly changing, that stars evolve from one sort to another, that the earth on which we live is very old, and that rock formations differ from age to age. As before, the Creation model points the other way. Given its premises, it follows that galaxies never change, that stars do not evolve but remain the same, that this earth of ours is probably very young, and that rock formations start and stay the same from beginning to end.

Then, most crucially, we have the different predictions between the two models about the organic world. The evolution model predicts that life evolved (and apparently is evolving) from non-life, that organisms today present a sort of continuum or unbroken spectrum, that the fossil record has lots of specimens bridging or intermediate between one distinct form and another, and that new "kinds" of organism are always coming into being. Additionally, with respect to mechanisms, the evolution model sees mutation as beneficial, and it sees natural selection as a creative force. Finally, in the context of our own species, the evolution model looks for ape-human forms in the fossil record, it regards man as superior only in degree to other animals (*i.e.*, we have no unique features), and it believes civilization to have developed in a slow and gradual fashion.

The Creation model opposes all of these inferences about organisms. Life comes only from other life: a living organism must have a living parent or parents. Organisms today are broken up into distinct kinds, there are many gaps in the fossil record, and no new kinds ever appear. Mutations are harmful and natural selection is, at best, a minor conservative force, eliminating the inadequate. And most pertinently, there are no ape-human intermediates, and there never have been, man is qualitatively different from the brutes, and as long as there has been man, there has always been a fully developed civilization.

We have, therefore, two completely different sets of predictions: those given by the evolution model and those given by the Creation model. It is claimed by the authors of the text that, judged purely by *scientific* criteria, we shall see that the Creation model wins. (For a full description of the different supposed predictions of the two models, see Morris, 1974, p. 13.)

With the stage set, we move to a discussion entitled: *Chaos or Cosmos?* Here, a number of rather metaphysical questions are raised, and it is concluded that they all imply great difficulties for evolutionism. Let me pick out three of the most important.

First, there is the fact that the evolutionist is committed to the constantly changing nature of law, even though there is absolutely no empirical evidence for such change. To the best of our knowledge, all of the great laws of science—the law of gravitational attraction, Newton's laws of motion, the laws of thermodynamics, and so forth—go on their ways, without any change or variation whatsoever. Clearly, this poses grave difficulties for the evolutionist, and it would seem that his beliefs are thus countered. Creationism, on the other hand, predicts what seems to occur, namely that laws remain constant (Morris, 1974, p. 18).

Second, there is the question of cause and effect. One cannot get more from effect, than one puts in as cause. Therefore, the cause of human love must be loving. Blind matter cannot be loving. Hence, it cannot be, as supposed by the evolutionist, that blind matter is the cause of human beings. The ultimate cause must be an All-loving Being. "An omnipotent Creator is an adequate First Cause for all observable effects in the universe, whereas evolution is *not* an adequate cause. The universe could not be its own cause" (Morris, 1974, p. 20).

Third, there is the question of purpose. The Creationists point out that the world is purposeful, and that evolution denies such purpose, whereas Creationism affirms it. "The creation model does include, quite explicitly, the concept of purpose. The Creator was purposive, not capricious or indifferent, as He planned and then created the universe" (Morris, 1974, p. 33). A point that should not go unmentioned is that, here, as indeed throughout the whole discussion, frequent reference is made to the writings of non-Creationist scientists to support or otherwise justify the arguments and conclusions of the text.

We come next to the chapter that raises the question of *Uphill or Downhill?* In fact, as before, a number of discussions are subsumed under this general heading. First, we get a much-favored argument of the Creationists, namely that all forms of evolutionism violate the second law of thermodynamics.

This law states essentially that physical processes always go from order to randomness—you can scramble an egg, but you cannot unscramble it. (More formally, the law states that entropy always increases, where entropy can be thought of as a measure of the energy which can no longer be used to get things done.) But, evolution supposedly is a process from randomness to order. Hence, the clash. The second law of thermodynamics says that the world is running down. Evolution says that it is not. There is, however, no clash between Creationism and physics. The Creationist agrees that all is downhill, since the beginning of the world.

As the Creationists realize, evolutionists have a standard reply to this objection. They distinguish between "closed systems," where no new usable energy can come in, and "open systems," where new usable energy can come in. The second law obviously applies only to closed systems. But, argue evolutionists, given the influx of usable energy from the sun, the organic world is an open system. Hence evolution is possible. Entropy may be increasing through the universe, taken as a whole, but it does not mean that, in small localized areas, entropy cannot decrease. The world of organic evolution is one such area. The sun shines down on the Earth. This makes the plants grow. Animals live and feed on the plants. And thus life goes forward.

Nevertheless, the Creationists are not convinced. They deny that the second law of thermodynamics could be broken in such a case as evolution. There is simply no way in which any kind of process of growth can or could occur, if one starts with and uses only randomly caused phenomena. All one can possibly get in such a situation is a diffuse heterogeneous mess. In order to get structured functioning, one *must* put in design: a "blueprint" or a pattern or something. Just as one must have the architect's plans before one can put up a building, so one must have the information encoded in the DNA molecule before an organism can be produced (Morris, 1974, pp. 43-44). Evolution denies the existence of such a blueprint for organisms. Hence, it offers no protection against the unfettered randomizing operation of the second law of thermodynamics.

Second, in this chapter, the problem of the origin of life is raised. The very thought of this phenomenon creates a difficulty for the evolutionist, who is "really" committed to the idea that life is constantly being created. Therefore, since life is obviously no longer being created naturally, before the evolutionist can even start to explain the origin of life, ad hoc hypotheses must be introduced, "proving" that life no longer can be created naturally. But then, of course, we still have the difficulty of explaining how life was produced naturally, at some point in the past. In fact, no one is anywhere close to solving this problem. Artificially created amino acids are just simply not genuinely living things (Morris, 1974, p. 49). And, in any case, even if life were produced, it proves nothing. The production of life through careful thought and precise experimentation is hardly analogous to life coming through random, blind law.

Third, we get discussion of variation and selection. Both of these concepts are inadequate for the tasks assigned to them by evolutionists. There is no example of variation leading to truly new characteristics of the kind required by evolutionists, and selection is a purely conservative force. Selection never leads to new kinds of features. It merely eliminates the inadequate, and other kinds of subnormal features and organisms. And obviously, this very limited power of selection is just what one is led to expect from the Creation model, which is thus confirmed (Morris, 1974, p. 54).

Finally, in this chapter, the topic of mutation is raised. The simple fact of the matter is that mutation is totally inadequate for the tasks supposed by the evolutionists. It is quite impossible that a complex, functioning organism could have been built, step by step, out of mutations. Indeed, the words of evolutionists are turned against themselves. Mutations are random (Waddington), very rare (Ayala), and almost invariably, very harmful to their possessors (Muller and Julian Huxley). They just cannot lead to new kinds of organisms.

Moving along, we come to the chapter dealing with the question of: *Accident or Plan?* A major discussion here is of the complexity of living organisms, and of the total improbability of such complexity arising instantly, by chance. Suppose we have an organism made of only 100 parts. Each of these parts must link up, in a unique way, with every other part. One can show, mathematically, that there are 10^{158} different possible ways for linkage. Hence, the chance of the organism being formed is only 1 in 10^{158}! This is totally improbable. If one considers all of the molecules in the universe, and all of the instants of time (10^{-9} seconds) since the beginning of the universe, and if one pretends that new combinations are constantly being formed, then at most one could get 10^{105} new combinations. Hence, there is only a 1 in 10^{53} chance of our 100 unit piece of life being formed, which means that in actual fact there is no chance whatsoever of life being formed (Morris, 1974, p. 61). And, a 100 unit piece of life is impossibly simple.

Would it help to suppose that life is formed on a gradual process? Not at all! Pretend that one needed 1500 steps to make primitive life. Pretend, also, that there was a 50:50 chance that random processes would bring about the required step at any stage. Were one to go on successively to the formation of higher and higher living order, every new step would have to be a success and to help the newly forming organism. One could not have or allow any steps which "failed," that is, which took the organism back to a more primitive form or which killed it entirely. Hence, the probability of new life is $(1/2)^{1500}$, or, approximately, one chance in 10^{450}. This is an impossible demand. Even if one took a new step every billionth of a second, there would only be time for 10^{107} steps! Life could not have been formed by chance (Morris, 1974, p. 63).

Next, in this chapter, we have discussion of things like homologies, which phenomena evolutionists take to be evidence of common ancestry. The Creationists make short work of this. Relying on the authority of Ernst Mayr, they point out that classification above the species' level is really all very arbitrary. Hence, homologies between organisms of different groups may indeed be indicative of common ancestry; but, this proves nothing very much really. By evolutionists' own admission, the groups are man made! There is no "real" breaking into separate classes. And, even if there were, this would be an embarrassment for the evolutionist. "If an evolutionary continuum existed, as the evolution model should predict, there would be no gaps, and thus it would be impossible to demark specific categories of life" (Morris, 1974, p. 72). The Creationists treat other biological similarities, like those of embryology and behavior, in the same way.

Following this, the question of vestigial organs and of recapitulation is raised. The "biogenetic law," namely that "Ontogeny recapitulates phylogeny," is presented. It is pointed out that this once popular rule is now discredited by most modern embryologists. Nevertheless, to their surprise the Creationists find that evolutionists—important evolutionists—go on citing the law as evidence for their evolutionary case. This is despite the fact that anyone who knows anything at all about embryos or fossils knows the law to be false (Morris, 1974, p. 77).

Concluding the chapter, we get a fairly extensive discussion of the fossil record. The space given to this topic is in line with Creationist claims elsewhere, that: "The only *solid* evidence in the discussion of origins is the fossil record— anything else is circumstantial evidence and conjecture" (advertisement blurb for *Evolution: The Fossils Say No!* by Duane T. Gish). Expectedly, much is made of gaps in the record. Even more expectedly, much is made of what the Creationists believe to be one of the most dreadful gaps of them all, namely that which occurs between the supposed micro-organisms of the pre-Cambrian era, and the very rich and complex organisms of the Cambrian: the gap between virtually nothing and the thriving and abundant trilobites (Morris, 1974, p. 80). Several evolutionists are quoted to show what a problem this is. Similar attention is drawn to other gaps in the record, including that between reptiles and birds. What about *Archaeopteryx*? Well, with respect, because it had feathers, it has to be classed as a bird. Hence, since *Archaeopteryx* is a bird, it cannot be "a reptile-bird transition"! The gap remains. Also noted in this discussion are the many "living fossils," which live today, but which occur only in ancient strata. These should not exist, given evolutionism. Given Creationism, they are expected. All of today's organisms go back to the beginning of time.

To this point, discussion has been more one of criticism (of Darwinian evolutionism) than one of positive affirmation (of Creationism). However, in the next chapter, *Uniformitarianism or Catastrophism?*, much of the positive, Creationist case is laid out. After noting how many geologists today feel unhappy with uniformitarianism (Gould is a leading authority here), we are given a major part of the Creationist thesis. Originally, all of the organisms that we find living today *and* all those organisms that we find represented in the fossil record lived together in the world, as it was created by the Creator during that initial time when everything was produced. Obviously, although organisms all lived together at the same time, this early life was subject to the ecological restraints we find in operation today. Everyone did not live on top of everyone, humans cheek to jowl with trilobites. As today, those organisms suited for water lived in water, and those suited for dry land lived on dry land.

Then, into this picture of harmony there came a monstrous flood. The heavens opened and, if this were not enough, at the same time, water poured forth from subterranean sources. This deluge went on for weeks on end, until the whole earth was submerged. Simultaneously, along with all of these watery calamities, there were earthquakes, volcanic eruptions, landslides, and just about every other possible natural disaster. Clearly, this all had horrendous implications for organisms, particularly for land animals. They ran hither and thither, trying to escape their awful fate; but, inevitably, the forces of nature eventually caught up with them. At long last, unless a few enterprising humans had managed to build super-strong boats to ride out the flood, all organisms (especially all land organisms) were drowned.

Finally, bringing everything to a conclusion, sediments settled down out of the waters, forming the various strata that we find around the whole world today. As a function of various physical and chemical processes and constraints, the layers of presently visible or discoverable rocks were formed. And, at the end, for some reason or reasons, the water receded and dry land appeared once more. (For details, see Morris, 1974, pp. 117-118.)

But what about the fossil record? Does not the gradual, progressive rise, that we see, support evolution? Not really! First of all, the record is truly not all that progressive. Indeed, many if not most plants and animals living today can be found in the fossil record. Conversely, a huge number of fossil animals and plants have similar living counterparts, and such differences as do exist between past and present are easily and totally explicable in terms of environmental effects (Morris, 1974, p. 116). Also, Mayan relief sculptures show that *Archaeopteryx* lived at the same time as man. And, in the cretaceous Glen Rose formation of central Texas, there are dinosaur and human footprints occurring together. There are even human traces in trilobite beds! (I had hoped to be able to show the reader photographs which the Creationists use to bolster their case for contemporaneous human and dinosaur prints. Unfortunately, the Creationists did not feel that they could let me use their pictures. Let me therefore draw your attention to (Milne, 1981), where the pictorial evidence is laid out in full.)

Second, in reply to evolutionist queries about progression, there is the simple fact that the Creationist expects a broadly progressive fossil record! As organisms rushed (unsuccessfully) to avoid the flood, they left their remains in a progressive fashion, because of their different original habitats and abilities. Thus, for instance, one expects to find mammals and birds higher up in the fossil record than reptiles and amphibians. Mammals and birds live at higher elevations than do reptiles and amphibians. Moreover, the former are more mobile than the latter, and thus would have scrambled to greater heights, in their desperate albeit futile efforts to avoid the rising flood waters (Morris, 1974, p. 119).

The penultimate chapter, *Old or Young?*, deals with the question of the age of the earth. It is pointed out that "no one can possibly *know* what happened before there were people to observe and record what happened. Science means knowledge and the essence of the scientific method is experimental observation" (Morris, 1974, p. 131, his italics). This means that, at best, we can really only go, with certainty, back to the beginning of written records, some 2000 to 3000 years before Christ.

But what about physicochemical methods of dating the age of the earth and of the fossils? What about so-called "radiometric dating"? The Creationists claim that every one of these methods is highly speculative, probably leading to quite erroneous results. Essentially, all of the methods work on the same principle: certain elements "decay" into other elements. Hence, by measuring the amount of decay that has gone on in a rock sample, knowing the rate of decay, one can calculate the absolute age of the rock. The Creationists argue that at least three things could go wrong, and probably do go wrong. First, all sorts of elements could come in and out of the rock, after it formed. This would obviously render the extant ratios totally meaningless. Second, one cannot know the initial composition of the rock. Hence, the final ratios could be distorted, because of initial ratios. Third, who can say that rates of decay are constant? Processes in nature go at all sorts of different rates, dependent on intervening factors. Change the facts and the rates change. In any case, rates are statistical. They are never constants (Morris, 1974, p. 139).

What, then, is the age of the earth? All the evidence points to a very recent origin. For instance, Dr. Thomas G. Barnes, Professor of Physics at the University of Texas in El Paso, has shown that the earth's magnetic field strength is decaying, with a most probably half-life of 1400 years. Immediately, this gives rise to some interesting implications. Only 1400 years ago the magnetic field was twice what it is today, and going back in such a geometric fashion we very soon get an incredibly strong field. Within 10,000 years we get a putative magnetic field which is at least as powerful as that which one finds on a magnetic star. Obviously, this is totally impossible, for the earth would be destroyed under such a force. "Thus, 10,000 years seems to be an outside limit for the age of the earth, based on the present decay of its magnetic field" (Morris, 1974, pp. 157-158). Interestingly, in this part of the discussion, the authors do not refer solely to external, non-Creationists, as their authorities. Dr. Barnes is, in fact, one of the writers and/or consultants of this text we are discussing.

Finally, in this examination of time-spans, a rather different, Malthusian argument is offered, to show that man is a recent phenomenon. Annual human population growth rate today is of the order of 2 percent. Which model, evolutionism with a human-span of at least a million years, or Creationism, with a human-span of 4000-5000 years, best fits with this figure? In fact, it is easy to show that it is the Creationist model. With an average growth rate of 1/2% (adjusted down to take account of war and disease), we can get today's numbers (3.5×10^9), in about 4000 years.

Conversely, under the burden of the demographic facts just mentioned, the evolution model collapses entirely. Suppose, with the evolutionists, that humans have been around for a million years. This gives us about 25,000 generations. Hence, one has to infer that all of these humans could have produced only 3.5 billion people by today. This is ridiculous! Pretend that population increased at only the above-supposed rate of $1/2\%$; that is, pretend that there were only 2.5 children per family. Given 25,000 generations, we would expect there to be 10^{2100} people today. This is an absolutely impossible figure. There are, after all, only 10^{130} electrons maximally possible in the entire known universe. Hence, evolutionary hypotheses come crashing down, salvable only through the addition of all sorts of implausible secondary *ad hoc* suppositions. (See Morris, 1974, p. 169, for full details of this argument.)

Finally, we come to the question everyone has been waiting for: *Apes or Men*? Many phenomena (languages, cultures, religions) point to recent human origins; but, of course, the most important information comes from the fossil record. There is absolutely no solid evidence, whatsoever, for Darwinian claims that humans evolved from the apes. Neanderthal man was probably an ordinary man, with arthritis or rickets. *Homo erectus,* also, seems to have been a true man, although hardly a prime specimen. Unfortunately, because of inbreeding, inadequate food, severe environmental forces, and the like, he was second-rate, both in size and in culture (Morris, 1974, p. 174). What about *Australopithecus*? With respect, he was no man at all, but a kind of ape. He had a small brain, was long-armed and short-legged, and probably walked on his knuckles, rather than in an upright fashion, like us. "In other words, *Australopithecus* not only had a brain like an ape, but he also looked like an ape and walked like an ape. He, the same as *Ramapithecus,* is no doubt simply an extinct ape" (Morris, 1974, p. 173).

Creationism wins; evolutionism loses!

REFERENCES

Medawar, P., P. S. Moorhead and M. M. Kaplan (eds.). 1967. *Mathematical Challenges to the Neo-Darwinism Interpretations of Evolution.* Wistar Institute Press; Philadelphia.

Milne, D. H. 1981. "How to Debate with Creationists and 'Win,' " *American Biology Teacher 43*, 235 245.

Morris, H. M. ed. 1974. *Scientific Creationism.* Creation-Life; San Diego.

17

Creation, Evolution, and the Historical Evidence

Duane T. Gish

For a clear understanding of the issues to be discussed in this paper, I must begin by defining evolution and creation. When the term evolution is used it will refer to the general theory of organic evolution, or the molecules-to-man theory of evolution. According to this theory all living things have arisen by naturalistic, mechanistic, evolutionary processes from a single living source, which itself had arisen by similar processes from inanimate matter. These processes are attributable solely to properties inherent in matter and are, therefore, still operative today. Creation theory postulates, on the other hand, that all basic animal and plant types (the created kinds) were brought into being by the acts of a preexisting Being by means of special processes that are not operative today. The variation that has occurred since creation has been restricted within the limits of each created kind.

Evolutionists adamently insist that special creation be excluded from any consideration as a possible explanation for origins, because it does not qualify as a scientific theory. The proponents of evolution theory at the same time would view as unthinkable the consideration of evolution as anything less than pure science; and indeed most of them insist that evolution must no longer be thought of as a theory, but must be considered to be a fact.

From *The American Biology Teacher* (March 1973): 132 40. Reprinted by permission of the publisher.

WHAT IS THEORY? WHAT IS FACT?

What criteria must be met for a theory to be considered scientific in the usually accepted sense? George Gaylord Simpson (1964) has stated, "It is inherent in any definition of science that statements that cannot be checked by observation are not really about anything . . . or at the very least, they are not *science.*" A definition of science in the *Oxford English Dictionary* is "a branch of study which is concerned either with a connected body of *demonstrated truths* or with *observed facts* systematically classified and more or less colligated by being brought under general laws, and which includes trustworthy methods for the discovery of new truth within its own domain" (emphasis added).

Thus, for a theory to qualify as a scientific theory, it must be supported by events or processes that can be observed to occur, and the theory must be useful in predicting the outcome of future natural phenomena or laboratory experiments. An additional limitation usually imposed is that the theory must be capable of falsification; that is, one must be able to conceive some experiment the failure of which would disprove the theory. It is on the basis of such criteria that most evolutionists insist that creation be refused consideration as a possible explanation for origins. Creation has not been witnessed by human observers, it cannot be tested scientifically, and as a theory it is non-falsifiable.

The general theory of evolution (molecules-to-man theory) also fails to meet all three of these criteria, however. Dobzhansky (1958), while seeking to affirm the factuality of evolution, actually admits that it does not meet the criteria of a scientific theory, when he says, "The occurrence of the evolution of life in the history of the earth is established about as well as events *not witnessed by human observers* can be" (emphasis added).

Goldschmidt, who has insisted that evolution is a fact for which no further proof is needed, also reveals its failure to meet the usually accepted criteria for a scientific theory. After outlining his postulated systemic-mutation, or "hopeful monster," mechanism for evolution, Goldschmidt (1952, p. 94) states, "Such an assumption is violently opposed by the majority of geneticists, who claim that the facts found on the sub-specific level must apply also to the higher categories. Incessant repetition of this *unproved claim,* glossing lightly over the difficulties, and the assumption of an arrogant attitude toward those who are not so easily swayed by fashions in science, are considered to afford scientific proof of the doctrine. It is true that nobody thus far has produced a new species or genus, etc., by macromutation. It is equally true that nobody has produced even a species by the selection of micromutations" (emphasis added). Later in the same paper (p. 97) he says, "Neither has anyone witnessed the production of a new specimen of a higher taxonomic category by selection of micromutants." Goldschmidt has thus affirmed that, in the molecules-to-man context, only the most trivial change, or that at the subspecies level, has actually ever been observed.

Furthermore, the architects of the modern synthetic theory of evolution have so skillfully constructed their theory that it is not capable of falsification. The theory is so plastic that it is capable of explaining anything. This is the complaint of Olson (1960, p. 580) and of several participants in the Wistar Symposium on Mathematical Challenges to the Neo-Darwinian Interpretation of Evolution (Moorhead and Kaplan, 1967)—even including Ernst Mayr, a leading exponent of the theory. Eden (1967, p. 71), one of the mathematicians, puts it this way, with reference to falsifiability: "This cannot be done in evolution, taking it in its broad sense, and this is really all I meant when I called it tautologous in the first place. It can, indeed, explain anything. You may be ingenious or not in proposing a mechanism which looks plausible to human beings and mechanisms which are consistent with other mechanisms which you have discovered, but it is still an unfalsifiable theory."

A RISING TIDE OF CRITICISM

In addition to scientists who are creationists, a growing number of other scientists have expressed doubts that modern evolution theory could explain more than trivial change. Eden (1967, p. 109) is so discouraged, after a thorough consideration of the modern theory from a probabilistic point of view, that he proclaims, "an adequate scientific theory of evolution must await the discovery and elucidation of new laws—physical, physico-chemical, and biological." Salisbury (1969, 1971) similarly expresses doubts based on probabilistic considerations.

The attack on the theory by French scientists has been intense in recent years. In a review of the French situation Litynski (1961) says, "This year saw the controversy rapidly growing, until recently it culminated in the title 'Should We Burn Darwin?' spread over two pages of the magazine *Science et Vie*. The article, by the science writer Aimé Michel, was based on the author's interviews with such specialists as Mrs. Andrée Tetry, professor at the famous Ecole des Hautes Etudes and a world authority on problems of evolution, Professor René Chauvin and other noted French biologists, and on his thorough study of some 600 pages of biological data collected, in collaboration with Mrs. Tetry, by the late Michael Cuenot, a biologist of international fame. Aimé Michel's conclusion is significant: the classical theory of evolution in its strict sense belongs to the past. Even if they do not publicly take a definite stand, almost all French specialists hold today strong mental reservations as to the validity of natural selection."

E. C. Olson (1960, p. 523), one of the speakers at the Darwinian Centennial Celebration at Chicago, made the following statement on that occasion: "There exists, as well, a generally silent group of students engaged in biological pursuits who tend to disagree with much of the current thought but say and write little because they are not particularly interested, do not see that controversy over evolution is of any particular importance, or are so strongly in disagreement that it seems futile to undertake the monumental task of controverting the immense body of information and theory that exists in the formulation of modern thinking. It is, of course, difficult to judge the size and composition of this silent segment, but there is no doubt that the numbers are not inconsiderable."

Fothergill (1961) refers to what he calls "the paucity of evolutionary theory as a whole." Ehrlich and Holm (1962) have stated their reservations in the following way: "Finally, consider the third question posed earlier: 'What accounts for the observed patterns in nature?' It has become fashionable to regard modern evolutionary theory as the *only* possible explanation of these patterns rather than just the best explanation that has been developed so far. It is conceivable, even likely, that what one might facetiously call a non-Euclidean theory of evolution lies over the horizon. Perpetuation of today's theory as dogma will not encourage progress toward more satisfactory explanations of observed phenomena."

Sometimes the attacks are openly critical. Such is Danson's letter that appeared recently in *New Scientist*. He states in part, "The Theory of Evolution is no longer with us, because neo-Darwinism is now acknowledged as being unable to explain anything more than trivial changes and in default of some other theory we have none. . . . despite the hostility of the witness provided by the fossil record, despite the innumerable difficulties, and despite the lack of even a credible theory, evolution survives. . . . Can there be any other area of science, for instance, in which a concept as intellectually barren as embryonic recapitulation could be used as evidence for a theory?" (Danson, 1971).

Macbeth (1971) has provided an especially incisive criticism of evolution theory. He points out that although evolutionists have abandoned classical Darwinism, the modern synthetic theory they have proposed as a substitute is equally inadequate to explain progressive change as the result of natural selection; as a matter of fact, they cannot even define natural selection in nontautologous terms. Inadequacies of the present theory and failure of the fossil record to substantiate its predictions leave macroevolution, and even microevolution, intractable mysteries, according to Macbeth. He suggests that no theory at all may be preferable to the existing one.

In view of the above considerations, it is incredible that leading scientists, including several who addressed the NABT convention in San Francisco, dogmatically insist that the molecules-to-man evolution theory be taught as a fact to the exclusion of all other postulates. Evolution in this broad sense is unproven and unprovable and thus cannot be considered to be fact. It is not subject to test by the ordinary methods of experimental science: observation and falsification. It thus does not, in a strict sense, even qualify as a scientific theory. It is a postulate, and it may serve as a model within which attempts may be made to explain and correlate the evidence from the historical record—that is, the fossil record—and to make predictions concerning the nature of future discoveries.

Creation is, of course, unproven and unprovable by the methods of experimental science. Neither can it qualify, according to the above criteria, as a scientific theory, because creation would have been unobservable and, as a theory, would be nonfalsifiable. Creation is therefore (like evolution) a postulate that may serve as a model to explain and correlate the evidence relating to origins. Creation is, in this sense, no more religious or less scientific than evolution. In fact, to many well-informed scientists creation seems to be far superior to the evolution model as an explanation for origins.

I strongly suspect that the dogmatic acceptance of evolution is not due, primarily, to the nature of the evidence but the philosophic bias peculiar to our times. Watson (1929), for example, has referred to the theory of evolution as "a theory universally accepted not because it can be proved by logically coherent evidence to be true but because the only alternative, special creation, is clearly incredible."

That this is the philosophy held by most biologists has been recently emphasized by Dobzhansky. In his review of Monod's book *Chance and Necessity* Dobzhansky (1972) says, "He has stated with admirable clarity, and eloquence often verging on pathos, the mechanistic materialist philosophy shared by most of the present 'establishment' in the biological sciences."

TWO MODELS TO BE TESTED

The exclusion of creation from science-teaching as a credible explanation of origins is unwarranted and undesirable on both philosophic and scientific grounds. Under the present system, whereby evolution is taught as an established fact to the exclusion of creation, the student is being indoctrinated in a philosophy of secular humanism rather than benefiting from an objective presentation of the evidence.

This situation could be remedied by (i) presenting creation and evolution in the form of models, (ii) making predictions based on each model, and (iii) comparing the actual scientific evidence with the predictions of the models. The students would then be able to make up their minds on the basis of this objective presentation. This is what I would like to do in the remainder of this paper. I will restrict myself to an examination of the fossil record.

Although various scientific fields could be investigated in attempts to determine which model appears to be the more plausible of the two, the fossil record offers the only source of scientific evidence that would allow a determination of whether living organisms actually did arise by creation or by evolutionary process. The case is well stated by Le Gros Clark (1955) when he says, "That evolution actually *did* occur can only be scientifically established by the discovery of the fossilized remains of representative samples of those intermediate types which have been postulated on the basis of the indirect evidence. In other words, the really crucial evidence for evolution must be provided by the paleontologist whose business it is to study the evidence of the fossil record." Gavin de Beer (1964) echoes this view when he states, "The last word on the credibility and course of evolution lies with the paleontologists."

In his revolutionary work *The Origin of Species,* Darwin (1859) says, "the number of intermediate and transitional links, between all living and extinct species, must have been inconceivably great." This conclusion seems inescapable, whether it be based either on the concepts of classical Darwinism or on those of the modern synthetic theory. Because the number of transitional forms predicted by evolution theory is inconceivably great, the number of such forms that would have become fossilized, according to this theory, would have been very great indeed, even though only a very minute fraction of all plants and animals that ever existed had become fossilized.

Sampling of the fossil record has now been so thorough that appeals to the imperfections in the record are no longer valid. George (1960, p. 1) has stated, "There is no need to apologize any longer for the poverty of the fossil record. In some ways it has become almost unmanageably rich and discovery is outpacing integration." It seems clear, then, that after 150 years of intense searching, a very large number of obvious transitional forms would have been discovered if the predictions of evolution theory are valid.

On the basis of the creation model, on the other hand, the visual absence of apparent transitional forms between the higher categories or created kinds would be predicted. The presence of apparent transitional forms could not be rigidly excluded, however, for two reasons: (1) tremendous diversity is exhibited within each major type of plant and animal and (ii) possession of similar modes of existence or activities would require similar structures or functions. On the basis of the creation model such pseudotransitional forms should be rare and would not be connected by intermediate types. Gaps in the fossil record, therefore, should be systematic and nearly universal between the higher categories or created kinds. The fossil record should permit a clear choice between the two models.

The two models may thus be constructed as follows:

Creation model	*Evolution model*
By acts of a Creator	By naturalistic, mechanistic processes due to properties inherent in inanimate matter
Creation of basic plant and animal kinds with ordinal characteristics complete in first representatives	Origin of all living things from a single living source, which itself arose from inanimate matter. Origin of each kind from an ancestral form by slow, gradual change
Variation and speciation limited within each kind	Unlimited variation. All forms genetically related

These two models would permit the following predictions to be made about the fossil record:

Creation model	Evolution model
Sudden appearance in great variety of highly complex forms	Gradual change of simplest forms into more and more complex forms
Sudden appearance of each created kind with ordinal characteristics complete. Sharp boundaries separating major taxonomic groups. No transitional forms between higher categories	Transitional arteries linking all categories. No systematic gaps

Let us now compare the known facts of the fossil record with the predictions of the two models.

ADVENT OF LIFE IN THE CAMBRIAN

The oldest rocks in which indisputable fossils are found are those of the Cambrian Period. In these sedimentary deposits are found billions and billions of fossils of highly complex forms of life. These include sponges, corals, jellyfish, worms, mollusks, and crustaceans; in fact, every one of the major invertebrate forms of life has been found in Cambrian rocks. These animals were so highly complex that, it is conservatively estimated, they would have required 1.5 billion years to evolve.

What do we find in rocks older than the Cambrian? Not a single, indisputable multicellular fossil has ever been found in Precambrian rocks. Certainly it can be said without fear of contradiction that the evolutionary ancestors of the Cambrian fauna, if they ever existed, have never been found (Simpson, 1960, p. 143; Cloud, 1968; Axelrod, 1968).

Concerning this problem, Axelrod (1968) has stated, "One of the major unsolved problems of geology and evolution is the occurrence of diversified, multicellular marine invertebrates in Lower Cambrian rocks on all the continents and their absence in rocks of greater age." After discussing the varied types that are found in the Cambrian, Axelrod goes on to say, "However, when we turn to examine the Precambrian rocks for the forerunners of these Early Cambrian fossils they are nowhere to be found. Many thick (over 5,000 feet) sections of sedimentary rock are now known to lie in unbroken succession below strata containing the earliest Cambrian fossils. These sediments apparently were suitable for the preservation of fossils because they are often identical with overlying rocks which are fossiliferous, yet no fossils are found in them."

From all appearances, then, based on the known facts of the historical record, there occurred a sudden great outburst of life at a high level of complexity. The fossil record gives no evidence that these Cambrian animals were derived from preceding, ancestral forms. Furthermore, not a single fossil has been found that can be considered to be a transitional form between the major groups, or phyla. At their earliest appearance these major invertebrate types were just as clearly and distinctly set apart as they are today.

How do these facts compare with the predictions of the evolution model? They are in clear contradiction to such predictions. This has been admitted, for instance, by George (1960, p. 5), who states, "Granted an evolutionary origin of the main groups of animals and not an act of special creation, the absence of any record whatsoever of a single member of any of the phyla in the Precambrian rocks remains as inexplicable on orthodox grounds as it was to Darwin." Simpson has struggled valiantly but not fruitfully with this problem and has been forced to concede (1949, p. 18) that the absence of Precambrian fossils (other than alleged fossil microorganisms) is the "major mystery of the history of life."

These facts, however, are in full agreement with the predictions of the creation model. The fossil record *does* reveal (i) a sudden appearance, in great variety, of high complex forms with no evolutionary ancestors and (ii) the absence of transitional forms between the major taxonomic groups, just as postulated on the basis of creation. Most emphatically, the known facts of the fossil record from the very outset support the predictions of the creation model but unquestionably contradict the predictions of the evolution model.

DISCRETE NATURE OF VERTEBRATE CLASSES

The remainder of the history of life reveals a remarkable absence of the many transitional forms demanded by the theory. There is, in fact, a *systematic* deficiency of transitional forms between the higher categories, just as predicted by the creation model.

The idea that the vertebrates are derived from the invertebrates is purely an assumption that cannot be documented from the fossil record. In the history of the study of the comparative anatomy and embryology of living forms almost every invertebrate group has been proposed, at one time or another, as the ancestor of the vertebrates (E. G. Conklin, as quoted in Allen, 1969; Romer, 1966, p. 12). The transition from invertebrate to vertebrate supposedly passed through a simple chordate stage. Does the fossil record provide evidence for such a transition? Not at all. Ommaney (1964) has stated, "How this earliest chordate stock evolved, what stages of development it went through to eventually give rise to truly fishlike creatures we do not know. Between the Cambrian when it probably originated, and the Ordovician when the first fossils of animals with really fishlike characteristics appeared, there is a gap of perhaps 100 million years which we will probably never be able to fill."

Incredible! 100 million years of evolution and no transitional forms! All hypotheses combined, no matter how ingeniously, could never pretend, on the basis of evolution theory, to account for a gap of such magnitude. Such facts, on the other hand, are in perfect accord with the predictions of the creation model.

A careful reading of Romer's *Vertebrate Paleontology* (1966) seems to allow no other conclusion than that the major classes of fish are clearly and distinctly set apart from one another with no transitional forms linking them.

The fossil record has not produced ancestors or transitional forms for these classes. Hypothetic ancestors and the required transitional forms must, on the basis of the known record, be merely the products of speculation. How then can it be argued that the evolution model's explanation of such evidence is more scientific than that of the creation model?

The fossil record has been diligently searched for a transitional series linking fish to amphibian, but as yet no such series has been found. The closest link that has been proposed is that allegedly existing between rhipidistian crossopterygian fish and the amphibians of the genus *Ichthyostega,* of the labyrinthodont family Ichthyostegidae. There is a tremendous gap, however, between the crossopterygians and the ichthyostegids—a gap that would have spanned many millions of years, during which innumerable transitional forms should have existed. These transitional forms should reveal a slow, gradual change of the pectoral and pelvic fins of the crossopterygian fish into the feet and legs of the amphibian, along with loss of other fins, and the accomplishments of other transformations required for adaptation to a terrestrial habitat.

What is the fact? Not a single transitional form has ever been found showing an intermediate stage between the fin of the crossopterygian and the foot of the ichthyostegid. The limb and the limb-girdle of *Ichthyostega* is already of the basic amphibian type, showing no vestige of a fin ancestry.

The extremely broad gap between fish and amphibians, as observed between the rhipidistian crossopterygians and the ichthyostegids; the sudden appearance, in fact, of all Paleozoic amphibian orders with diverse ordinal characteristics complete in the first representatives; the absence of any transitional forms between these Paleozoic orders; and the absence of transitional forms between the Paleozoic orders and the three living orders—all these conditions are contradictory to the predictions of the evolution model. These facts, however, are just as predicted by the creation model.

It is at the amphibian-reptilian and the reptilian-mammalian boundaries that the strongest claims have been advanced for transitional types bridging classes. But these are just those classes that are most closely similar in skeletal features; that is, the parts that are preserved in the fossil record.

The conversion of an invertebrate into a vertebrate, a fish into a tetrapod with feet and legs, and a nonflying animal into a flying animal are a few examples of changes that would require a revolution in structure. Such transformations should provide readily recognizable transitional series in the fossil record if they occurred through evolutionary processes. On the other hand, if the creation model is the true model, it is at just such boundaries that the absence of transitional forms would be most evident.

The opposite is true at the amphibian-reptilian and reptilian-mammalian boundaries—particularly the former. Although it is feasible to distinguish between living reptiles and amphibians on the basis of skeletal features, they are much more readily distinguishable by means of their soft parts; and, in fact, the major definitive characteristic that separates reptiles from amphibians is the possession by the reptile, in contrast with the amphibian, of an amniote egg.

Many of the diagnostic features of mammals, of course, reside in their soft anatomy or their physiology. These include their mode of reproduction, warm-bloodedness, mode of breathing due to possession of a diaphragm, suckling of the young, and possession of hair.

The two most easily distinguishable osteologic differences between reptiles and mammals, however, have never been bridged by transitional series. All mammals, living or fossil, have a single bone, the dentary, on each side of the lower jaw; and all mammals, living or fossil, have three auditory ossicles, or ear bones: the malleus, incus, and stapes. In some fossil reptiles the number and size of the bones of the lower jaw are reduced, by comparison with those of living reptiles. Every reptile, living or fossil, however, has at least four bones in the lower jaw and only one auditory ossicle, the stapes. There are no transitional forms showing, for instance, three or two jaw bones or two ear bones. No one has explained yet, for that matter, how the transitional form would have managed to chew while its jaw was being unhinged and rearticulated or how it would hear while dragging two of its jaw bones up into its ear.

SPECIAL FEATURES OF FLYING ANIMALS

The origin of flight should provide excellent case histories for testing the evolution model vs. the creation model. Almost every structure in a nonflying animal would require modification for flight, and resultant transitional forms should be easily detectable in the fossil record. Flight is supposed to have evolved four times, separately and independently: in insects, birds, mammals (bats), and reptiles (pterosaurs, now extinct). In each case the origin of flight is supposed to have required many millions of years, and almost innumerable transitional forms would have been required in each case. Yet not in a single case can anything even approaching a transitional series be produced.

E. C. Olson, an evolutionist and geologist, in his book *The Evolution of Life* (1965) states that "As far as flight is concerned there are some very big gaps in the record" (p. 180). Concerning insects he says, "There is almost nothing to give any information about the history of the origin of flight in insects" (p. 180). Concerning flying reptiles, Olson reports that "True flight is first recorded among the reptiles by the pterosaurs in the Jurassic Period. Although the earliest of these were rather less specialized for flight than in the later ones, there is absolutely no sign of intermediate stages" (p. 181). As for birds: Olson refers to *Archaeopteryx* as "reptile-like" but says that in possession of feathers "it shows itself to be a bird" (p. 182). Finally, with reference to mammals Olson states that "The first evidence of flight in mammals is in *fully developed* bats of the Eocene epoch" (p. 182; emphasis added).

Thus, in not a single investigation of the origin of flight has a transitional series been documented. In the case of *Archaeopteryx*—a so-called intermediate—all paleontologists now acknowledge that it was a true bird. It had wings; it was completely feathered; it flew. It was not a half-way bird; it *was* a bird. No transitional form with part-wings and part-feathers has ever been found.

The alleged reptilian features of *Archaeopteryx* consist of the clawlike appendages on the leading edges of its wings and the possession of teeth and of vertebrae that extend out along the tail. It is believed to have been a poor flier, with a small keel on the sternum. Although such features might be expected if birds had evolved from reptiles, in no sense of the word do they constitute proof that *Archaeopteryx* was an intermediate between reptile and bird. For example, there is a bird living today in South America—the hoatzin, *Opisthocomus hoazin*—which in the juvenile stage possesses two claws. Furthermore, it is a poor flier, with an astonishingly small keel (Grimmer, 1962). This bird is unquestionably 100% bird, yet it possesses two of the characteristics that are used to impute a reptilian ancestry to *Archaeopteryx*.

Modern birds do not possess teeth; but certain ancient birds, unquestionably 100% birds, possessed teeth. Does the possession of teeth denote a reptilian ancestry for birds, or does it simply prove that some ancient birds had teeth and others did not? Some reptiles have teeth and some do not; some amphibians have teeth and some do not. In fact, this is true throughout the entire range of the vertebrate subphylum. On the principle that toothed birds are primitive and that toothless birds are more advanced, the Monotremata (the duck-billed platypus and the spiny anteater), which are mammals that do not possess teeth, should be considered more "advanced" than humans. Yet in every other respect these egg-laying mammals are considered to be the most primitive of all mammals (although they are among the last mammals to appear in the fossil record). Just what phylogenetic value, then, can be assigned to the possession or absence of teeth?

Concerning the status of *Archaeopteryx*, Lecomte du Noüy (1947, p. 58) has stated, "Unfortunately, the greater part of the fundamental types in the animal realm are disconnected from a paleontological point of view. In spite of the fact that it is undeniably related to the two classes of reptiles and birds (a relation which the anatomy and physiology of actually living specimens demonstrates), we are not even authorized to consider the exceptional case of the *Archaeopteryx* as a true link. By link, we mean a necessary stage of transition between classes such as reptiles and birds, or between smaller groups. An animal displaying characters belonging to two different groups cannot be treated as a true link as long as the intermediary stages have not been found, and as long as the mechanisms of transition remain unknown."

What seems to be the most reasonable conclusion? I believe that the fossil record would permit no better assessment of the facts than that voiced by Swinton (1960): "The origin of birds is largely a matter of deduction. There is no fossil of the stages through which the remarkable change from reptile to bird was achieved."

The absence of any indication whatsoever from the fossil record that feathers gradually evolved is usually excused by the allegation that such delicate structures are not likely to be preserved in fossils. No such explanation is admissible, however, in the case of flying reptiles and the bats.

There are many significant differences between nonflying reptiles and flying reptiles. Again I refer to Romer's *Vertebrate Paleontology.* On p. 140 is shown a reconstruction of *Saltoposuchus* (fig. 214), which was a representative of the Triassic thecodonts—a group that Romer believes gave rise to flying reptiles (pterosaurs), dinosaurs, and birds. Comparison of this form with reconstructions of the earliest representatives among the two suborders of pterosaurs (pp. 144 and 146) reveals the vast gulf between them—a gulf not bridged by fossil intermediates. A similar gulf also exists, of course, between this creature and *Archaeopteryx.*

Almost every structure in *Rhamphorhynchus,* a long-tailed pterosaur (fig. 222, p. 144), was unique to this creature. Especially obvious (as in all pterosaurs) is the enormous length of the fourth finger, in contrast with the other three fingers possessed by this reptile. This fourth finger provided support for the wing membrane. It is certainly not a delicate structure; and if the pterosaurs evolved from the thecodonts or some other earth-bound reptile, transitional forms should have been found showing a gradual lengthening of this fourth finger. Not even a hint of such a transitional form has ever been discovered.

Even more unusual was the pterodactyloid group of pterosaurs (Romer, fig. 225, p. 146). *Pteranodon* not only had a large, toothless beak and a long, rearward-extending bony crest, but its fourth fingers supported a wingspan of 25 feet. Where are the transitional forms documenting an evolutionary origin of these and other structures unique to the pterosaurs?

The bat is presumed to have evolved from nonflying insectivores—although, as stated earlier, the oldest-known bat to appear in the fossil record is 100% bat, and no trace of a transitional form can be found (Jepsen, 1966). In the bat four of the five fingers support the membrane of the wing and are extremely long, compared with the normal hand. These and other unique structures are solid bone and are anything but delicate structures. Transitional forms, if they ever existed, should certainly have been preserved. The absence of such forms leaves unanswered, on the basis of the evolution model, such questions as when, from what, where, and how bats originated.

Now let me ask this question: concerning the origin of flight, does the creation model or the evolution model have greater support from the fossil record? To me the answer seems obvious. Not a single fact contradicts the predictions of the creation model; but the actual evidence fails miserably to support the predictions of the evolution model. Here, where transitional forms should be the most obvious and easiest to find if evolution really accounts for the origin of these highly adapted and unique creatures, *none* is found. Could the fossil record really be that cruel and capricious to evolutionary paleontologists? The historical record inscribed in the rocks literally cries "Creation!"

SYSTEMIC DISCONTINUITY IS PERVASIVE

The examples cited in this paper are in no way exceptional; rather, they serve to illustrate what is characteristic of the fossil record. Although transitions at the subspecies level are observable and those at the species level may be inferred, the absence of transitional forms between higher categories (the created kinds of the creation model) is regular and systematic.

Simpson, in his book *Tempo and Mode in Evolution* (1944), under the heading "Major Systematic Discontinuities of Record" states that nowhere in the world is there any trace of a fossil that would close the considerable gap between *Hyracotherium* and its supposed ancestral order, Condylarthra. He then goes on to say (p. 106), "This is true of all the thirty-two orders of mammals. . . . The earliest and most primitive known members of every order already have the basic ordinal characters, and in no case is an approximately continuous sequence from one order to another known. In most cases the break is so sharp and the gap so large that the origin of the order is speculative and much disputed." Later (p. 107), Simpson states, "this regular absence of transitional forms is not confined to mammals, but is an almost universal phenomenon, as has long been noted by paleontologists. It is true of almost all orders of all classes of animals, both vertebrate and invertebrate. A fortiori, it is also true of the classes themselves, and of the major animal phyla, and it is apparently also true of analogous categories of plants."

In his book *The Meaning of Evolution* (1949) Simpson, with reference to the appearance of new phyla, classes, and other major groups, states (p. 231), "The process by which such *radical events* occur in evolution is the subject of one of the most serious remaining disputes among qualified professional students of evolution. The question is whether such *major events* take place *instantaneously*, by some process essentially unlike those involved in lesser or more gradual evolutionary change, or whether all of evolution, including these major changes, is explained by the same principles and processes throughout, their results being greater or less according to the time involved, the relative intensity of selection, and other material variables in any given situation." He continues: "Possibility for such dispute exists because transitions between major grades of organization are seldom well recorded by fossils. There is in this respect a tendency toward *systematic deficiency* in the record of the history of life. It is thus possible to claim that such transitions are not recorded because they did not exist, that the changes were not by transition but by sudden leaps in evolution" (emphasis added).

If phyla, classes, orders, and other major groups were connected by transitional forms rather then appearing suddenly in the fossil record with basic characteristics complete, it would not be necessary, of course, to refer to their appearances in the fossil record as "radical events." Furthermore, it cannot be emphasized too strongly that even evolutionists are arguing among themselves as to whether these major categories appeared *instantaneously* or not.

It is precisely the argument of creationists that these forms *did* arise instantaneously and that the transitional forms are not recorded because they never existed. Creationists thus would reword Simpson's statement to read, "It is thus possible to claim that such transitions are not recorded because they did not exist—that these major types arose by creation rather than by a process of gradual evolution."

In a more recent work, Simpson (1960, p. 149) says, "It is a feature of the known fossil record that most taxa appear abruptly." In the same paragraph he states further, "Gaps among known species are sporadic and often small. Gaps among known orders, classes, and phyla are systematic and almost always large."

It would hardly be necessary to document further the nature of the fossil record. It seems obvious that if the above statements of Simpson were stripped of all presuppositions and presumed evolutionary mechanisms to leave the bare record, they would describe exactly what is required by the creation model. This record is woefully deficient, however, in the light of the predictions of the evolution model.

No one has devoted himself more wholeheartedly than Simpson to what Dobzhansky (1972) has called the "mechanistic materialist philosophy shared by most of the present 'establishment' in the biological sciences." Simpson (1953, p. 360) therefore asserts that most paleontologists "find it logical, if not scientifically required, to assume that the sudden appearance of a new systematic group is not evidence for creation. . . ." He has expended considerable effort (1944, p. 105-124; 1953, p. 360-376; 1960, p. 140-152) in attempts to bend and twist every facet of evolution theory to explain away the deficiencies of the fossil record. One needs to be reminded, however, that if evolution is adopted as an a-priori principle, it is always possible to imagine auxiliary hypotheses—unproved and by nature unprovable—to make it work in any specific case. By this process biological evolution degenerates into what Thorpe (1969) calls one of his "four pillars of unwisdom": mental evolution that is the result of random tries preserved by reinforcements.

Concerning the plant kingdom, the following remark of E. J. H. Corner (1961), of the Cambridge University botany school, is refreshingly candid: "Much evidence can be adduced in favor of the theory of evolution—from biology, biogeography and paleontology, but I still think that to the unprejudiced, the fossil record of plants is in favor of special creation."

Even in the famous horse "series," which has been so highly touted as proof of evolution within an order, transitional forms between major types are missing. Lecomte du Noüy (1947, p. 74) has stated in reference to horses, "But each one of these intermediaries seems to have appeared 'suddenly,' and it has not yet been possible, because of the lack of fossils, to reconstitute the passage between those intermediaries. Yet it must have existed. The known forms remain separated like the piers of a ruined bridge. We know that the bridge has been built, but only vestiges of the stable props remain. The continuity we surmise may never be established by facts." Goldschmidt (1952, p. 97) has said, "Moreover, within the slowly evolving series, like the famous horse series, the decisive steps are abrupt."

THE "HOPEFUL MONSTER" ALTERNATIVE

Goldschmidt (1940; 1952, p. 84-98), in contrast with Simpson and the majority of evolutionists, accepts the discontinuities in the fossil record at face value. He rejects the neo-Darwinian interpretation of evolution (the modern synthesis), which is accepted by almost all evolutionists, at least among those who accept any theory concerning mechanisms at all. Goldschmidt instead has proposed that major categories (phyla, classes, orders, families) arose instantaneously by major saltations or systemic mutations. Goldschmidt terms this the "hopeful monster" mechanism. He has proposed, for instance, that at one time a reptile laid an egg and a bird was hatched from the egg. All major gaps in the fossil record are accounted for, according to Goldschmidt, by similar events: something laid an egg, and something else got born. Neo-Darwinists prefer to believe that Goldschmidt is the one who laid the egg; they maintain that there is not a shred of evidence to support his "hopeful monster" mechanism. Goldschmidt insists just as strongly that there is no evidence for the postulated neo-Darwinian mechanism (major transformations by the accumulation of micromutations). Creationists agree with both the neo-Darwinists and Goldschmidt: they are *both* wrong. However, Goldschmidt's publications do offer cogent arguments against the neo-Darwinian view of evolution, from both genetics and paleontology.

No one was more wholly committed to evolutionary philosophy than was Goldschmidt. If anybody wanted to find transitional forms, he did. If anybody would have admitted that a transitional form was a transitional form, if indeed that's what it was, he would have. But, concerning the fossil record, this is what Goldschmidt (1952, p. 97) says: "The facts of greatest general importance are the following. When a new phylum, class, or order appears, there follows a quick, explosive (in terms of geological time) diversification so that practically all orders or families known appear suddenly and without any apparent transitions."

Now, creationists ask: what better description of the fossil record could one expect, based on the predictions of the creation model? On the other hand, unless one accepts Goldschmidt's "hopeful monster" mechanism of evolution, this description contradicts the most critical prediction of the evolution model: the presence in the fossil record of the intermediates demanded by the theory.

AGAINST AUTHORITARIAN MATERIALISM

Kerkut (1960), although not a creationist, wrote a notable little volume to expose the weaknesses and fallacies in the usual evidence used to support evolution theory. In the concluding paragraph of the book this author states that "there is the theory that all the living forms in the world have arisen from a single source which itself came from an inorganic form. This theory can be called the 'General Theory of Evolution' and the evidence that supports it is not sufficiently strong to allow us to consider it as anything more than a working hypothesis." There is a world of difference, of course, between a working hypothesis and established scientific fact. If one's philosophic presuppositions lead him to accept evolution as his working hypothesis, he should restrict it to that use, rather than force it on others as an established fact.

If, without the philosophic presuppositions of either the materialist or the theist, creation and evolution are used as models to predict the nature of the historical evidence, it can be seen that the creation model is just as credible as the evolution model (and, I believe, much more credible). And I reiterate: the one model is no more religious or any less scientific than the other.

No less convinced an evolutionist than Thomas H. Huxley (as quoted in L. Huxley, 1903) acknowledged that " 'creation', in the ordinary sense of the word is perfectly conceivable. I find no difficulty in conceiving that, at some former period, this universe was not in existence, and that it made its appearance in six days (or instantaneously, if that is preferred), in consequence of the volition of some pre-existing Being. Then, as now, the so-called *a priori* arguments against Theism and, given a Deity, against the possibility of creative acts, appeared to me to be devoid of reasonable foundation."

The majority in the scientific community and educational circles are using the cloak of "science" to force the teaching of their view of life upon all. The authoritarianism of the medieval church has been replaced by the authoritarianism of rationalistic materialism. Constitutional guarantees are violated and free scientific inquiry is stifled under this blanket of dogmatism. It is time for a change.

REFERENCES

Allen, G. E. 1969. "T. H. Morgan and the Emergence of a New American Biology," *Quarterly Review of Biology 44*: 168-188.

Axelrod, D. I. 1956. "Early Cambrian Marine Fauna," *Science 123*: 7-9

Cloud, P. E. 1955. "Significance of the Gunflint (Precambrian) Microflora," *Science 143*: 27-35.

Corner, E. J. H. 1961. "Evolution," In *Contemporary Botanical Thought,* ed. by A. M. McLeod and L. S. Cobley. Quadrangle Books; Chicago, pp. 93-114.

Danson, R. 1971. "Evolution," *New Scientist 49*: 25.

Darwin, C. 1869. *Origin of Species.* Reprinted 1956, J. M. Dent & Sons; London, p. 304.

de Beer, G. 1964. "The World of an Evolutionist." *Science 143*: 1911-1917.

Dobzhansky, T. 1936. "Evolution at Work," *Science 137*: 1891-1903.

———. 1972. "A Biologist's World View," *Science 175*: 49.

Eden, M. 1967. "Inadequacies of Neo-Darwinian Evolution as a Scientfic Theory." In *Mathematical Challenges to the Neo-Darwinian Interpretation of Evolution,* ed. by P. S. Moorhead and M. M. Kaplan. Wistar Institute Press; Philadelphia.

Ehrlich, P. R., and Holm, R. W. 1962. "Patterns and Populations," *Science 137*: 655.

Fothergill, P. G. 1961. "Issues in Evolution," *Nature 189*: 425.

George, T. N. 1960. "Fossils in Evolutionary Perspective," *Science Progress 48*: 1-5.

Goldschmidt, R. B. 1940. *The Material Basis of Evolution.* Yale University Press; New Haven, Conn.

———. 1952. "Evolution as Viewed by One Geneticist," *American Scientist 40*: 84-98.

Grimmer, J. L. 1962. "Strange Little World of the Hoatzin," *National Geographic 122*: 391-400.

Huxley, L., ed. 1903. *Life and Letters of Thomas Henry Huxley.* D. Appleton & Co.; New York. Vol. 2, p. 439.

Jepsen, G. L. 1966. "Early Eocene Bat from Wyoming," *Science 154*: 1333-1339.

Kerkut, G. A. 1960. *Implications of Evolution.* Pergamon Press; New York, p. 157.

Lecomte du Noüy, P. 1947. *Human Destiny.* New American Library; New York.

Le Gros Clark, W. E. 1955. *Discovery* Jan.: 7.

Litynski, L. 1961. "Should We Burn Darwin?" *Science Digest 50*: 61-63.

Macbeth, N. 1971. *Darwin Retried.* Gambit, Inc.; Boston.

Moorhead, P. S., and Kaplan, M. M., eds. 1937. *Mathematical Challenges to the Neo-Darwinian Interpretation of Evolution.* Wistar Institute Press; Philadelphia.

Olson, E. C. 1960. "Morphology, Paleontology, and Evolution," In *Evolution after Darwin; Volume 1: The Evolution of Life,* ed. by Sol Tax. University of Chicago Press; Chicago.

———. 1965. *The Evolution of Life.* New American Library; New York.

Ommaney, F. D. 1964. *The Fishes.* Life Nature Library, Time-Life, Inc.; New York. p. 60.

Romer, A. S. 1966. *Vertebrate Paleontology,* 3rd ed. University of Chicago Press; Chicago.

Salisbury, F. B. 1969. "Natural Selection and the Complexity of the Gene," *Nature 224*: 342-343.

———. 1971. "Doubts about the Modern Synthetic Theory of Evolution," *American Biology Teacher 33*: 335-338.

Simpson, G. G. 1944. *Tempo and Mode in Evolution.* Columbia University Press; New York.

———. 1949. *The Meaning of Evolution.* Yale University Press; New Haven, Conn.

———. 1953. *The Major Features of Evolution.* Columbia University Press; New York.

———. 1960. "The History of Life." In *Evolution after Darwin; Volume 1: The Evolution of Life,* ed. by Sol Tax. University of Chicago Press; Chicago.

———. 1964. "The Non-prevalence of Humanoids," *Science 143*: 709.

Swinton, W. E. 1980. "The Origin of Birds." In *Biology and Comparative Physiology of Birds,* ed. by A. J. Marshall, Academia Press; New York, vol. 1, p. 1.

Thorpe, W. 1969, "Reductionism v. Organicism," *New Scientist 43*: 635-638.

Watson, D. M. S. 1929. "Adaptation," *Nature 194*: 233.

18

Act 590 of 1981,
General Assembly, State of Arkansas

AN ACT TO REQUIRE BALANCED TREATMENT OF CREATION-SCIENCE AND EVOLUTION-SCIENCE IN PUBLIC SCHOOLS; TO PROTECT ACADEMIC FREEDOM BY PROVIDING STUDENT CHOICE; TO ENSURE FREEDOM OF RELIGIOUS EXERCISE; TO GUARANTEE FREEDOM OF BELIEF AND SPEECH; TO PREVENT ESTABLISHMENT OF RELIGION; TO PROHIBIT RELIGIOUS INSTRUCTION CONCERNING ORIGINS; TO BAR DISCRIMINATION ON THE BASIS OF CREATIONIST OR EVOLUTIONIST BELIEF; TO PROVIDE DEFINITIONS AND CLARIFICATIONS; TO DECLARE THE LEGISLATIVE PURPOSE AND LEGISLATIVE FINDINGS OF FACT; TO PROVIDE FOR SEVERABILITY OF PROVISIONS: TO PROVIDE FOR REPEAL OF CONTRARY LAWS; AND TO SET FORTH AN EFFECTIVE DATE.

Be It Enacted by the General Assembly of the State of Arkansas:

SECTION 1. *Requirement for Balanced Treatment.* Public Schools within this State shall give balanced treatment to creation-science and to evolution-science. Balanced treatment to these two models shall be given in classroom lectures taken as a whole for each course, in textbook materials taken as a whole for each course, in library materials taken as a whole for the sciences and taken as a whole for the humanities, and in other educational programs in public schools, to the extent that such lectures, textbooks, library materials, or educational programs deal in any way with the subject of the origin of man, life, the earth, or the universe.

SECTION 2. *Prohibition against Religious Instruction.* Treatment of either evolution-science or creation-science shall be limited to scientific evidence for each model and inferences from those scientific evidences, and must not include any religious instruction or references to religious writings.

SECTION 3. *Requirement for Nondiscrimination.* Public schools within this State, or their personnel, shall not discriminate, by reducing a grade of a student or by singling out and making public criticism, against any student who demonstrates a satisfactory understanding of both evolution-science and creation-science and who accepts or rejects either model in whole or part.

SECTION 4. *Definitions.* As used in this Act:

(a) "Creation-science" means the scientific evidences for creation and inferences from those scientific evidences. Creation-science includes the scientific evidences and related inferences that indicate: (1) Sudden creation of the universe, energy, and life from nothing; (2) The insufficiency of mutation and natural selection in bringing about development of all living kinds from a single organism; (3) Changes only within fixed limits of originally created kinds of plants and animals; (4) Separate ancestry for man and apes; (5) Explanation of the earth's geology by catastrophism, including the occurrence of a worldwide flood; and (6) A relatively recent inception of the earth and living kinds.

(b) "Evolution-science" means the scientific evidences for evolution and inferences from those scientific evidences. Evolution-science includes the scientific evidences and related inferences that indicate (1) Emergence by naturalistic processes of the universe from disordered matter and emergence of life from nonlife; (2) The sufficiency of mutation and natural selection in bringing about development of present living kinds from simple earlier kinds; (3) Emergence by mutation and natural selection of present living kinds from simple earlier kinds; (4) Emergence of man from a common ancestor with apes; (5) Explanation of the earth's geology and the evolutionary sequence by uniformitarianism; and (6) An inception several billion years ago of the earth and somewhat later of life.

(c) "Public schools" mean secondary and elementary schools.

SECTION 5. *Clarification.* This Act does not require or permit instruction in any religious doctrine or materials. This Act does not require any instruction in the subject of origins, but simply requires instruction in both scientific models (of evolution-science and creation-science) if public schools choose to teach either. This Act does not require each individual textbook or library book to give balanced treatment to the models of evolution-science and creation-science; it does not require any school books to be discarded. This Act does not require each individual classroom lecture in a course to give such balanced treatment, but simply requires the lectures as a whole to give balanced treatment; it permits some lectures to present evolution-science and other lectures to present creation-science.

SECTION 6. *Legislative Declaration of Purpose.* This Legislature enacts this Act for public schools with the purpose of protecting academic freedom for students' differing values and beliefs; ensuring neutrality toward students' diverse religious convictions, ensuring freedom of religious exercise for students and their parents; guaranteeing freedom of belief and speech for students; preventing establishment of Theologically Liberal, Humanist, Nontheist, or Atheist religions; preventing discrimination against students on the basis of their personal beliefs concerning creation and evolution; and assisting students in their search for truth. This Legislature does not have the purpose of causing instruction in religious concepts or making an establishment of religion.

SECTION 7. *Legislative Findings of Fact.* This Legislature finds that:

(a) The subject of the origin of the universe, earth, life, and man is treated within many public school courses, such as biology, life science, anthropology, sociology, and often also in physics, chemistry, world history, philosophy, and social studies.

(b) Only evolution-science is presented to students in virtually all of those courses that discuss the subject of origins. Public schools generally censor creation-science and evidence contrary to evolution.

(c) Evolution-science is not an unquestionable fact of science, because evolution cannot be experimentally observed, fully verified, or logically falsified, and because evolution-science is not accepted by some scientists.

(d) Evolution-science is contrary to the religious convictions or moral values or philosophical beliefs of many students and parents, including individuals of many different religious faiths and with diverse moral values and philosophical beliefs.

(e) Public school presentation of only evolution-science without any alternative model of origins abridges the United States Constitution's protections of freedom of religious exercise and of freedom of belief and speech for students and parents, because it undermines their religious convictions and moral or philosophical values, compels their unconscionable professions of belief and hinders religious training and moral training by parents.

(f) Public school presentation of only evolution-science furthermore abridges the Constitution's prohibition against establishment of religion, because it produces hostility toward many Theistic religions and brings preference to Theological Liberalism, Humanism, Nontheistic religions, and Atheism, in that those religious faiths generally include religious belief in evolution.

(g) Public school instruction in only evolution-science also violates the principle of academic freedom, because it denies students a choice between scientific models and instead indoctrinates them in evolution-science alone.

(h) Presentation of only one model rather than alternative scientific models of origins is not required by any compelling interest of the State, and exemption of such students from a course or class presenting only evolution-science does not provide an adequate remedy because of teacher influence and student pressure to remain in that course or class.

(i) Attendance of those students who are at public schools is compelled by law, and school taxes from their parents and other citizens are mandated by law.

(j) Creation-science is an alternative scientific model of origins and can be presented from a strictly scientific standpoint without any religious doctrine just as evolution-science can, because there are scientists who conclude that scientific data best support creation-science and because scientific evidences and inferences have been presented for creation-science.

(k) Public school presentation of both evolution-science and creation-science would not violate the Constitution's prohibition against establishment of religion, because it would involve presentation of the scientific evidences and related inferences for each model rather than any religious instruction.

(l) Most citizens, whatever their religious beliefs about origins, favor balanced treatment in public schools of alternative scientific models of origins for better guiding students in their search for knowledge, and they favor a neutral approach towards subjects affecting the religious and moral and philosophical convictions of students.

SECTION 8. *Short Title.* This Act shall be known as the "Balanced Treatment for Creation-Science and Evolution-Science Act."

SECTION 9. *Severability of Provisions.* If any provision of this Act is held invalid, that invalidity shall not affect other provisions that can be applied in the absence of the invalidated provisions, and the provisions of this Act are declared to be severable.

SECTION 10. *Repeal of Contrary Laws.* All State laws or parts of State laws in conflict with this Act are hereby repealed.

SECTION 11. *Effective Date.* The requirements of the Act shall be met by and may be met before the beginning of the next school year if that is more than six months from the date of enactment, or otherwise one year after the beginning of the next school year, and in all subsequent school years.

Signed on 19 March 1981 by Governor Frank White.

19

Witness Testimony Sheet
McLean v. Arkansas

Michael Ruse

Question: Please state your name.
Answer: Michael Escott Ruse.

*   *   *

Q: Are you a Canadian citizen?
A: Yes.

Q: What is your occupation?
A: I am a professor of philosophy at the University of Guelph in Ontario, Canada.

Q: What is your educational background?
A: [Attached vita.]

Q: Are you a member of any professional organizations?
A: [Attached vita.]

Q: Have you authored any books or articles?
A: Yes, I have authored the following books: *The Philosophy of Biology; Sociobiology: Sense or Nonsense?; The Darwinian Revolution: Science Red in Tooth and Claw; Is Science Sexist? And Other Problems in the Biomedical Sciences;* and *Darwinism Defended,* which is now in manuscript form and about to be published. In addition I have published well over seventy articles in professional journals. [See vita.]

Q: Professor Ruse, will you please describe your academic specialties at the present time.

A: My major academic interest at this time is the history and philosophy of science, in particular the history and philosophy of biology. I teach courses in the philosophy of science, ethics, logic, introductory philosophy, and the history of science. I also teach a course in the philosophy of religion.

THE HAPPENINGS AND MECHANISMS OF EVOLUTION

Q: Dr. Ruse, you mentioned that your latest book is titled *Darwinism Defended*. Does the title of that book suggest that evolution is in question and in need of defense?

A: No. That evolution happened is not questioned by any credible scientist. My book is, in large part, a defense of the Darwinian theory of about *how* evolution happened.

Q: Would you please explain that distinction.

A: Yes, that distinction is very important because creation scientists often confuse the two ideas. What is properly known as "evolution" has two major components. The first is the *happening* of evolution. That is, did one or a few organisms develop by naturalistic processes through a succession of forms which changed over long periods of time into the organisms alive today. I do not know of a single credible scientist, other than the people who call themselves "creation-scientists," who question that evolution actually happened.

The second component of evolution is *how* evolution happened. When scientists today speak of "evolutionary theory," they are usually referring to a theory explaining the mechanics of evolution; that is, the "how" of evolution.

Q: You say that scientists agree that evolution happened. Why is that?

A: Because the evidence is absolutely overwhelming. It convinces the unbiased observer beyond any reasonable doubt.

Q: What is the history of that consensus? When did scientists come to agree about the happening of evolution?

A: Scientists have been considering ideas about evolution since ancient Greece —although those early notions were rather primitive. In modern times, evolution has been seriously considered by scientists since the scientific revolution in the sixteenth and seventeenth centuries. Apart from the pressure that Copernicanism put upon literal readings of the Bible, the heliocentric world view inspired all sorts of thoughts about inorganic evolution. In the eighteenth century, an idea which enjoyed much popularity was the "nebular hypothesis," formulated by Kant, Laplace, and William Herschel, supposing that universes are formed out of gaseous clouds. It was not long before people started to think analogically about the organic world. Developments in geology had started to convince people that the earth is far older than the traditionally supposed, biblically based 6000 years.

The first great organic evolutionist was the French scientist Jean Baptiste de Lamarck, whose major work *Philosophie Zoologique* was published in 1809. The next important figure, certainly in influence, was the then-anonymous Scottish publisher Robert Chambers, whose *Vestiges of the Natural History of Creation* appeared in 1844. And this leads to Charles Darwin and his *Origin of Species,* arriving on the scene in 1859. (In fact, Darwin's seminal work was performed in the 1830s, but he did not publish for twenty years.) The publication of *Origin* was the conclusive scholarship necessary to convince the science community of the day that evolution happened. That consensus has never been threatened since because every new piece of data, from every relevant discipline of science, confirms that evolution happened.

Of course, Darwin also proposed a theory about *how* evolution happened. That theory—natural selection—has been far more controversial.

Q: Do scientists generally agree now about how evolution happened?
A: No, not at all. With respect to this issue of how evolution happened there is still much debate.

Darwin suggested *natural selection* as the most important, but not the only, mechanism of evolution. For the most part, I think he was right and that accounts for the title of my book *Darwinism Defended.* However, other scientists think that natural selection may not be the most important element of evolutionary change. Some emphasize *speciation.* Others emphasize *genetic drift. Pure chance* may be another important factor. Still other scientists, such as Stephen Gould (who I understand will also be a witness in this trial), propose a theory of *punctuated equilibria*—that evolution happened more abruptly than the slow, gradual change suggested by Darwinian theory.

In other words, though scientists are all agreed that evolution happened, there is a vigorous and, I believe, a healthy debate among proponents of very many different ideas about how evolution happened.

Q: Dr. Ruse, you testified earlier that creation scientists often confuse the difference between the *happening* of evolution and the *how* of evolution. Would you please explain what you meant?
A: Creation scientists frequently extract quotes from scientific literature which seem to challenge, or question the sufficiency of, evolution. They use those quotes to advance the creationist theory that evolution did not happen. However, in virtually every instance I have examined, the authors of those quotations were discussing or criticizing a particular theory about *how* evolution happened, not whether evolution happened. Indeed, the scientists quoted are often among the foremost evolutionists. However, by taking their words out of context, the creation scientists camouflage the fact that

those same authorities are unanimously in agreement that evolution actually happened.

Q: Do you know of any specific examples of such out of context quotations?

A: Yes. [Insert appropriate e.g.'s re Theodosius Dobzhansky, Ernst Mayr, or Stephen Gould.]

Q: In your book *Darwinism Defended* do you also say anything about creation science?

CREDENTIALS AS EXPERT IN CREATION SCIENCE

A: Yes. I devote two chapters to creation science. In the first of those chapters, chapter 14 in the manuscript (which will be chapter 13 in the published edition), I describe the creation-science arguments in some detail. In the next chapter I analyze those contentions and conclude that philosophically and methodologically the creationists do not act like scientists, and that substantively the creationists' contentions are without scientific merit.

Q: Have you read much creation-science literature?

A: Yes.

Q: Would you please describe the kinds of material you are familiar with.

A: In *Darwinism Defended* I analyzed in great detail a book titled *Scientific Creationism, Public School Edition*. The book is published by the Institute for Creation Research, a leading creation science organization. The book is edited by Henry Morris and was written by twenty-three creation scientists on the staff of ICR. Inasmuch as it is denoted a public school edition, it purports to be the best attempt to eliminate religion from creation science. I consider *Scientific Creationism* to be an authoritative statement of creation-science doctrine.

In addition to *Scientific Creationism* I have read [Ruse offers list].

CONNECTION: ACT 590 AND CREATION SCIENCE GENERALLY

Q: Dr. Ruse, I would like to show you a copy of Act 590. I direct your attention to the references to creation science, especially in Section 4(a) that defines that term. As a philosopher and historian of science, and someone who has read extensively in creation-science literature, what are your thoughts about how Act 590 relates to creation science generally?

A: When I read Act 590, and Section 4 in particular, I am struck by the similarities between the act and the principal creation-science claims found throughout all of the creation-science literature that I have read and that I know about.[1] Of course, some creation-science books focus on specific issues. However, taken as a whole, the elements of creation science found in Section 4(a) track the major themes of creation science generally.

Q: What are the similarities that you see?

A: First of all, the act establishes the so-called "dual model" approach to the teaching of what creation scientists call "origins."

Secondly, the six elements of creation science listed in 4(a) are the same claims repeatedly made throughout creation-science literature.

Finally, creation scientists are the only ones I know of who talk about evolution as a single body of knowledge relating to everything from the origin of the universe, through the origin of the world, the origin of life, and the origin of human life, to flood geology. That typical creationist distortion of evolutionary theory is also incorporated into the statute.

ACT 590: DUAL MODEL

Q: Dr. Ruse, I would like to explore each of those elements with you one at a time.

First you mentioned the dual-model approach to teaching origins. Would you please describe what you meant by that?

A: Creation scientists like to say that there are only two models of origins: evolution and creation. This so-called "dual model" approach to origins is a critical ingredient of the creation-science model for two reasons. First, it enables creation scientists to argue that if evolution is taught, then *only* creation need also be taught—without having to include other possible explanations of what they call "origins." In other words, the "dual model" is an attempt to distinguish creation from other unconventional theories about how life on this planet began. Second, creation scientists use the dual model as a rationale for arguing that disproof of evolution is proof of creation.

Q: What are the various theories about how life began?

A: Of course, there is the creation theory and the theory of abiogenesis—that life developed from nonlife. Another theory is that life on this planet began as a result of the intelligent intervention of beings from elsewhere in the universe. That theory has been proposed by Von Däniken, in *Chariots of the Gods,* and more recently by Francis Crick.

Still another theory is that life began on this planet as a result of passing through a cloud of organic material, some of which took root here. In fact, I think one of the defendants' witnesses in this case, Chandra Wickramasinghe, is a proponent of this theory.

Q: As a philosopher of science, what is your professional opinion about the logic of the creation-science contention that disproof of Darwinian evolution is proof of creation?

A: The contention is fallacious because, as I just demonstrated, creation and abiogenesis are not the only alternatives. The creation scientists try to invoke the following proposition of logic: If A or B, then "not A" equals

B. That statement is true—but only if A and B are the only alternatives. If the statement read: If A, B, or C, then "not A" would not equal B. In fact, "not A" would not tell you anything about the truth value of B or C.

Similarly, there are many theories about how life began on this planet. Disproving one cannot prove any other. Also keep in mind that I am only speaking about theories of how life began on this planet. Creation science is also a theory about how the universe began; how the earth originated, how human life developed, and flood geology. And, in regard to each of those subjects scientists have proposed a great many alternative theories.

Q: Dr. Ruse, Act 590 does not contain the words "dual model." Would you tell us precisely what in the statute you are referring to?

A: The dual-model approach is incorporated throughout Act 590. For example, in the very first sentences of Section 1 the statute talks about creation science and evolution science as the two models that must be given balanced treatment in the classroom. Similarly, Section 3 talks about discrimination with respect to "either model," meaning the two models of evolution science and creation science. However, the clearest example of the two so-called models of origin is apparent in Section 4. Section 4 defines creation science and evolution science. As you can see from looking at that section, the definitions have been written in a way to establish a stark contrast between every element of creation science and corresponding elements of evolution science.

Q: What is your understanding of creation theory?

A: That the universe, the earth, all life, and especially human life, were created by a supernatural Creator—spelled with a capital C—and that the Creator also inundated the world with a great flood.

ACT 590: CREATION THEORY

Q: Is the creation theory part of Act 590?

A: Yes, Act 590 cannot be understood, in my opinion, unless you realize that the explanations of science are being contrasted with the explanations of supernatural creation.

Q: Where do you see that contrast?

A: In the first sentence of 4(a) and the first sentence of 4(b).

Looking first at 4(b) ["Evolution-science means the scientific evidences for evolution and inferences from those scientific evidences"]. Scientific evidences and inferences from evidences do not mean anything independent of a theory that joins them together. In Section 4(b) the unifying theory for those evidences and the inferences is evolution.

Now look at 4(a) ["Creation-science means the scientific evidences for creation and inferences from those scientific evidences"]. Section 4(a) is identical to 4(b) word for word except that where 4(b) says "evolution," 4(a) speaks of "creation." Thus, in 4(a) the various evidences and inferences are meant to support the theory of creation. And that theory is that the Creator created the universe, the world, life, human life, and a worldwide flood.

Q: Do you see the word Creator in Act 590?

A: No.

Q: Why then do you refer to the intervention of a creator in analyzing the statute?

A: It is true that the act does not mention "Creator." It is also true that the act even tries to conceal that creation science is the theory of creation. And the theory of creation is that the Creator did it.

For example, look at sections 4(a)(1) and 4(b)(1). Section 4(b)(1) ["Emergence by naturalistic processes of the Universe from disordered matter and emergence of life from non-life"], which is supposed to be the evolution-science explanation of the origin of matter and life, refers to emergence by *naturalistic* processes. Now I do not like the word "emergence"—it is not scientific—and I do not think that evolutionary theory has anything to say about the origin of matter or life. However, the important point of comparison here is that 4(b)(1) pertains to "naturalistic processes."

Contrast that with 4(a)(1) ["Sudden creation of the universe, energy and life from nothing"]. In the statute, "sudden creation" in 4(a)(1) is contrasted with "naturalistic processes" in 4(b)(1). Therefore, "sudden creation" must mean by nonnaturalistic processes. And since we know that the unifying theory of 4(a) is the theory of creation, it follows that the nonnaturalistic process referred to in 4(a)(1) must be the intervention of a Creator.

Similarly, compare 4(b)(3) with 4(a)(3). 4(b)(3) ["Emergence by mutation and natural selection of present living kinds from simple earlier kinds"] refers to "mutation and natural selection." Both are naturalistic processes that contribute to the evolution of present living organisms from simpler earlier organisms.

In contrast, 4(a)(3) ["Changes only within fixed limits of originally created kinds of plants and animals"] refers to "originally created kinds of plants and animals." Since "originally created" must mean something other than "by naturalistic processes," it can only be understood with reference to the theory of creation—which is that the Creator created these original kinds of plants and animals.

I might add that there is another reason to believe that the reference to "originally created kinds" refers to the acts of a Creator. The word "kinds" is not a scientific word. The science of taxonomy—which involves classifying organisms—divides living organisms into categories of species, genus, family, order, class, and phylum. There is no taxonomic classification of "kind." In fact the only place I know where the word "kind" is used in this sense is in the Bible. In the very first chapter of Genesis, describing the creation, the Bible says, "and God made the beast of the earth after its kind and the cattle after their kind and everything that creepeth upon the ground after its kind and God saw that it was good."

ACT 590: SIMILARITIES WITH CREATION SCIENCE GENERALLY

Q: Dr. Ruse, I believe you testified earlier that the six elements of creation science identified in sections 4(a)(1) through 4(a)(6) were similar to the elements of creation science as explained in creation-science literature, is that so?

A: Yes.

Q: Would you please elaborate a bit on those similarities.

A: In my reading of creation-science literature, I find that it constantly emphasizes each of the elements that are also incorporated into Section 4(a) of Act 590.

A good example of this is clear from chapter 14 of my book *Darwinism Defended.* In that chapter, as I mentioned earlier, I analyzed the creation-science model as articulated in a basic text, *Scientific Creationism.* That chapter was written long before I ever saw Act 590; yet the elements of creation science as articulated in Section 4(a) are precisely the same as the elements of creation science articulated in *Scientific Creationism.*

In *Scientific Creationism* the authors refer to a "cause-and-effect" analysis and a "purpose" analysis to conclude that an omnipotent Creator is the first cause of matter, energy, and life (see, *Darwinism Defended,* at p. 430), just as in 4(a)(1).

The book *Scientific Creationism* then argues at some length that mutation and selection are not adequate explanations for the variable life forms found today (*Darwinism Defended,* at p. 433), which is directly comparable to the contention in 4(a)(2).

Next the book *Scientific Creationism* discusses homologies—which are structural similarities of different organisms—and concludes that homologies *are not* evidence of common ancestry (*Darwinism Defended,* at p. 434). The authors similarly dismiss embryological similarities and behavorial similarities as evidences of common ancestry. That creationist position is incorporated into the statute in sections 4(a)(2) and 4(a)(3).

Creation scientists generally also refer to the absence of transitional forms in the fossil record to suggest the lack of common ancestry generally and, more specifically, that man did not descend from apes (*Darwinism Defended,* at p. 435, 441). These creationists' contentions are incorporated in 4(a)(3) and 4(a)(4).

Continuing with the analysis of parallels, *Scientific Creationism* next argues that a catastrophic worldwide flood once enveloped the earth, destroyed all living things, and profoundly affected earth's geology. Of course, this notion of a worldwide flood is incorporated into 4(a)(5) of Act 590.

Finally, *Scientific Creationism* devotes a whole chapter to its contention that the earth is relatively young (*Darwinism Defended*, at p. 438). And that creation science is also in Act 590 in Section 4(a)(6).

ACT 590: EVOLUTION SCIENCE

Q: I would now like to direct your attention to Section 4(b) of the statute, which purports to define evolution science. What is your professional opinion of the definition of evolution science?

A: The definition does not make sense in any way that scientists talk about evolution.

Q: Would you please explain?

A: The statute adopts a common and quite distorted creation-science characterization of evolution. Creation science attempts to create the impression that evolution pertains to the same things as creation science. That is not so. There is no scientific discipline known as evolution science. One cannot get a degree in evolution science.

In fact, the body of knowledge implicated by creation science includes a dozen different scientific disciplines including cosmology, astronomy, geology, biology and all of its many subdisciplines, paleontology, chemistry, physics, botany, and the technologies of radiocarbon and radiometric dating.

Therefore, I find that incorporating many aspects of all these disciplines under the rubric of "evolution science" is both typical of creation science and thoroughly nonsensical.

Q: What then do scientists usually mean by "evolution"?

A: Scientists commonly think of evolutionary theory as attempting to explain how life developed *after* it was formed. Evolutionary theory does not focus on how life began, but only on what happened to life after it began. For example, one of the defendants' own experts—Chandra Wickrama-singhe—believes that life on this planet took hold after earth passed through a cloud of organic material. Even if that were so, evolutionary theory would begin at that point and describe what happened to that life after it started to grow on the planet.

Q: Understanding that scientists do not generally recognize this thing called "evolution science," let me nonetheless direct your attention to Section 4(b)(1). What is your professional judgment of that provision as a scientific statement?

A: Of course, I do not know, and scientists generally do not know, how the universe originated or even how life on this planet originated.

Section, 4(b)(1) may or may not be right. It is not the only naturalistic explanation of how the universe began. For example, the universe might not have ever "begun." Though it is a very hard concept to grasp, the universe may have always existed. Moreover, even if it began at some point, the so-called big bang may not have happened from "disordered matter," but rather from the most highly ordered matter imaginable: all of the energy and matter of the universe compressed into a single mass for an instant before explosion. Similarly, the theory of abiogenesis—that life developed through naturalistic processes from nonlife—is only a working hypothesis that scientists are presently attempting to confirm or deny. It will probably be quite some time before science has any clear understanding of how life began.

Thus the elements of 4(b)(1) are by no means the only scientific theories being examined.

The most important thing about 4(b)(1), however, is its reliance on naturalistic processes. I believe that the first and the most important characteristic of science is that it relies exclusively on blind, undirected natural laws and naturalistic processes. Therefore, whether Section 4(b)(1) is right or wrong, at least it is science.

Q: What is your professional assessment of 4(b)(2)?

A: Section 4(b)(2) is a good example of the confusion and distortion that results from the dual-model comparison of evolution science with creation science.

Section 4(b)(2) ["The sufficiency of mutation and natural selection in bringing about development of present living kinds from single earlier kinds"] is probably not accepted by any scientist as an accurate statement about evolution. In other words, scientists do *not* believe that mutation and natural selection are sufficient to bring about the development of present living kinds from simple earlier kinds. Other additional factors are surely also at work. For example, in addition to mutation and natural selection scientists recognize the importance of genetic drift, speciation, and pure chance in the development of present living organisms from simple earlier organisms. Therefore, Section 4(a)(2), which pertains to the *in*sufficiency of mutation and natural selection, is an accurate reflection of current scientific learning about evolution.

However, as used in the statute, 4(a)(2) is supposed to be an indication for creation. In that context, the section obviously means that mutation and natural selection are insufficient, not because other naturalistic processes are also operative, but because the Creator brought about the development of all living kinds.

Q: Dr. Ruse, do you understand the meaning of Section 4(b)(3)?

A: Standing alone, 4(b)(3) does not seem to mean very much in addition to 4(b)(2). There is the same problem with limiting the operative mechanisms of evolution to mutation and natural selection. However, the point of 4(b)(3) becomes evident by reference to 4(a)(3).

Section 4(a)(3) adopts the general creation-science concession that some evolutionary changes have occurred. Creation science contends however that evolution has occurred only within kinds and not between kinds.

The reason creation science has to admit the happening of some evolution is that some evolutionary changes cannot be denied. To use an oft-cited example, [use moths, flies, Darwin's finches, etc.]. In order to account for these evolutionary changes creation science resorts to the 4(b)(3) contention: that the Creator created an original moth kind; although that moth may experience evolutionary adaptations to deal with changing environmental circumstances, it cannot change into any other kind.

By contrast, Section 4(b)(3) must mean that evolutionary changes do not know any fixed limits of original kinds, and that organisms existing today have descended from other organisms of very different species. Of course, 4(b)(4)—that man descended from apes—is the most troublesome example of that evolutionary change for the creation scientists.

Q: What is your professional opinion about the contrast between unrestrained evolutionary change and evolutionary changes only within fixed limits of created kinds.?

A: I believe that these subsections are quite revealing about the creation-science/evolution-science dual model. In section 4(b), although subsections (3) and (4) are incorrectly limited to mutation and natural selection, they reflect a reliance on natural processes that are not constrained by any supernatural limitations.

On the other hand, sections 4(a)(3) and (a)(4) impose the limitation of "originally created kinds." However, creation scientists offer no explanation of why admittedly ongoing evolutionary changes should stop at originally created kinds. In the creation model the only explanation is that the Creator fixed those limits of originally created kinds and that His will cannot be altered.

Q: Dr. Ruse, do you understand the use of the words "catastrophism" and "uniformitarianism" in subsection 5 of sections 4(a) and (b).

A: No. These two words are used in the statute very loosely, without any sensitivity to the historical meanings of these terms.

In Section 4(a)(5) the reference to a worldwide flood suggests that catastrophism is used to mean that earth geology can be explained by catastrophes of a different order of magnitude than we know today. However, by the time the word "catastrophism" was introduced in the 1830s, it was meant to apply to naturalistic phenomena and its proponents, such as Whewell, specifically denied that it applied to a worldwide flood as described in the Bible. Indeed, one element of catastrophism was that the earth was very old and that the process of cooling from its molten state caused these catastrophes.

Q: And what about "uniformitarianism" as used in Section 4(b)(5)?

A: Standing alone, 4(b)(5) gives very little clue to the meaning of the word "uniformitarianism" as used in the statute. Once again, its content is understood only by reference to 4(b)(5). In that regard, it seems to me that uniformitarianism was intended to mean that the processes acting on the world's geology have been subject to the same natural laws in the past as in the present. Of course, conditions in the past may have different from what they are today, and that would mean that the operations of the same laws might have different effects in the past from those they do today. However, the laws of nature would remain the same.

If that is the definition, then I think most scientists would agree with it. However, uniformitarianism then also does not say very much because all science is premised on the same natural laws operating in the past as they do today. Moreover, that meaning of uniformitarianism takes absolutely no account of the varied meanings attached to that word throughout the last hundred years.

Q: Dr. Ruse, do you find much reference to these words "uniformitarianism" and "catastrophism" in the creationist literature generally?

A: Yes. Indeed, it is quite striking that the statute embodies the same loose and historically meaningless use of the two words.

Q: What is your professional opinion about the significance of the worldwide flood contention as it relates to creation science?

A: In my opinion, the worldwide flood idea further highlights the nonscience of creation science. As used in 590, "uniformitarianism" refers to scientific reliance on blind, unchanging natural laws. Contrarily, since catastrophism is cited as an element of *creation* theory, and as an alternative to the naturalistic processes that are presumed to govern evolution theory, the word can only mean that catastrophes occurred in the past in a way that we do not see today because a Creator intervened and disrupted the natural processes. The most prominent example of that supernatural intervention is the worldwide flood described in Genesis: the Noachian flood.

I think the worldwide flood is especially revealing because it has nothing to do with *"origins"*—which is supposed to be the subject matter of creation science. To use the words of the statute, the worldwide flood is not directly related to the origin of man, life, the earth, or the universe. A cataclysmic worldwide flood is, of course, a biblical notion. Creation science generally has incorporated the flood into the creation model. That the worldwide flood found its way into the statute, even though it does not have anything to do with origins, further highlights the direct connection between the existing body of literature defining creation science and the creation science defined in the statute.

Q: Do you have an opinion about why creation science incorporates reference to the flood?

A: Yes. I think there are two reasons.

First, because it is an essential element of fundamentalist religious belief, and creation science is a manifestation of those religious views.

Second, I think creation scientists grasped at the flood to explain away a very difficult problem for them—the worldwide uniform ordering of the fossil record. In other words, if as the creationists contend, the Creator created all organisms at once, then the fossil remains of man and dinosaurs and fish, etc., should be found uniformly throughout the geologic column. Of course, as we know, that is not so. The creation scientists fabricate a farcically ridiculous explanation based on the flood. According to creation science this catastrophic catastrophe that destroyed the earth also had the side effect of neatly ordering the fossil record.

Q: And finally, Dr. Ruse, do you have any professional observations with respect to subsection 6?

A: Section of 4(b) is rather vague for a scientific statement: the earth was created "*several* billion years ago" and "*life was created somewhat* later." However, at least the point of this provision is clear: the earth and life on the earth are very old.

In contrast, 4(a)(6), though similarly vague, identifies a young age for youth and life on the earth. In the creation-science literature I've read, a standard tenet of the theory is that the world is only six to ten thousand years old.

Interestingly, a creation theory—that a Creator created the earth and life on the earth—does not necessarily require a young earth. The Creator could have acted many billions of years ago. Indeed, many of the government witnesses at this trial apparently believe that the earth is very old. Since a young earth is not a necessary prerequisite for creation by a Creator, it must have some other explanation. Obviously, that explanation is the Bible. Creation science attempts to prove that the world is six to ten thousand years old because the Bible, if literally read, would have the earth only that old.

Q: Dr. Ruse, having reviewed the definition and elements of creation science and evolution science, the dual models, do you have a professional opinion about the dual-model approach adopted in the statute?

A: Yes. The central premise of the dual-model approach is that an argument against evolution science must be an argument in support of creation science. That premise is without any logical basis whatsoever. The reason is that the so-called scientific evidences against evolution are not really against evolution at all. At very best, each scientific evidence cited by creation science suggests no more than some uncertainty about some small part of one discipline in an interrelated scheme of about a dozen scientific disciplines that comprise the content of what is defined in the statute as evolution science.

Q: Can you give us any example of a creation-science argument that illustrates your point?

A: Yes. I think the issue of abiogenesis is a good example because science does not yet know how life on this planet began. Abiogenesis is one working theory that some scientists are considering.

Creation scientists take great comfort in the present inability of science to fabricate life in a laboratory. They dismiss the whole effort for its present failure, conclude that naturalistic processes were not sufficient to create life, and cite that present failure as scientific evidence in support of creation.

The silliness of that conclusion is apparent. First, that scientists cannot *now* create life in the laboratory from naturalistic processes does not mean that, as knowledge and technology advances, life will not be so created in the future. And even if life could never be created in the laboratory, that does not mean that creation is the only alternative. Even today there are other theories. And advanced scientific learning may suggest still more other alternative explanations through naturalistic mechanisms for the origin of life. Science does not know the *right* answer to most things. But that ignorance does not prove creation, rather it proves only that current knowledge is incomplete.

Q: Dr. Ruse, does your learning in the history of science suggest any parallels with the example you have been discussing about the scientific attempt to understand the origins of life?

A: Yes, absolutely. We can see throughout the history of science that when people do not understand the world around them, some always attribute the unknown to the workings of a god.

The ancient Greeks' ignorance of celestial mechanics led them to believe that the sun was pulled across the sky by a god. And indeed that all natural phenomena were directed by gods in the sky at the mere mortals that inhabited the earth.

Similarly, for thousands of years, up until only about a century ago, people believed that disease was inflicted by God: divine retribution for sin.

Virtually all scientific learning—astronomy, biology, chemistry, physics, of course, psychology—was at one time profoundly influenced by the belief that what could not be understood must reflect the intervention of God. Consequently, perhaps the single most important element of our modern understanding of science is that science is limited to naturalistic processes that do not rely on, or permit, the intervention of supernatural forces. If something is not now understood, then that means only that more work must be done. When faced with the unknown, scientists today do not just throw up their hands and say the Creator must have done it.

WHAT IS SCIENCE?

Q: Dr. Ruse, you refer to a modern understanding of what science is. Will you please describe to the court your understanding, as a historian and philosopher of science, of what science is today?

A: Mr. Novik, that is a misleading question, in part because science is not one thing. For thousands of years, beginning with the Greek philosophers, people have been seriously thinking about that question. Needless to say, many different notions have been posited, considered, and criticized. Therefore, I cannot give you one single definition of science. Instead, I can describe what philosophers of science generally consider some of the attributes that characterize scientific thinking and methodology today.

Q: Would you please do that for the court.

A: To begin with, as I have mentioned earlier, the most important characteristic of modern science is that it depends entirely on the operation of blind, unchanging regularities in nature. We call those regularities "natural laws." Thus, scientists seek to understand the empirical world by reference to natural law and naturalistic processes. Therefore, any reliance on a supernatural force, a Creator intervening in a natural world by supernatural processes, is necessarily not science.

Q: If a theory involving the supernatural intervention of a Creator is not science, then what is it?

A: It is religion. In my opinion, reliance on the acts of a Creator is inherently religious. It is not necessarily wrong. It is just a different perspective. It has its place, just as science has its place, but it is not science.

Q: In addition to exclusive reliance on natural law, are there any other characteristics commonly recognized among philosophers of science as distinguishing science and scientific thought?

A: Yes, in my opinion science must also be at least explanatory, testable, and tentative.

Q: Tell the court what you mean in saying that science must be explanatory.

A: Science attempts to explain the empirical world in terms of natural law and naturalistic processes. A scientific explanation may try to explain how one phenomenon follows in a tight and definite way, as a result of the working of natural law. A scientific explanation may also attempt to explain how two different phenomena are related to each other by the operation of natural law.

   For example, a scientist might attempt to explain the path of a basketball from player to hoop. A scientific explanation would focus on the physical laws of motion as they apply to the variables in the situation: in this case the force and direction of the throw as well as the resistance of gravity and friction. A scientific explanation would also involve a consistency that would transcend the one-of-a-kind throw of a basketball and relate that one throw to other basketball shots as well as to football passes and the flight of a rocket.

Q: Please explain what you mean by testable.

A: Testability is related to explanation.

Genuine science must lay itself open to testing against empirical reality. A scientific explanation leads to certain inferences—predictions, if you will—about what to expect the next time the explanation is applied to the empirical evidence. That ability to compare the evidence against the proposed explanation is the essence of scientific testability.

This characteristic of testability also suggests another consideration that does not usually come up but is peculiarly relevant to creation science. If a theory is to be taken seriously as science, it must have some positive confirming evidence in support of it that can be tested. If a theory does not have any empirical evidence in support of it, it cannot be tested and, therefore, cannot be science.

Q: In connection with your explanation of testable, can you explain the meaning of falsifiability?

A: Falsifiability is another way of looking at what I have called testability.

Falsifiability means that a theory is scientific only if there is some fact or observation that, if true, would tend to disprove, or falsify, the theory. In other words, science must be subject to falsification.

Falsifiability was a philosophical analysis developed by Karl Popper. The reason Popper focused on falsification, rather than verification, is that it is theoretically never possible to verify a theory with absolute certainty. In order for a theory to be absolutely verified it would have to be proven in every possible circumstance. Inasmuch as there are an infinite number of circumstances in which any theory could be tested, it would be impossible ever to verify a theory fully and unequivocally.

However, if a theory were false in any one circumstance, the theory would be falsified. (Of course, the theory might be modified or the theory might be discarded in favor of a new theory, but the original theory tested would be false.) Therefore, Popper argued that the true method of scientific inquiry is the attempt to falsify. And unless something was subject to falsification it was not scientific.

Q: Can you give us an example of how falsification works?

A: Yes, using an example in paleontology, evolutionary theory predicts an ordering of fossils in fossil record: less complex to more complex. If the fossil record contained evidences of all species—from trilobites to dinosaurs to man—at every level of the geologic column, that finding would tend to falsify evolution. That has not been the case, and therefore evolution has not been falsified. But it is subject to falsification.

Similarly, if modern technological advances in microbiology had resulted in the discovery that there were absolutely no similarities between the genetic material of different organisms, that finding would tend to falsify the evolutionary theory of common descent. Once again, that theory of common descent has not been falsified by the evidence. But the theory is nonetheless falsifiable.

Q: How does falsifiability relate to testability?

A: In attempting to falsify a theory, scientists test that theory in every imaginable circumstance. If such tests are negative, then the theory is falsified. But if the tests are positive then the theory is confirmed, at least in the circumstance of that test. And, as a theory is confirmed in more and more circumstances, it is given increasing weight by science. Of course, as I explained earlier, as a theoretical matter such verification will never prove the theory with absolute certainty. However, as a practical matter such confirming evidence may be so overwhelming, the theory may have been tested and validated under so many different circumstances, that scientists generally accept it as true. Such is the state of the theory that evolution has actually happened. There is not yet such a consensus with respect to the theories about how evolution has happened.

Q: The last characteristic you mentioned was that science is tentative. Would you please explain that concept.

A: The tentativeness of science follows directly from its testability. Science knows no ultimate truth not subject to revision. If a theory fails the test, then it is rejected.

Q: You testified earlier with respect to testability that sometimes a theory is confirmed to such a degree that it is accepted as true, for practical purposes. If that is so, how can science be tentative?

A: By saying that science is tentative, I do not mean that scientists must start from the beginning every Monday morning. For purposes of continuing their research and expanding their knowledge, scientists regularly accept many theories that have been proven sufficiently valid to warrant day-to-day reliance on their validity. That is not to say that such accepted theories can never be challenged. If accumulated evidence cannot be explained by the theory, then scientists regularly reassess their initial underlying theoretical assumptions.

For example, up until the early part of this century, Newtonian physics enjoyed virtually universal acceptance. Within a very short period of time, however, it was displaced by Einstein's theories, which better explained the empirical world.

Of course, a scientist does not have to discard the whole theoretical framework of his discipline merely because he finds some aberration. First, the peculiarity will be studied and explored and exposed and questioned, and certainly an effort will be made to fix it within the existing theoretical framework. Only then, if it is found hopelessly inconsistent with existing theory, will new ideas be formulated to take account of the discrepancy. That is precisely what happened with Newton's laws of motion. In certain respects they were seen to be inadequate, they were not able to explain the empirical evidence. Scientists did not immediately throw out Newtonian theory. They spent a considerable amount of time attempting to fit the facts within that theory. It was only after Einstein formulated a new theory, which was an alternative to Newtonian physics and which did explain the evidence, that scientists accepted the displacement of Newton's laws.

I like to think of tentativeness in science as analogous to the tentativeness in the judicial system. When a person is accused of crime, he is assumed innocent until proven guilty. The judicial system spends a considerable amount of time and energy examining and questioning the evidence and being very tentative about the guilt of the defendant. At some point, however, the case is presented to the jury, the evidence is assessed, and if the jury is convinced beyond a reasonable doubt, the defendant is convicted. At that point the judicial system ceases to look at the evidence; it does not reassess guilt every day. Of course, for sufficient cause, if the evidence is compelling, the conviction can be reexamined even though the jury was convinced beyond a reasonable doubt. But the extent to which the jury's initial judgment is subject to reexamination is a function of the strength of the contrary evidence. If a person is convicted, and then the very next week an identical crime is committed by another person actually caught doing it who happens to look just like the person convicted for the first crime, that is much stronger reason to question the determination than a bit of tangential evidence found twenty years after the fact.

You change your mind much more quickly in one case than the other. Thus, "tentative" means that a working conclusion or a theory is always open to change, depending upon the strength of the contrary evidence. Science must strive to be an honest, rigorous, and objective enterprise that constantly self-corrects to overcome personal bias and misconception. In pursuit of those goals, scientists search out new facts and generate new information to test and challenge their theories. Scientists subject their work to the logical and critical assessment of their colleagues. In other words, science rejects reliance on a priori assumptions that are not subject to scientific investigation.

Q: Dr. Ruse, having examined the creationist literature at great length, do you have a professional opinion about whether creation science measures up to the standards and characteristics of science that you have just been describing?

A: Yes, I do.

In my opinion creation science does not have those attributes that distinguish science from other endeavors.

Q: Would you please explain why you think it does not.

A: First, and most importantly, creation science necessarily looks to the supernatural acts of a Creator. According to creation-science theory, the Creator has intervened in supernatural ways using supernatural forces. Moreover, because the supernatural forces are the acts of a Creator, that is, the acts of God, they are not subject to scientific investigation or understanding. This nonscientific aspect of creation science emerges quite clearly from the creation-science literature I have read. [Appropriate quotes.]

Q: Do you think that creation science is explanatory?

A: No, because creation science relies on the acts of a Creator, whose purposes cannot be understood by us mortals, creation science has no explanatory power.

Let me give you an example in the evolution/creation controversy. The example pertains to homologies. Homologies are structural similarities between different organisms of widely different species. For example [as can be seen in these two pictures] the wing of a bat and the arm of a man are strikingly similar. These similarities cannot be explained by function: obviously, a man's arm and a bat's wing have different functions. Evolutionary theory gives an explanation for homologies. That explanation is: Organisms with homologies are descended from common ancestors. Creationism, on the other hand, does not give any explanation for homologies. The theory of creation is that a Creator created the bat's wing and man's arm independently. But that theory does not give any explanation of why they are similar. Nor can any explanation be provided because the purpose of the Creator can never be understood. The Creator could have made them similar; He could have made them different. Creation theory does not explain why He chose similarity rather than dissimilarity.

Q: But doesn't creation-science theory explain some things? For example, the eye is a remarkable structure. Creation science would say it was made by the Creator. Isn't that an explanation?

A: No, not in the scientific sense. In science an explanation must explain more than that for which it was invented. Saying God created the eye of a dog, for example, does not explain anything about why it is structured the way it is, why it works the way it does, and why it is similar to the eye of a cat in some respects and different in others.

Q: Do you think that creation science is testable?

A: Creation science is neither testable nor tentative. Indeed, an attribute of creation science that distinguishes it quite clearly from science is that it is absolutely certain about all of the answers. And considering the magnitude of the questions it addresses—the origins of man, life, the earth, and the universe—that certainty is all the more revealing. Whatever the contrary evidence, creation science never accepts that its theory is falsified. This is just the opposite of tentativeness and makes a mockery of testing. [Quotes to illustrate.]

Q: Do you find that creation science measures up to the methodological considerations you described?

A: Creation science is woefully lacking in this regard as well. First of all, the dogmatic certainty of creation science I have described earlier necessarily precludes the requisite objectivity. For example, members of the Creation Research Society take an oath that is an outright admission that their religious views supersede scientific objectivity. [Read oath.]

Second, and most regrettably, I have found innumerable instances of outright dishonesty, deception, and distortion used to advance creation-science arguments. [Quotes to illustrate.]

Q: Dr. Ruse, do you have an opinion to a reasonable degree of professional certainty about whether creation science is science?
A: Yes.

Q: What is your opinion?
A: In my opinion creation science is not science.

Q: What do you think it is?
A: As someone also trained in the philosophy of religion, in my opinion creation science is religion.

NOTE

1. In order to provide a constant reminder of the connection, all our witnesses should refer *only* to creation science and creation scientists. I think we make up what we lose in propaganda by enhancing the relevance of creation-science literature. If the witness is uncomfortable, a general disclaimer at the outset ("just because they call it 'science' does not mean I concede it is") should suffice.

20

United States District Court Opinion
McLean v. Arkansas

Judge William R. Overton

JUDGMENT

Pursuant to the Court's Memorandum Opinion filed this date, judgment is hereby entered in favor of the plaintiffs and against the defendants. The relief prayed for is granted.

Dated this January 5, 1982.

INJUNCTION

Pursuant to the Court's Memorandum Opinion filed this date, the defendants and each of them and all their servants and employees are hereby permanently enjoined from implementing in any manner Act 590 of the Acts of Arkansas of 1981.

It is so ordered this January 5, 1982.

MEMORANDUM OPINION

Introduction

On March 1, 1981, the Governor of Arkansas signed into law Act 590 of 1981, entitled the "Balanced Treatment for Creation-Science and Evolution-Science Act." The Act is codified as Ark. Stat. Ann. §80-1663, *et seq.* (1981 Suppl.). Its essential mandate is stated in its first sentence: "Public schools within this State shall give balanced treatment to creation-science and to evolution-science." On May 27, 1981, this suit was filed[1] challenging the constitutional validity of Act 590 on three distinct grounds.

First, it is contended that Act 590 constitutes an establishment of religion prohibited by the First Amendment to the Constitution, which is made applicable to the states by the Fourteenth Amendment. Second, the plaintiffs argue the Act violates a right to academic freedom which they say is guaranteed to students and teachers by the Free Speech clause of the First Amendment. Third, plaintiffs allege the Act is impermissibly vague and thereby violates the Due Process Clause of the Fourteenth Amendment.

The individual plaintiffs include the resident Arkansas Bishops of the United Methodist, Episcopal, Roman Catholic and African Methodist Episcopal Churches, the principal official of the Presbyterian Churches in Arkansas, other United Methodist, Southern Baptist and Presbyterian clergy, as well as several persons who sue as parents and next friends of minor children attending Arkansas public schools. One plaintiff is a high school biology teacher. All are also Arkansas taxpayers. Among the organizational plaintiffs are the American Jewish Congress, the Union of American Hebrew Congregations, the American Jewish Committee, the Arkansas Education Association, the National Association of Biology Teachers and the National Coalition for Public Education and Religious Liberty, all of which sue on behalf of members living in Arkansas.[2]

The defendants include the Arkansas Board of Education and its members, the Director of the Department of Education, and the State Textbooks and Instructional Materials Selecting Committee.[3] The Pulaski County Special School District and its Directors and Superintendent were voluntarily dismissed by the plaintiffs at the pre-trial conference held October 1, 1981.

The trial commenced December 7, 1981, and continued through December 17, 1981. This Memorandum Opinion constitutes the Court's findings of fact and conclusions of law. Further orders and judgment will be in conformity with this opinion.

I

There is no controversy over the legal standards under which the Establishment Clause portion of this case must be judged. The Supreme Court has on a number of occasions expounded on the meaning of the clause, and the pronouncements are clear. Often the issue has arisen in the context of public education, as it has here. In *Everson v. Board of Education,* Justice Black stated:

> The "establishment of religion" clause of the First Amendment means at least this: Neither a state nor the Federal Government can set up a church. Neither can pass laws which aid one religion, aid all religions, or prefer one religion over another. Neither can force nor influence a person to go to or to remain

away from church against his will or force him to profess a belief or disbelief in any religion. No person can be punished for entertaining or professing religious beliefs or disbeliefs, for church attendance or non-attendance. No tax, large or small, can be levied to support any religious activities or institutions, whatever they may be called, or whatever form they may adopt to teach or practice religion. Neither a state nor the Federal Government can, openly or secretly, participate in the affairs of any religious organizations or groups and *vice versa*. In the words of Jefferson, the clause . . . was intended to erect "a wall of separation between church and State." (*Everson* at 15-16.)

The Establishment Clause thus enshrines two central values: voluntarism and pluralism. And it is in the area of the public schools that these values must be guarded most vigilantly.

Designed to serve as perhaps the most powerful agency for promoting cohesion among a heterogeneous democratic people, the public school must keep scrupulously free from entanglement in the strife of sects. The preservation of the community from divisive conflicts, of Government from irreconcilable pressures by religious groups, of religion from censorship and coercion however subtly exercised, requires strict confinement of the State to instruction other than religious, leaving to the individual's church and home, indoctrination in the faith of his choice. [*McCollum v. Board of Education* at 216-217 (Opinion of Justice Frankfurter, joined by Justices Jackson, Burton, and Rutledge).]

The specific formulation of the establishment prohibition has been refined over the years, but its meaning has not varied from the principles articulated by Justice Black in *Everson*. In *Abbington School District v. Schempp*, Justice Clark stated that "to withstand the strictures of the Establishment Clause there must be a secular legislative purpose and a primary effect that neither advances nor inhibits religion." The Court found it quite clear that the First Amendment does not permit a state to require the daily reading of the Bible in public schools, for "[s]urely the place of the Bible as an instrument of religion cannot be gainsaid" (*Idem* at 224). Similarly, in *Engel v. Vitale*, the Court held that the First Amendment prohibited the New York Board of Regents from requiring the daily recitation of a certain prayer in the schools. With characteristic succinctness, Justice Black wrote, "Under [the First] Amendment's prohibition against governmental establishment of religion, as reinforced by the provisions of the Fourteenth Amendment, government in this country, be it state or federal, is without power to prescribe by law any particular form of prayer which is to be used as an official prayer in carrying on any program of governmentally sponsored religious activity" (*Idem* at 430). Black also identified the objective at which the Establishment Clause was aimed: "Its first and most immediate purpose rested on the belief that a union of governent and religion tends to destroy government and to degrade religion" (*Idem* at 431).

Most recently, the Supreme Court has held that the clause prohibits a state from requiring the posting of the Ten Commandments in public school classrooms for the same reasons that officially imposed daily Bible reading is prohibited (*Stone v. Graham*). The opinion in *Stone* relies on the most recent formulation of the Establishment Clause test, that of *Lemon v. Kurtzman*:

> First, the statute must have a secular legislative purpose; second, its principal or primary effect must be one that neither advances nor inhibits religion . . . ; finally, the statute must not foster "an excessive government entanglement with religion." (*Stone* at 40.)

It is under this three-part test that the evidence in this case must be judged. Failure on any of these grounds is fatal to the enactment.

II

The religious movement known as Fundamentalism began in nineteenth-century America as part of evangelical Protestantism's response to social changes, new religious thought, and Darwinism. Fundamentalists viewed these developments as attacks on the Bible and as responsible for a decline in traditional values.

The various manifestations of Fundamentalism have had a number of common characteristics,[4] but a central premise has always been a literal interpretation of the Bible and a belief in the inerrancy of the Scriptures. Following World War I, there was again a perceived decline in traditional morality, and Fundamentalism focused on evolution as responsible for the decline. One aspect of their efforts, particularly in the South, was the promotion of statutes prohibiting the teaching of evolution in public schools. In Arkansas, this resulted in the adoption of Initiated Act 1 of 1929.[5]

Between the 1920s and early 1960s, anti-evolutionary sentiment had a subtle but pervasive influence on the teaching of biology in public schools. Generally, textbooks avoided the topic of evolution and did not mention the name of Darwin. Following the launch of the Sputnik satellite by the Soviet Union in 1957, the National Science Foundation funded several programs designed to modernize the teaching of science in the nation's schools. The Biological Sciences Curriculum Study (BSCS), a nonprofit organization, was among those receiving grants for curriculum study and revision. Working with scientists and teachers, BSCS developed a series of biology texts which, although emphasizing different aspects of biology, incorporated the theory of evolution as a major theme. The success of the BSCS effort is shown by the fact that fifty percent of American school children currently use BSCS books directly and the curriculum is incorporated in directly in virtually all biology texts (Testimony of Mayer, Nelkin; Plaintiffs' Exhibit 1).[6]

In the early 1960s, there was again a resurgence of concern among Fundamentalists about the loss of traditional values and a fear of growing secularism in society. The Fundamentalist movement became more active and has steadily grown in numbers and political influence. There is an emphasis among current Fundamentalists on the literal interpretation of the Bible and the Book of Genesis as the sole source of knowledge about origins.

The term "scientific creationism" first gained currency around 1965 following publication of *The Genesis Flood* in 1961 by Whitcomb and Morris. There is undoubtedly some connection between the appearance of the BSCS texts emphasizing evolutionary thought and efforts by Fundamentalists to attack the theory (Testimony of Mayer).

In the 1960s and early 1970s, several Fundamentalist organizations were formed to promote the idea that the Book of Genesis was supported by scientific data. The terms "creation science" and "scientific creationism" have been adopted by these Fundamentalists as descriptive of their study of creation and the origins of man. Perhaps the leading creationist organization is the Institute for Creation Research (ICR), which is affiliated with the Christian Heritage College and supported by the Scott Memorial Baptist Church in San Diego, California. The ICR, through the Creation-Life Publishing Company, is the leading publisher of creation science material. Other creation science organizations include the Creation Science Research Center (CSRC) of San Diego and the Bible Science Association of Minneapolis, Minnesota. In 1963, the Creation Research Society (CRS) was formed from a schism in the American Scientific Affiliation (ASA). It is an organization of literal Fundamentalists[7] who have the equivalent of a master's degree in some recognized area of science. A purpose of the organization is "to reach all people with the vital message of the scientific and historic truth about creation" (Nelkin, *The Science Textbook Controversies and the Politics of Equal Time,* p. 66). Similarly, the CSRC was formed in 1970 from a split in the CRS. Its aim has been "to reach the 63 million children of the United States with the scientific teaching of Biblical creationism" (*Idem* at 69).

Among creationist writers who are recognized as authorities in the field by other creationists are Henry M. Morris, Duane Gish, G. E. Parker, Harold S. Slusher, Richard B. Bliss, John W. Moore, Martin E. Clark, W. L. Wysong, Robert E. Kofahl, and Kelly L. Segraves. Morris is Director of ICR, Gish is Associate Director, and Segraves is associated with CSRC.

Creationists view evolution as a source of society's ills, and the writings of Morris and Clark are typical expressions of that view.

> Evolution is thus not only anti-Biblical and anti-Christian, but it is utterly unscientific and impossible as well. But it has served effectively as the pseudo-scientific basis of atheism, agnosticism, socialism, fascism, and numerous other false and dangerous philosophies over the past century. [Morris and Clark, *The Bible Has the Answer* (Plaintiffs' Exhibit 31 and Plaintiffs' Pretrial Exhibit 89).][8]

Creationists have adopted the view of Fundamentalists generally that there are only two positions with respect to the origins of the earth and life: belief in the inerrancy of the Genesis story of creation and of a worldwide flood as fact, or belief in what they call evolution.

Henry Morris has stated, "It is impossible to devise a legitimate means of harmonizing the Bible with evolution" [Morris, "Evolution and the Bible," *ICR Impact Series,* Number 5 (undated, unpaged), quoted in Mayer, Plaintiffs' Exhibit 8, p. 3]. This dualistic approach to the subject of origins permeates the creationist literature.

The creationist organizations consider the introduction of creation science into the public schools part of their ministry. The ICR has published at least two pamphlets[9] containing suggested methods for convincing school boards, administrators and teachers that creationism should be taught in public schools. The ICR has urged its proponents to encourage school officials to voluntarily add creationism to the curriculum.[10]

Citizens For Fairness In Education is an organization based in Anderson, South Carolina, formed by Paul Ellwanger, a respiratory therapist who is trained in neither law nor science. Mr. Ellwanger is of the opinion that evolution is the forerunner of many social ills, including Nazism, racism and abortion (Ellwanger Deposition, pp. 32-34). About 1977, Ellwanger collected several proposed legislative acts with the idea of preparing a model state act requiring the teaching of creationism as science in opposition to evolution. One of the proposals he collected was prepared by Wendell Bird, who is now a staff attorney for ICR.[11] From these various proposals, Ellwanger prepared a "model act" which calls for "balanced treatment" of "scientific creationism" and "evolution" in public schools. He circulated the proposed act to various people and organizations around the country.

Mr. Ellwanger's views on the nature of creation science are entitled to some weight since he personally drafted the model act which became Act 590. His evidentiary deposition with exhibits and unnumbered attachments (produced in response to a subpoena *duces tecum*) speaks to both the intent of the Act and the scientific merits of creation science. Mr. Ellwanger does not believe creation science is a science. In a letter to Pastor Robert E. Hays he states, "While neither evolution nor creation can qualify as a scientific theory, and since it is virtually impossible at this point to educate the whole world that evolution is not a true scientific theory, we have freely used these terms—the evolution theory and the theory of scientific creationism—in the bill's text" (Unnumbered attachment to Ellwanger Deposition, p. 2). He further states in a letter to Mr. Tom Bethell, "As we examine evolution (remember, we're not making any scientific claims for creation, but we are challenging evolution's claim to be scientific) . . ." (Unnumbered attachment to Ellwanger Deposition, p. 1).

Ellwanger's correspondence on the subject shows an awareness that Act 590 is a religious crusade, coupled with a desire to conceal this fact. In a letter to State Senator Bill Keith of Louisiana, he says, "I view this whole battle as one between God and anti-God forces, though I know there are a large number of evolutionists who believe in God." And further, ". . . it behooves

Satan to do all he can to thwart our efforts and confuse the issue at every turn." Yet Ellwanger suggests to Senator Keith, "If you have a clear choice between having grassroots leaders of this statewide bill promotion effort to be ministerial or non-ministerial, be sure to opt for the non-ministerial. It does the bill effort no good to have ministers out there in the public forum and the adversary will surely pick at this point . . . Ministerial persons can accomplish a tremendous amount of work from behind the scenes, encouraging their congregations to take the organizational and P.R. initiatives. And they can lead their churches in storming Heaven with prayers for help against so tenacious an adversary" (Unnumbered attachment to Ellwanger Deposition, p. 1).

Ellwanger shows a remarkable degree of political candor, if not finesse, in a letter to State Senator Joseph Carlucci of Florida:

> 2. It would be very wise, if not actually essential, that all of us who are engaged in this legislative effort be careful not to present our position and our work in a religious framework. For example, in written communications that might somehow be shared with those other persons whom we may be trying to convince, it would be well to exclude our own personal testimony and/or witness for Chirst, but rather, if we are so moved, to give that testimony on a separate attached note. (Unnumbered attachment to Ellwanger Deposition, p. 1.)

The same tenor is reflected in a letter by Ellwanger to Mary Ann Miller, a member of FLAG (Family, Life, America under God) who lobbied the Arkansas Legislature in favor of Act 590:

> . . . . we'd like to suggest that you and your co-workers be very cautious about mixing creation-science with creation-religion . . . Please urge your co-workers not to allow themselves to get sucked into the "religion" trap of mixing the two together, for such mixing does incalculable harm to the legislative thrust. It could even bring public opinion to bear adversely upon the higher courts that will eventually have to pass judgment on the constitutionality of this new law. [Exhibit 1 to Miller Deposition.]

Perhaps most interesting, however, is Mr. Ellwanger's testimony in his deposition as to his strategy for having the model act implemented:

> Q. You're trying to play on other people's religious motives.
> A. I'm trying to play on their emotions, love, hate, their likes, dislikes, because I don't know any other way to involve, to get humans to become involved in human endeavors. I see emotions as being a healthy and legitimate means of getting people's feelings into action, and . . . I believe that the predominance of population in America that represents the greatest potential for taking some kind of action in this area is a Christian community. I see the Jewish community as far less potential in taking action . . . but I've seen a lot of interest among Christians and I feel, why not exploit that to get the bill going if that's what it takes. (Ellwanger Deposition, pp. 146—147.)

Mr. Ellwanger's ultimate purpose is revealed in the closing of his letter to Mr. Tom Bethell: "Perhaps all this is old hat to you, Tom, and if so, I'd appreciate your telling me so and perhaps where you've heard it before— the idea of killing evolution instead of playing these debating games that we've been playing for nigh over a decade already" (Unnumbered attachment of Ellwanger Deposition, p. 3).

If was out of this milieu that Act 590 emerged. The Reverend W. A. Blount, a Biblical literalist who is pastor of a church in the Little Rock area and was, in February, 1981, chairman of the Greater Little Rock Evangelical Fellowship, was among those who received a copy of the model act from Ellwanger.[12]

At Reverend Blount's request, the Evangelical Fellowship unanimously adopted a resolution to seek introduction of Ellwanger's act in the Arkansas Legislature. A committee composed of two ministers, Curtis Thomas and W. A. Young, was appointed to implement the resolution. Thomas obtained from Ellwanger a revised copy of the model act which he transmitted to Carl Hunt, a business associate of Senator James L. Holsted, with the request that Hunt prevail upon Holsted to introduce the act.

Holsted, a self-described "born again" Christian Fundamentalist, introduced the act in the Arkansas Senate. He did not consult the State Department of Education, scientists, science educators, or the Arkansas Attorney General.[13] The Act was not referred to any Senate committee for hearing and was passed after only a few minutes' discussion on the Senate floor. In the House of Representatives, the bill was referred to the Education Committee which conducted a perfunctory fifteen minute hearing. No scientist testified at the hearing, nor was any representative from the State Department of Education called to testify.

Ellwanger's model act was enacted into law in Arkansas as Act 590 without amendment or modification other that minor typographical changes. The legislative "findings of fact" in Ellwanger's act and Act 590 are identical, although no meaningful fact-finding process was employed by the General Assembly.

Ellwanger's efforts in preparation of the model act and campaign for its adoption in the states were motivated by his opposition to the theory of evolution and his desire to see the Biblical version of creation taught in the public schools. There is no evidence that the pastors, Blount, Thomas, Young, or the Greater Little Rock Evangelical Fellowship were motivated by anything other than their religious convictions when proposing its adoption or during their lobbying efforts in its behalf. Senator Holsted's sponsorship and lobbying efforts in behalf of the Act were motivated solely by his religious beliefs and desire to see the Biblical version of creation taught in the public schools.[14]

The State of Arkansas, like a number of states whose citizens have relatively homogeneous religious beliefs, has a long history of official opposition to evolution which is motivated by adherence to Fundamentalist beliefs in the inerrancy of the Book of Genesis. This history is documented in Justice Fortas' opinion in *Epperson v. Arkansas,* which struck down Initiated Act 1 of 1929, Ark. Stat. Ann. §§80-1627-1628, prohibiting the teaching of the theory of evolution.

To this same tradition may be attributed Initiated Act 1 of 1930, Ark. Stat. Ann. §80-1606 (Repealed 1980), requiring "the reverent daily reading of a portion of the English Bible" in every public school classroom in the State.[15]

It is true, as defendants argue, that courts should look to legislative statements of a statute's purpose in Establishment Clause cases and accord such pronouncements great deference (See, e.g., *Committee for Public Education & Religious Liberty v. Nyquist* at 773 and *McGowan v. Maryland* at 445). Defendants also correctly state the principle that remarks by the sponsor or author of a bill are not considered controlling in analyzing legislative intent (See, e.g., *United States v. Emmons* and *Chrysler Corporation v. Brown*).

Courts are not bound, however, by legislative statements of purpose or legislative disclaimers *(Stone v. Graham, Abbington School District v. Schempp)*. In determining the legislative purpose of a statute, courts may consider evidence of the historical context of the Act *(Epperson v. Arkansas)*, the specific sequence of events leading up to passage of the Act, departures from normal procedural sequences, substantive departures from the normal *(Village of Arlington Heights v. Metropolitan Housing Corp.)*, and contemporaneous statements of the legislative sponsor *(Federal Energy Administration v. Algonquin SNG, Inc.* at 564).

The unusual circumstances surrounding the passage of Act 590, as well as the substantive law of the First Amendment, warrant an inquiry into the stated legislative purposes. The author of the Act had publicly proclaimed the sectarian purpose of the proposal. The Arkansas residents who sought legislative sponsorship of the bill did so for a purely sectarian purpose. These circumstances alone may not be particularly persuasive, but when considered with the publicly announced motives of the legislative sponsor made contemporaneously with the legislative process; the lack of any legislative investigation, debate or consultation with any educators or scientists; the unprecedented intrusion in school curriculum;[16] and official history of the State of Arkansas on the subject, it is obvious that the statement of purposes has little, if any, support in fact. The State failed to produce any evidence which would warrant an inference or conclusion that at any point in the process anyone considered the legitimate educational value of the Act. It was simply and purely an effort to introduce the Biblical version of creation into the public school curricula. The only inference which can be drawn from these circumstances is that the Act was passed with the specific purpose by the General Assembly of advancing religion. The Act therefore fails the first prong of the three-pronged test, that of secular legislative purpose, as articulated in *Lemon v. Kurtzman* and *Stone v. Graham*.

III

If the defendants are correct and the Court is limited to an examination of the language of the Act, the evidence is overwhelming that both the purpose and effect of Act 590 is the advancement of religion in the public schools.

Section 4 of the Act provides:

Definitions. As used in this Act:

(a) "Creation-science" means the scientific evidences for creation and inferences from those scientific evidences. Creation-science includes the scientific evidences and related inferences that indicate: (1) Sudden creation of the universe, energy, and life from nothing; (2) The insufficiency of mutation and natural selection in bringing about development of all living kinds from a single organism; (3) Changes only within fixed limits of originally created kinds of plants and animals; (4) Separate ancestry for man and apes; (5) Explanation of the earth's geology by catastrophism, including the occurrence of a worldwide flood; and (6) A relatively recent inception of the earth and living kinds.

(b) "Evolution-science" means the scientific evidences for evolution and inferences from those scientific evidences. Evolution science includes the scientific evidences and related inferences that indicate: (1) Emergence by naturalistic processes of the universe from disordered matter and emergence of life from nonlife; (2) The sufficiency of mutation and natural selection in bringing about development of present living kinds from simple earlier kinds; (3) Emergence by mutation and natural selection of present living kinds from simple earlier kinds; (4) Emergence of man from a common ancestor with apes; (5) Explanation of the earth's geology and the evolutionary sequence by uniformitarianism; and (6) An inception several billion years ago of the earth and somewhat later of life.

(c) "Public schools" means public secondary and elementary schools.

The evidence establishes that the definition of "creation science" contained in 4(a) has as its unmentioned reference the first 11 chapters of the Book of Genesis. Among the many creation epics in human history, the account of sudden creation from nothing, or *creatio ex nihilo,* and subsequent destruction of the world by flood is unique to Genesis. The concepts of 4(a) are the literal Fundamentalists' view of Genesis. Section 4(a) is unquestionably a statement of religion, with the exception of 4(a)(2) which is a negative thrust aimed at what the creationists understand to be the theory of evolution.[17]

Both the concepts and wording of Section 4(a) convey an inescapable religiosity. Section 4(a)(1) describes "sudden creation of the universe, energy and life from nothing." Every theologian who testified, including defense witnesses, expressed the opinion that the statement referred to a supernatural creation which was performed by God.

Defendants argue that: (1) the fact that 4(a) conveys ideas similar to the literal interpretation of Genesis does not make it conclusively a statement of religion; (2) that reference to a creation from nothing is not necessarily a religious concept since the act only suggests a creator who has power, intelligence and a sense of design and not necessarily the attributes of love, compassion and justice;[18] and (3) that simply teaching about the concept of a creator is not a religious exercise unless the student is required to make a commitment to the concept of a creator.

The evidence fully answers these arguments. The ideas of 4(a)(1) are not merely similar to the literal interpretation of Genesis; they are identical and parallel to no other story of creation.[19]

The argument that creation from nothing in 4(a)(1) does not involve a supernatural deity has no evidentiary or rational support. To the contrary, "creation out of nothing" is a concept unique to Western religions. In traditional Western religious thought, the conception of a creator of the world is a conception of God. Indeed, creation of the world "out of nothing" is the ultimate religious statement because God is the only actor. As Dr. Langdon Gilkey noted, the Act refers to one who has the power to bring all the universe into existence from nothing. The only "one" who has this power is God.[20]

The leading creationist writers, Morris and Gish, acknowledge that the idea of creation described in 4(a)(1) is the concept of creation by God and make no pretense to the contrary.[21] The idea of sudden creation from nothing, or *creatio ex nihilo*, is an inherently religious concept (Testimony of Vawter, Gilkey, Geisler, Ayala, Blount, and Hicks).

The argument advanced by defendants' witness Dr. Norman Geisler, that teaching the existence of God is not religious unless the teaching seeks a commitment, is contrary to common understanding and contradicts settled case law *(Stone v. Graham, Abbington School District v. Schempp).*

The facts that creation science is inspired by the Book of Genesis and that Section 4(a) is consistent with a literal interpretation of Genesis leave no doubt that a major effect of the Act is the advancement of particular religious beliefs. The legal impact of this conclusion will be discussed further at the conclusion of the Court's evaluation of the scientific merit of creation science.

IV(A)

The approach to teaching "creation science" and "evolution science" found in Act 590 is identical to the two-model approach espoused by the Institute for Creation Research and is taken almost verbatim from ICR writings. It is an extension of Fundamentalists' view that one must either accept the literal interpretation of Genesis or else believe in the godless system of evolution.

The two-model approach of the creationists is simply a contrived dualism[22] which has no scientific factual basis or legitimate educational purpose. It assumes only two explanations for the origins of life and existence of man, plants, and animals: It was either the work of a creator or it was not. Application of these two models, according to creationists, and the defendants, dictates that all scientific evidence which fails to support the theory of evolution is necessarily scientific evidence in support of creationism and is, therefore, creation science "evidence" in support of Section 4(a).

IV(B)

The emphasis on origins as an aspect of the theory of evolution is peculiar to creationist literature. Although the subject of origins of life is within the province of biology, the scientific community does not consider origins of life a part of evolutionary theory. The theory of evolution assumes the existence of life and is directed to an explanation of *how* life evolved. Evolution does not presuppose the absence of a creator or God and the plain inference conveyed by Section 4 is erroneous.[23]

As a statement of the theory of evolution, Section 4(b) is simply a hodge-podge of limited assertions, many of which are factually inaccurate.

For example, although 4(b)(2) asserts, as a tenet of evolutionary theory, "the sufficiency of mutation and natural selection in bringing about the existence of present living kinds from simple earlier kinds," Drs. Ayala and Gould both stated that biologists know that these two processes do not account for all significant evolutionary change. They testified to such phenomena as recombination, the founder effect, genetic drift and the theory of punctuated equilibrium, which are believed to play important evolutionary roles. Section 4(b) omits any reference to these. Moreover, 4(b) utilizes the term "kinds" which all scientists said is not a word of science and has no fixed meaning. Additionally, the Act presents both evolution and creation science as "package deals." Thus, evidence critical of some aspect of what the creationists define as evolution is taken as support for a theory which includes a worldwide flood and a relatively young earth.[24]

IV(C)

In addition to the fallacious pedagogy of the two model approach, Section 4(a) lacks legitimate educational value because "creation science" as defined in that section is simply not science. Several witnesses suggested definitions of science. A descriptive definition was said to be that science is what is "accepted by the scientific community" and is "what scientists do." The obvious implication of this description is that, in a free society, knowledge does not require the imprimatur of legislation in order to become science.

More precisely, the essential characteristics of science are:
(1) It is guided by natural law;
(2) It has to be explanatory by reference to natural law;
(3) It is testable against the empirical world;
(4) Its conclusions are tentative, i.e., are not necessarily the final word; and
(5) It is falsifiable (Testimony of Ruse and other science witnesses).

Creation science as described in Section 4(a) fails to meet these essential characteristics. First, the section revolves around 4(a)(1) which asserts a sudden creation "from nothing." Such a concept is not science because it depends upon a supernatural intervention which is not guided by natural law. It is not explanatory by reference to natural law, is not testable, and is not falsifiable.[25]

If the unifying idea of supernatural creation by God is removed from Section 4, the remaining parts of the section explain nothing and are meaningless assertions.

Section 4(a)(2), relating to the "insufficiency of mutation and natural selection in bringing about development of all living kinds from a single organism," is an incomplete negative generalization directed at the theory of evolution.

Section 4(a)(3) which describes "changes only within fixed limits of originally created kinds of plants and animals" fails to conform to the essential characteristics of science for several reasons. First, there is no scientific definition of "kinds" and none of the witnesses was able to point to any scientific authority which recognized the term or knew how many "kinds" existed. One defense witness suggested there may be 100 to 10,000 different "kinds." Another believes there were "about 10,000, give or take a few thousand." Second, the assertion appears to be an effort to establish outer limits of changes within species. There is no scientific explanation for these limits which is guided by natural law and the limitations, whatever they are, cannot be explained by natural law.

The statement in 4(a)(4) of "separate ancestry of man and apes" is a bald assertion. It explains nothing and refers to no scientific fact or theory.[26]

Section 4(a)(5) refers to "explanation of the earth's geology by catastrophism, including the occurrence of a worldwide flood." This assertion completely fails as science. The Act is referring to the Noachian flood described in the Book of Genesis.[27] The creationist writers concede that *any* kind of Genesis Flood depends upon supernatural intervention. A worldwide flood as an explanation of the world's geology is not the product of natural law, nor can its occurrence be explained by natural law.

Section 4(a)(6) equally fails to meet the standards of science. "Relatively recent inception" has no scientific meaning. It can only be given meaning by reference to creationist writings which place the age at between 6,000 and 20,000 years because of the genealogy of the Old Testament. See, e.g., Plaintiffs' Exhibit 78, Gish (6,000 to 10,000); Plaintiffs' Exhibit 87, Segraves (6,000 to 20,000). Such a reasoning process is not the product of natural law; not explainable by natural law; nor is it tentative.

Creation science, as defined in Section 4(a), not only fails to follow the canons of defining scientific theory, it also fails to fit the more general descriptions of "what scientists think" and "what scientists do." The scientific community consists of individuals and groups, nationally and internationally, who work independently in such varied fields as biology, paleontology, geology and astronomy. Their work is published and subject to review and testing by their peers. The journals for publication are both numerous and varied. There is, however, not one recognized scientific journal which has published an article espousing the creation science theory described in Section 4(a). Some of the State's witnesses suggested that the scientific community was "close-minded" on the subject of creationism and that explained the lack of acceptance of the creation science arguments. Yet no witness produced a scientific article for which publication had been refused. Perhaps some members of the scientific community are resistant to new ideas. It is, however, inconceivable that such a loose knit group of independent thinkers in all the varied fields of science could, or would, so effectively censor new scientific thought.

The creationists have difficulty maintaining among their ranks consistency in the claim that creationism is science. The author of Act 590, Ellwanger, said that neither evolution nor creationism was science. He thinks both are religion. Duane Gish recently responded to an article in *Discover* critical of creationism by stating:

> Stephen Jay Gould states that creationists claim creation is a scientific theory. This is a false accusation. Creationists have repeatedly stated that neither creation nor evolution is a scientific theory (and each is equally religious). (Gish, letter to the editor of *Discover*, July 1981, Appendix 30 to Plaintiffs' Pretrial Brief.)

The methodology employed by creationists is another factor which is indicative that their work is not science. A scientific theory must be tentative and always subject to revision or abandonment in light of facts that are inconsistent with, or falsify, the theory. A theory that is by its own terms dogmatic, absolutist and never subject to revision is not a scientific theory.

The creationists' methods do not take data, weigh it against the opposing scientific data, and thereafter reach the conclusions stated in Section 4(a). Instead, they take the literal wording of the Book of Genesis and attempt to find scientific support for it. The method is best explained in the language of Morris in his book *Studies in the Bible and Science* at page 114 (Plaintiffs' Exhibit 31):

> . . . it is . . . quite impossible to determine anything about Creation through a study of present processes, because present processes are not creative in character. If man wishes to know anything about Creation (the time of Creation, the duration of Creation, the order of Creation, the methods of Creation, or anything else) his sole source of true information is that of divine revelation. God was there when it happened. We were not there . . . Therefore, we are completely limited to what God has seen fit to tell us, and this information is in His written Word. This is our textbook on the science of Creation! (Morris, Plaintiffs' Exhibit 312.)

The Creation Research Society employs the same unscientific approach to the issue of creationism. Its applicants for membership must subscribe to the belief that the Book of Genesis is "historically and scientifically true in all of the original autographs."[28] The Court would never criticize or discredit any person's testimony based on his or her religious beliefs. While anybody is free to approach a scientific inquiry in any fashion they choose, they cannot properly describe the methodology used as scientific, if they start with a conclusion and refuse to change it regardless of the evidence developed during the course of the investigation.

IV(D)

In efforts to establish "evidence" in support of creation science, the defendants relied upon the same false premise as the two model approach contained in Section 4, i.e., all evidence which criticized evolutionary theory was proof in support of creation science. For example, the defendants established that the mathematical probability of a chance chemical combination resulting in life from non-life is so remote that such an occurrence is almost beyond imagination. Those mathematical facts, the defendants argue, are scientific evidences that life was the product of a creator. While the statistical figures may be impressive evidence against the theory of chance chemical combinations as an explanation of origins, it requires a leap of faith to interpret those figures so as to support a complex doctrine which includes a sudden creation from nothing, a worldwide flood, separate ancestry of man and apes, and a young earth.

The defendants' argument would be more persuasive if, in fact, there were only two theories or ideas about the origins of life and the world. That there are a number of theories was acknowledged by the State's witnesses, Dr. Wickramasinghe and Dr. Geisler. Dr. Wickramasinghe testified at length in support of a theory that life on earth was "seeded" by comets which delivered genetic material and perhaps organisms to the earth's surface from interstellar dust far outside the solar system. The "seeding" theory further hypothesizes that the earth remains under the continuing influence of genetic material from space which continues to affect life. While Wickramasinghe's theory[29] about the origins of life on earth has not received general acceptance within the scientific community, he has, at least, used scientific methodology to produce a theory of origins which meets the essential characteristics of science.

Perhaps Dr. Wickramasinghe was called as a witness because he was generally critical of the theory of evolution and the scientific community, a tactic consistent with the strategy of the defense. Unfortunately for the defense, he demonstrated that the simplistic approach of the two model analysis of the origins of life is false. Furthermore, he corroborated the plaintiffs' witnesses by concluding that "no rational scientist" would believe the earth's geology could be explained by reference to a worldwide flood or that the earth was less than one million years old.

The proof in support of creation science consisted almost entirely of efforts to discredit the theory of evolution through a rehash of data and theories which have been before the scientific community for decades. The arguments asserted by creationists are not based upon new scientific evidence or laboratory data which has been ignored by the scientific community.

Robert Gentry's discovery of radioactive polonium haloes in granite and coalified woods is, perhaps, the most recent scientific work which the creationists use as argument for a "relatively recent inception" of the earth and a "worldwide flood." The existence of polonium haloes in granite and coalified wood is thought to be inconsistent with radiometric dating methods based upon constant radioactive decay rates. Mr. Gentry's findings were published almost ten years ago and have been the subject of some discussion in the scientific community. The discoveries have not, however, led to the formulation of any scientific hypothesis or theory which would explain a relatively recent inception of the earth or a worldwide flood. Gentry's discovery has been treated as a minor mystery which will eventually be explained. It may deserve further investigation, but the National Science Foundation has not deemed it to be of sufficient import to support further funding.

The testimony of Marianne Wilson was persuasive evidence that creation science is not science. Ms. Wilson is in charge of the science curriculum for Pulaski County Special School District, the largest school district in the State of Arkansas. Prior to the passage of Act 590, Larry Fisher, a science teacher in the District, using materials from the ICR, convinced the School Board that it should voluntarily adopt creation science as part of its science curriculum. The District Superintendent assigned Ms. Wilson the job of producing a creation science curriculum guide. Ms. Wilson's testimony about the project was particularly convincing because she obviously approached the assignment with an open mind and no preconceived notions about the subject. She had not heard of creation science until about a year ago and did not know its meaning before she began her research.

Ms. Wilson worked with a committee of science teachers appointed from the District. They reviewed practically all of the creationist literature. Ms. Wilson and the committee members reached the unanimous conclusion that creationism is not science; it is religion. They so reported to the Board. The Board ignored the recommendation and insisted that a curriculum guide be prepared.

In researching the subject, Ms. Wilson sought the assistance of Mr. Fisher who initiated the Board action and asked professors in the science departments of the University of Arkansas at Little Rock and the University of Central Arkansas[30] for reference material and assistance, and attended a workshop conducted at Central Baptist College by Dr. Richard Bliss of the ICR staff. Act 590 became law during the course of her work so she used Section 4(a) as a format for her curriculum guide.

Ms. Wilson found all available creationists' materials unacceptable because they were permeated with religious references and reliance upon religious beliefs.

It is easy to understand why Ms. Wilson and other educators find the creationists' textbook material and teaching guides unacceptable. The materials misstate the theory of evolution in the same fashion as Section 4(b) of the Act, with emphasis on the alternative mutually exclusive nature of creationism and evolution. Students are constantly encouraged to compare and make a choice between the two models, and the material is not presented in an accurate manner.

A typical example is *Origins* (Plaintiffs' Exhibit 76) by Richard B. Bliss, Director of Curriculum Development of the ICR. The presentation begins with a chart describing "preconceived ideas about origins" which suggests that some people believe that evolution is atheistic. Concepts of evolution, such as "adaptive radiation," are erroneously presented. At page 11, Figure 1.6, of the text, a chart purports to illustrate this "very important" part of the evolution model. The chart conveys the idea that such diverse mammals as a whale, bear, bat and monkey all evolved from a shrew through the process of adaptive radiation. Such a suggestion is, of course, a totally erroneous and misleading application of the theory. Even more objectionable, especially when viewed in light of the emphasis on asking the student to elect one of the models, is the chart presentation at page 17, Figure 1.6. That chart purports to illustrate the evolutionists' belief that man evolved from bacteria to fish to reptile to mammals and, thereafter, into man. The illustration indicates, however, that the mammal from which man evolved was *a rat.*

Biology, A Search For Order in Complexity[31] is a high school biology text typical of creationists' materials. The following quotations are illustrative:

> Flowers and roots do not have a mind to have purpose of their own; therefore, this planning must have been done for them by the Creator. (p. 12)

> The exquisite beauty of color and shape in flowers exceeds the skill of poet, artist, and king. Jesus said (from Matthew's gospel), "Consider the lilies of the field, how they grow; they toil not, neither do they spin . . .". (p. 363)

The "public school edition" texts written by creationists simply omit Biblical references but the content and message remain the same. For example, *Evolution? The Fossils Say No!*[32] contains the following:

> Creation. By creation we mean the bringing into being by a supernatural Creator of the basic kinds of plants and animals by the process of sudden, or fiat, creation.
>   We do not know how the Creator created, what processes He used, *for He used processes which are not now operating anywhere in the natural universe.* This is why we refer to creation as Special Creation. We cannot discover by scientific investigation anything about the creative processes used by the Creator. (p. 40)

Gish's book also portrays the large majority of evolutionists as "materialistic atheists or agnostics."

Scientific Creationism (Public School Edition) by Morris is another text reviewed by Ms. Wilson's committee and rejected as unacceptable. The following quotes illustrate the purpose and theme of the text:

> *Foreword*

> Parents and youth leaders today, and even many scientists and educators, have become concerned about the prevalence and influence of evolutionary philosophy in modern curriculum. Not only is this system inimical to orthodox Christianity and Judaism, but also, as many are convinced, to a healthy society and true science as well. (p. iii)

*   *   *

The rationalist of course finds the concept of special creationism insufferably naive, even "incredible." Such a judgment, however, is warranted only if one categorically dismisses the existence of an omnipotent God. (p. 17)

Without using creationist literature, Ms. Wilson was unable to locate one genuinely scientific article or work which supported Section 4(a). In order to comply with the mandate of the Board she used such materials as an article from *Reader's Digest* about "atomic clocks" which inferentially suggested that the earth was less than 4-1/2 billion years old. She was unable to locate any substantive teaching material for some parts of Section 4 such as the worldwide flood. The curriculum guide which she prepared cannot be taught and has no educational value as science. The defendants did not produce any text or writing in response to this evidence which they claimed was usable in the public school classroom.[33]

The conclusion that creation science has no scientific merit or educational value as science has legal significance in light of the Court's previous conclusion that creation science has, as one major effect, the advancement of religion. The second part of the three-pronged test for establishment reaches only those statutes having as their *primary* effect the advancement of religion. Secondary effects which advance religion are not constitutionally fatal. Since creation science is not science, the conclusion is inescapable that the *only* real effect of Act 590 is the advancement of religion. The Act therefore fails both the first and second portions of the test in *Lemon v. Kurtzman.*

IV(E)

Act 590 mandates "balanced treatment" for creation science and evolution science. The Act prohibits instruction in any religious doctrine or references to religious writings. The Act is self-contradictory and compliance is impossible unless the public schools elect to forego significant portions of subjects such as biology, world history, geology, zoology, botany, psychology, anthropology, sociology, philosophy, physics and chemistry. Presently, the concepts of evolutionary theory as described in 4(b) permeate the public school textbooks. There is no way teachers can teach the Genesis account of creation in a secular manner.

The State Department of Education, through its textbook selection committee, school boards, and school administrators, will be required to constantly monitor materials to avoid using religious references. The school boards, administrators and teachers face an impossible task. How is the teacher to respond to questions about a creation suddenly and out of nothing? How will a teacher explain the concept of a relatively recent age of the earth? The answer is obvious because the only source of this information is ultimately contained in the Book of Genesis.

References to the pervasive nature of religious concepts in creation science texts amply demonstrate why State entanglement with religion is inevitable under Act 590. Involvement of the State in screening texts for impermissible religious references will require State officials to make delicate religious judgments. The need to monitor classroom discussion in order to uphold the Act's prohibition against religious instruction will necessarily involve administrators in questions concerning religion. These continuing involvements of State officials in questions and issues of religion create an excessive and prohibited entanglement with religion *(Brandon v. Board of Education* at 1230).

V

These conclusions are dispositive of the case and there is no need to reach legal conclusions with respect to the remaining issues. The plaintiffs raised two other issues questioning the constitutionality of the Act and, insofar as the factual findings relevant to these issues are not covered in the preceding discussion, the Court will address these issues. Additionally, the defendants raised two other issues which warrant discussion.

V(A)

First, plaintiff teachers argue the Act is unconstitutionally vague to the extent that they cannot comply with its mandate of "balanced" treatment without jeopardizing their employment. The argument centers around the lack of a precise definition in the Act for the word "balanced." Several witnesses expressed opinions that the word has such meanings as equal time, equal weight, or equal legitimacy. Although the Act could have been more explicit, "balanced" is a word subject to ordinary understanding. The proof is not convincing that a teacher using a reasonably acceptable understanding of the word and making a good faith effort to comply with the Act will be in jeopardy of termination. Other portions of the Act are arguably vague, such as the "relatively recent" inception of the earth and life. The evidence establishes, however, that relatively recent means from 6,000 to 20,000 years, as commonly understood in creation science literature. The meaning of this phrase, like Section 4(a) generally, is, for purposes of the Establishment Clause, all too clear.

V(B)

The plaintiffs' other argument revolves around the alleged infringement by the defendants upon the academic freedom of teachers and students. It is contended this unprecedented intrusion in the curriculum by the State prohibits teachers from teaching what they believe should be taught or requires them to teach that which they do not believe is proper. The evidence reflects that traditionally the State Department of Education, local school boards, and administration officials exercise little, if any, influence upon the subject matter taught by classroom teachers. Teachers have been given freedom to teach and emphasize those portions of subjects the individual teacher considered important. The limits to this discretion have generally been derived from the approval of textbooks by the State Department and preparation of curriculum guides by the school districts.

Several witnesses testified that academic freedom for the teacher means, in substance, that the individual teacher should be permitted unlimited discretion subject only to the bounds of professional ethics. The court is not prepared to adopt such a broad view of academic freedom in the public schools.

In any event, if Act 590 is implemented, many teachers will be required to teach material in support of creation science which they do not consider academically sound. Many teachers will simply forego teaching subjects which might trigger the "balanced treatment" aspects of Act 590 even though they think the subjects are important to a proper presentation of a course.

Implementation of Act 590 will have serious and untoward consequences for students, particularly those planning to attend college. Evolution is the cornerstone of modern biology, and many courses in public schools contain subject matter relating to such varied topics as the age of the earth, geology, and relationships among living things. Any student who is deprived of instruction as to the prevailing scientific thought on these topics will be denied a significant part of science education. Such a deprivation through the high school level would undoubtedly have an impact upon the quality of education in the State's colleges and universities, especially including the pre-professional and professional programs in the health sciences.

V(C)

The defendants argue in their brief that evolution is, in effect, a religion, and that by teaching a religion which is contrary to some students' religious views, the State is infringing upon the student's free exercise rights under the First Amendment. Mr. Ellwanger's legislative findings, which were adopted as a finding of fact by the Arkansas Legislature in Act 590, provides:

> Evolution-science is contrary to the religious convictions or moral values or philosophical beliefs of many students and parents, including individuals of many different religious faiths and with diverse moral and philosophical beliefs [Act 590, §7(d)].

The defendants argue that the teaching of evolution alone presents both a free exercise problem and an establishment problem which can only be redressed by giving balanced treatment to creation science, which is admittedly consistent with some religious beliefs. This argument appears to have its genesis in a student note written by Mr. Wendell Bird, "Freedom of Religion and

Science Instruction in Public Schools," 87 *Yale Law Journal* 515 (1978). The argument has no legal merit.

If creation science is, in fact, science and not religion, as the defendants claim, it is difficult to see how the teaching of such a science could "neutralize" the religious nature of evolution.

Assuming for the purposes of argument, however, that evolution is a religion or religious tenet, the remedy is to stop the teaching of evolution; not establish another religion in opposition to it. Yet it is clearly established in the case law, and perhaps also in common sense, that evolution is not a religion and that teaching evolution does not violate the Establishment Clause *(Epperson v. Arkansas, Willoughby v. Stever, Wright v. Houston Independent School District).*

V(D)

The defendants presented Dr. Larry Parker, a specialist in devising curricula for public schools. He testified that the public school's curriculum should reflect the subjects the public wants taught in schools. The witness said that polls indicated a significant majority of the American public thought creation science should be taught if evolution was taught. The point in this testimony was never placed in a legal context. No doubt a sizeable majority of Americans believe in the concept of a Creator or, at least, are not opposed to the concept and see nothing wrong with teaching school children about the idea.

The application and content of First Amendment principles are not determined by public opinion polls or by a majority vote. Whether the proponents of Act 590 constitute the majority or the minority is quite irrelevant under a constitutional system of government. No group, no matter how large or small, may use the organs of government, of which the public schools are the most conspicuous and influential, to foist its religious beliefs on others.

The Court closes this opinion with a thought expressed eloquently by the great Justice Frankfurter:

> We renew our conviction that "we have staked the very existence of our country on the faith that complete separation between the state and religion is best for the state and best for religion" *(Everson v. Board of Education* at 59). If nowhere else, in the relation between Church and State, "good fences make good neighbors" *(McCollum v. Board of Education* at 232).

An injunction will be entered permanently prohibiting enforcement of Act 590.

It is so ordered this January 5, 1982.

NOTES

1. The complaint is based on 42 U.S.C. §1983, which provides a remedy against any person who, acting under color of state law, deprives another of any right, privilege or immunity guaranteed by the United States Constitution or federal law. This Court's jurisdiction arises under 28 U.S.C. §§1331, 1343(3) and 1343(4). The power to issue declaratory judgments is expressed in 28 U.S.C. §§2201 and 2202.

2. The facts necessary to establish the plaintitfs' standing to sue are contained in the joint stipulation of facts, which is hereby adopted and incorporated herein by reference. There is no doubt that the case is ripe for adjudication.

3. The State of Arkansas was dismissed as a defendant because of its immunity from suit under the Eleventh Amendment. *Hans v. Louisiana*, 134 U.S. 1 (1890).

4. The authorities differ as to generalizations which may be made about Fundamentalism. For example, Dr. Geisler testified to the widely held view that there are five beliefs characteristic of all Fundamentalist movements, in addition, of course, to the inerrancy of Scripture: (1) belief in the virgin birth of Christ, (2) the belief in the deity of Christ, (3) belief in the substitutional atonement of Christ, (4) belief in the second coming of Christ, and (5) belief in the physical resurrection of all departed souls. Dr. Marsden, however, testified that this generalization, which had been common in religious scholarship, is now thought to be historical error. There is no doubt, however, that all Fundamentalists take the Scriptures as inerrant and probably most take them as literally true.

5. Initiated Act 1 of 1929, Arkansas Statutes Ann. § 80-1627 *et seq.*, which prohibited the teaching of evolution in Arkansas schools is discussed *infra* at text accompanying note 15 (below).

6. References to documentary exhibits are by the name of the author and the [trial or pre-trial] exhibit number.

7. Applicants for membership in the CRS must subscribe to the following statement of belief: "(1) The Bible is the written Word of God, and because we believe it to be inspired thruout (sic), all of its assertions are historically and scientifically true in all of the original autographs. To the student of nature, this means that the account of origins in Genesis is a factual presentation of simple historical truths. (2) All basic types of living things, including man, were made by direct creative acts of God during Creation Week as described in Genesis. Whate er biological changes have occurred since Creation have accomplished only changes within the original created kinds. (3) The great Flood described in Genesis, commonly referred to as the Noachian Deluge, was an historical event, worldwide in its extent and effect. (4) Finally, we are an organization of Christian men of science, who accept Jesus Christ as our Lord and Savior. The account of the special creation of Adam and Eve as one man and one woman, and their subsequent Fall into sin, is the basis for our belief in the necessity of a Savior for all mankind. Therefore, salvation can come only thru (sic) accepting Jesus Christ as our Savior." (Plaintiffs' Exhibit 115)

8. Because of the voluminous nature of the documentary exhibits, the parties were directed by pre-trial order to submit their proposed exhibits for the court's convenience prior to trial. The numbers assigned to the pre-trial submissions do not correspond with those assigned to the same documents at trial and, in some instances, the pre-trial submissions are more complete.

9. Plaintiffs' Exhibit 130, Morris, *Introducing Scientific Creationism Into the Public Schools* (1975); and Bird, "Resolution for Balanced Presentation of Evolution and Scientific Creationism," ICR Impact Series No. 71, Appendix 14 to Plaintiffs' Pretrial Brief.

10. The creationists often show candor in their proselytization. Henry Morris has stated, "Even if a favorable statute or court decision is obtained, it will probably be declared unconstitutional, especially if the legislation or injunction refers to the Bible account of creation." In the same vein he notes, "The only effective way to get creationism taught properly is to have it taught by teachers who are both willing and able to do it. Since most teachers

now are neither willing nor able, they must first be persuaded and instructed themselves." [Plaintiffs' Exhibit 130, Morris, *Introducing Scientific Creationism Into the Public Schools* (1975) (unpaged).]

11. Mr. Bird sought to particiate in this litigation by representing a number of individuals who wanted to intervene as defendants. The application for intervention was denied by this Court. . . .

12. The model act had been revised to insert "creation science" in lieu of creationism because Ellwanger had the impression people thought creationism was too religious a term (Ellwanger Deposition, p. 79).

13. The original model act had been introduced in the South Carolina Legislature, but had died without action after the South Carolina Attorney General had opined that the act was unconstitutional.

14. Specifically, Senator Holsted testified that he holds to a literal interpretation of the Bible; that the bill was compatible with his religious beliefs; that the bill does favor the position of literalists; that his religious convictions were a factor in his sponsorship of the bill; and that he stated publicly to the *Arkansas Gazette* (although not on the floor of the Senate) contemporaneously with the legislative debate that the bill does presuppose the existence of a divine creator. There is no doubt that Senator Holsted knew he was sponsoring the teaching of a religious doctrine. His view was that the bill did not violate the First Amendment because, as he saw it, it did not favor one denomination over another.

15. This statute is, of course, clearly unconstitutional under the Supreme Court's decision in *Abbington School District v. Schempp.*

16. The joint stipulation of facts establishes that the following areas are the only *information* specifically required by statute to be taught in all Arkansas schools: (1) The effects of alcohol and narcotics on the human body, (2) Conservation of natural resources, (3) Bird Week, (4) Fire Prevention, and (5) Flag etiquette. Additionally, certain specific courses, such as American history and Arkansas history, must be completed by each student before graduation from high school.

17. Paul Ellwanger stated in his deposition that he did not know why Section 4(a)(2) (insufficiency of mutation and natural selection) was included as an evidence supporting creation science. He indicated that he was not a scientist, "but these are the postulates that have been laid down by creation scientists" (Ellwanger Deposition, p. 136).

18. Although defendants must make some effort to cast the concept of creation in non-religious terms, this effort surely causes discomfort to some of the Act's more theologically sophisticated supporters. The concept of a creator God distinct from the God of love and mercy is closely similar to the Marcion and Gnostic heresies, among the deadliest to threaten the early Christian church. These heresies had much to do with development and adoption of the Apostle's Creed as the official creedal statement of the Roman Catholic Church in the West. (Testimony of Gilkey.)

19. The parallels between Section 4(a) and Genesis are quite specific: (1) "sudden creation from nothing" is taken from Genesis 1:1-10 (Testimony of Vawter and Gilkey); (2) destruction of the world by a flood of divine origin is a notion peculiar to Judeo-Christian tradition and is based on Chapters 7 and 8 of Genesis (Vawter); (3) the term "kinds" has no fixed scientific meaning, but appears repeatedly in Genesis (all scientific witnesses); (4) "relatively recent inception" means an age of the earth from 6,000 to 10,000 years and is based on the genealogy of the Old Testament using the rather astronomical ages assigned to the patriarchs (Gilkey and several of defendants' scientific witnesses); (5) Separate ancestry of man and ape focuses on the portion of the theory of evolution which Fundamentalists find most offensive *(Epperson v. Arkansas)*.

20. "[C]oncepts concerning . . . a supreme being of some sort are manifestly religious . . . These concepts do not shed that religiosity merely because they are presented as philosophy or as science . . ." *(Malnak v. Yogi* at 1322).

21. See, e.g., Plaintiffs' Exhibit 76, Morris *et al., Scientific Creationism,* p. 203 (1980)

("If creation really is a fact, this means there is a *Creator,* and the universe is His creation"). Numerous other examples of such admissions can be found in the many exhibits which represent creationist literature, but no useful purpose would be served here by a potentially endless listing.

22. Morris, the Director of ICR and one who first advocated the two-model approach, insists that a true Christian cannot compromise with the theory of evolution and that the Genesis version of creation and the theory of evolution are mutually exclusive (Plaintiffs' Exhibit 31, Morris, *Studies in the Bible & Science,* pp. 102-103). The two-model approach was the subject of Dr. Richard Bliss's doctoral dissertation (Defense Exhibit 35). It is presented in Bliss, *Origins: Two Models—Evolution, Creation* (1978). Moreover, the two-model approach merely casts in educational language the dualism which appears in all creationist literature— creation (i.e., God) and evolution are presented as two alternative and mutually exclusive theories [See, e.g., Plaintiffs' Exhibit 75, Morris, *Scientific Creationism* (1974) (public school edition); Plaintiffs' Exhibit 59, Fox, *Fossils: Hard Facts from the Earth*]. Particularly illustrative is Plaintiffs' Exhibit 61, Boardman, *et al., Worlds Without End* (1975), a CSRC publication: "One group of scientists, known as creationists, believe that God, in a miraculous manner, created all matter and energy . . .

"Scientists who insist that the universe just grew, by accident, from a mass of hot gases without the direction or help of a Creator are known as evolutionists."

23. The idea that belief in a creator and acceptance of the scientific theory of evolution are mutually exclusive is a false premise and offensive to the religious views of many (Testimony of Hicks). Dr. Francisco Ayala, a geneticist of considerable renown and a former Catholic priest who has the equivalent of a Ph.D. in theology, pointed out that many working scientists who subscribe to the theory of evolution are devoutly religious.

24. This is so despite the fact that some of the defense witnesses do not subscribe to the young earth or flood hypotheses. Dr. Geisler stated his belief that the earth is several billion years old. Dr. Wickramasinghe stated that no rational scientist would believe the earth is less than one million years old or that all the world's geology could be explained by a worldwide flood.

25. "We do not know how God created, what processes He used, for *God used processes which are not now operating anywhere in the natural universe.* This is why we refer to divine creation as Special Creation. We cannot discover by scientific investigation anything about the creative processes used by God" [Plaintiffs' Exhibit 78, Gish, *Evolution? The Fossils Say No!,* p. 42 (3rd ed., 1979) (emphasis in original)].

26. The evolutionary notion that man and some modern apes have a common ancestor somewhere in the distant past has consistently been distorted by anti-evolutionists to say that man descended from modern monkeys. As such, this idea has long been most offensive to Fundamentalists (See, *Epperson v. Arkansas*).

27. Not only was this point acknowledged by virtually all the defense witnesses, it is patent in the creationist literature. See, e.g., Plaintiffs' Exhibit 89, Kofahl & Segraves, *The Creation Explanation,* p. 40: "The Flood of Noah brought about vast changes in the earth's surface, including vulcanism, mountain building, and the deposition of the major part of sedimentary strata. This principle is called 'biblical catastophism.'"

28. See note 7, *supra,* for the full text of the CRS creed.

29. The theory is detailed in Wickramasinghe's book with Sir Fred Hoyle, *Evolution From Space* (1981), which is Defense Exhibit 79.

30. Ms. Wilson stated that some professors she spoke with sympathized with her plight and tried to help her find scientific materials to support Section 4(a). Others simply asked her to leave.

31. Plaintiffs' Exhibit 129, published by Zonderman Publishing House (1974), states that it was "prepared by the Textbook Committee of the Creation Research Society." It has a disclaimer pasted inside the front cover stating that it is not suitable for use in public schools.

32. Plaintiffs' Exhibit 77, by Duane Gish.

33. The passage of Act 590 apparently caught a number of its supporters off guard as much as it did the school district. The Act's author, Paul Ellwanger, stated in a letter to "Dick" (apparently Dr. Richard Bliss at ICR): "And finally, if you know of any textbooks at any level and for any subjects that you think are acceptable to you and also constitutionally admissible, these are things that would be of *enormous* to these bewildered folks who may be caught, as Arkansas now has been, by the sudden need to implement a whole new ball game with which they are quite unfamiliar." (sic) (Unnumbered attachment to Ellwanger Deposition.)

CASES

Abbington School District v. Schempp, 374 U.S. 203 (1963).

Brandon v. Board of Education, 487 F. Supp. 1219 (N.D.N.Y.), affirmed, 635 F.2d 971 (2nd Circuit 1980).

Chrysler Corporation v. Brown, 441 U.S. 281 (1979).

Committee for Public Education & Religious Liberty v. Nyquist, 413 U.S. 756 (1973).

Engel v. Vitale, 370 U.S. 421 (1962).

Epperson v. Arkansas, 393 U.S. 97 (1968).

Everson v. Board of Education, 330 U.S. 1 (1947).

Federal Energy Administration v. Algonquin SNG, Inc., 426 U.S. 548 (1976).

Hans v. Louisiana, 134 U.S. 1 (1890).

Lemon v. Kurtzman, 403 U.S. 602 (1971).

McCollum v. Board of Education, 333 U.S. 203 (1948).

McGowan v. Maryland, 366 U.S. 420 (1961).

Malnak v. Yogi, 440 F. Supp. 1284 (D.N.J. 1977); *affirmed per curiam,* 592 F.2d 197 (3rd Circuit 1979).

Stone v. Graham, 449 U.S. 39 (1980).

United States v. Emmons, 410 U.S. 396 (1979).

Village of Arlington Heights v. Metropolitan Housing Corp., 429 U.S. 252 (1977).

Willoughby v. Stever, No. 15574-75 (Denver District Court, 18 May 1973), *affirmed* 504 F.2d 271 (D.C. Circuit 1974), *certiorari denied,* 420 U.S. 927 (1975).

Wright v. Houston Independent School District, 366 F. Supp. 1208 (Southern District of Texas, 1978), *affirmed* 486 F.2d 137 (5th Circuit 1973), *certiorari denied* 417 U.S. 969 (1974).

Part Four

The Philosophical Aftermath

Introduction

The contributions to this section need little introduction, even apart from the fact that I as editor am loathe to write much about that in which I was greatly involved as disputant. The arguments must stand or fall on their own merits.

First, we have a general discussion by Larry Laudan, in which he argues both that all existent attempts to separate science from nonscience have failed and that there is not much hope of a satisfactory resolution to the problem in the future. Then, in the next article, Laudan criticizes Overton's claims about the nature of science, arguing that although creationism should be rejected, such rejection should be on the basis of its being bad science rather than nonscience. I defend Overton (and by implication myself), pointing out that, apart from anything else, it is not bad science that the First Amendment bans from classrooms, but nonscience in the form of religion.

Philip L. Quinn pushes the critique somewhat further. Like Laudan he rejects creationism, although like Laudan he believes that the arguments of Overton (and me) are inadequate and that different reasoning should have been used. He then relates the discussion to the broader context, pointing out that what we have at stake here is more than a narrow scholastic dispute, but an event where, in a way, the very integrity of the philosophy of science is on trial. In my exchange with Quinn, we clarify this point, and I think it is fair to say that for all our philosophical differences, we come to agree that there must be some way in which philosophers can be allowed to give an opinion. (I might add that if there were not, I for one would want to change subjects.)

I am glad to let my critic have the final response. However, I hope that his will not be the final word. It seems to me that the issues that I and others have raised in this volume are important, interesting, and merit much more discussion. I am sure I speak for all of the philosophers and all of the scientists when I say that we are less interested in making converts than in stimulating others to take up these issues themselves, to run with them a way on their own. Toward such an end, I hope you will now want to go on to some of the material mentioned in "Further Reading."

21

The Demise of the Demarcation Problem*

Larry Laudan

1. INTRODUCTION

We live in a society which sets great store by science. Scientific "experts"
play a privileged role in many of our institutions, ranging from the courts
of law to the corridors of power. At a more fundamental level, most of us
strive to shape our beliefs about the natural world in the "scientific" image.
If scientists say that continents move or that the universe is billions of years
old, we generally believe them, however counterintuitive and implausible their
claims might appear to be. Equally, we tend to acquiesce in what scientists
tell us not to believe. If, for instance, scientists say that Velikovsky was a
crank, that the biblical creation story is hokum, that UFOs do not exist,
or that acupuncture is ineffective, then we generally make the scientist's
contempt for these things our own, reserving for them those social sanctions
and disapprobations which are the just deserts of quacks, charlatans and con-
men. In sum, much of our intellectual life, and increasingly large portions
of our social and political life, rests on the assumption that we (or, if not
we ourselves, then someone whom we trust in these matters) can tell the
difference between science and its counterfeit.

For a variety of historical and logical reasons, some going back more
than two millennia, that "someone" to whom we turn to find out the difference
usually happens to be the philosopher. Indeed, it would not be going too
far to say that, for a very long time, philosophers have been regarded as

From *Physics, Philosophy, and Psychoanalysis*, ed. R. S. Cohen and L. Laudan (Dordrecht:
Reidel, 1983). pp. 111 27. Copyright © 1983 D. Reidel Publishing Company. D. Dordrecht,
Holland. Reprinted by permission of the publisher.

the gatekeepers to the scientific estate. They are the ones who are supposed to be able to tell the difference between real science and pseudo-science. In the familiar academic scheme of things, it is specifically the theorists of knowledge and the philosophers of science who are charged with arbitrating and legitimating the claims of any sect to "scientific" status. It is small wonder, under the circumstances, that the question of the nature of science has loomed so large in Western philosophy. From Plato to Popper, philosophers have sought to identify those epistemic features which mark off science from other sorts of belief and activity.

Nonetheless, it seems pretty clear that philosophy has largely failed to deliver the relevant goods. Whatever the specific strengths and deficiencies of the numerous well-known efforts at demarcation (several of which will be discussed below), it is probably fair to say that there is no demarcation line between science and non-science, or between science and pseudo-science, which would win assent from a majority of philosophers. Nor is there one which *should* win acceptance from philosophers or anyone else; but more of that below.

What lessons are we to draw from the recurrent failure of philosophy to detect the epistemic traits which mark science off from other systems of belief? That failure might conceivably be due simply to our impoverished philosophical imagination; it is conceivable, after all, that science really is *sui generis,* and that we philosophers have just not yet hit on its characteristic features. Alternatively, it may just be that there are no epistemic features which all and only the disciplines we accept as "scientific" share in common. My aim in this paper is to make a brief excursion into the history of the science/non-science demarcation in order to see what light it might shed on the contemporary viability of the quest for a demarcation device.

2. THE OLD DEMARCATIONIST TRADITION

As far back as the time of Parmenides, Western philosophers thought it important to distinguish knowledge (*episteme*) from mere opinion (*doxa*), reality from appearance, truth from error. By the time of Aristotle, these epistemic concerns came to be focused on the question of the nature of *scientific* knowledge. In his highly influential *Posterior Analytics,* Aristotle described at length what was involved in having scientific knowledge of something. To be scientific, he said, one must deal with causes, one must use logical demonstrations, and one must identify the universals which "inhere" in the particulars of sense. But above all, to have science one must have *apodictic certainty.* It is this last feature which, for Aristotle, most clearly distinguished the scientific way of knowing. What separates the sciences from other kinds of beliefs is the infallibility of their foundations and, thanks to that infallibility, the incorrigibility of their constituent theories. The first principles of nature are directly intuited from sense; everything else worthy of the name of science follows demonstrably from these first principles. What characterizes the whole enterprise is a degree of certainty which distinguishes it most crucially from mere opinion.

But Aristotle sometimes offered a second demarcation criterion, orthogonal to this one between science and opinion. Specifically, he distinguished between know-how (the sort of knowledge which the craftsman and the engineer possess) and what we might call "know-why" or demonstrative understanding (which the scientist alone possesses). A shipbuilder, for instance, knows how to form pieces of wood together so as to make a seaworthy vessel; but he does not have, and has no need for, a syllogistic, causal demonstration based on the primary principles or first causes of things. Thus, he needs to know that wood, when properly sealed, floats; but he need not be able to show by virtue of what principles and causes wood has this property of buoyancy. By contrast, the scientist is concerned with what Aristotle calls the "reasoned fact"; until he can show why a thing behaves as its does by tracing its causes back to first principles, he has no scientific knowledge of the thing.

Coming out of Aristotle's work, then, is a pair of demarcation criteria. Science is distinguished from opinion and superstition by the certainty of its principles; it is marked off from the crafts by its comprehension of first causes. This set of contrasts comes to dominate discussions of the nature of science throughout the later Middle Ages and the Renaissance, and thus to provide a crucial backdrop to the re-examination of these issues in the seventeenth century.

It is instructive to see how this approach worked in practice. One of the most revealing examples is provided by pre-modern astronomy. By the time of Ptolemy, mathematical astronomers had largely abandoned the (Aristotelian) tradition of seeking to derive an account of planetary motion from the causes or essences of the planetary material. As Duhem and others have shown in great detail,[1] many astronomers sought simply to correlate planetary motions, independently of any causal assumptions about the essence or first principles of the heavens. Straightaway, this turned them from scientists into craftsmen.[2] To make matters worse, astronomers used a technique of *post hoc* testing of their theories. Rather than deriving their models from directly-intuited first principles, they offered hypothetical constructions of planetary motions and positions and then compared the predictions drawn from their models with the observed positions of the heavenly bodies. This mode of theory testing is, of course, highly fallible and non-demonstrative; and it was known at the time to be so. The central point for our purposes is that, by abandoning a demonstrative method based on necessary first principles, the astronomers were indulging in mere opinion rather than knowledge, putting themselves well beyond the scientific pale. Through virtually the whole of the Middle Ages, and indeed up until the beginning of the seventeenth century, the predominant view of mathematical astronomy was that, for the reasons indicated, it did not qualify as a "science." (It is worth noting in passing that much of the furor caused by the astronomical work of Copernicus and Kepler was a result of the fact that they were claiming to make astronomy "scientific" again.)

More generally, the seventeenth century brought a very deep shift in demarcationist sensibilities. To make a long and fascinating story unconscionably brief, we can say that most seventeenth-century thinkers accepted Aristotle's first demarcation criterion (viz., between infallible science and fallible opinion), but rejected his second (between know-how and understanding). For instance, if we look to the work of Galileo, Huygens or Newton, we see a refusal to prefer know-why to know-how; indeed, all three were prepared to regard as entirely scientific, systems of belief which laid no claim to an understanding grounded in primary causes or essences. Thus Galileo claimed to know little or nothing about the underlying causes responsible for the free fall of bodies, and in his own science of kinematics he steadfastly refused to speculate about such matters. But Galileo believed that he could still sustain his claim to be developing a "science of motion" because the results he reached were, so he claimed, infallible and demonstrative. Similarly, Newton in *Principia* was not indifferent to causal explanation, and freely admitted that he would like to know the causes of gravitational phenomena; but he was emphatic that, even without a knowledge of the causes of gravity, one can engage in a sophisticated and *scientific* account of the gravitational behavior of the heavenly bodies. As with Galileo, Newton regarded his non-causal account as "scientifical" because of the (avowed) certainty of its conclusions. As Newton told his readers over and again, he did not engage in hypotheses and speculations: he purported to be deriving his theories directly from the phenomena. Here again, the infallibility of results, rather than their derivability from first causes, comes to be the single touchstone of scientific status.

Despite the divergence of approach among thinkers of the seventeenth and eighteenth centuries, there is widespread agreement that scientific knowledge is apodictically certain. And this consensus cuts across most of the usual epistemological divides of the period. For instance, Bacon, Locke, Leibniz, Descartes, Newton and Kant are in accord about this way of characterizing science.[3] They may disagree about how precisely to certify the certainty of knowledge, but none quarrels with the claim that science and infallible knowledge are co-terminous.

As I have shown elsewhere,[4] this influential account finally and decisively came unraveled in the nineteenth century with the emergence and eventual triumph of a *fallibilistic* perspective in epistemology. Once one accepts, as most thinkers had by the mid-nineteenth century, that science offers no apodictic certainty, that all scientific theories are corrigible and may be subject to serious emendation, then it is no longer viable to attempt to distinguish science from non-science by assimilating that distinction to the difference between knowledge and opinion. Indeed, the unambiguous implication of fallibilism is that there is no difference between knowledge and opinion: within a fallibilist framework, scientific belief turns out to be just a species of the genus opinion. Several nineteenth-century philosophers of science tried to take some of the sting out of this *volte-face* by suggesting that scientific opinions were more probable or more reliable than non-scientific ones; but even they conceded that it was no longer possible to make infallibility the hallmark of scientific knowledge.

With certainty no longer available as the demarcation tool, nineteenth-century philosophers and scientists quickly forged other tools to do the job. Thinkers as diverse as Comte, Bain, Jevons, Helmholtz and Mach (to name only a few) began to insist that what really marks science off from everything else is its *methodology*. There was, they maintained, something called "the scientific method"; even if that method was not fool-proof (the acceptance of fallibilism demanded that concession), it was at least a better technique for testing empirical claims than any other. And if it did make mistakes, it was sufficiently self-corrective that it would soon discover them and put them right. As one writer remarked a few years later: "if science lead us astray, more science will set us straight."[5] One need hardly add that the nineteenth century did not invent the idea of a logic of scientific inquiry; that dates back at least to Aristotle. But the new insistence in this period is on a fallible method which, for all its fallibility, is nonetheless superior to its non-scientific rivals.

This effort to mark science off from other things required one to show two things. First, that the various activities regarded as science utilized essentially the same repertoire of methods (hence the importance in the period of the so-called thesis of the "unity of method"); secondly, the epistemic credentials of this method had to be established. At first blush, this program of identifying science with a certain technique of inquiry is not a silly one; indeed, it still persists in some respectable circles even in our time. But the nineteenth century could not begin to deliver on the two requirements just mentioned because there was no agreement about what the scientific method was. Some took it to be the canons of inductive reasoning sketched out by Herschel and Mill. Others insisted that the basic methodological principle of science was that its theories must be restricted to observable entities (the nineteenth-century requirement of *"vera causa"*).[6] Still others, like Whewell and Peirce, rejected the search for *verae causae* altogether and argued that the only decisive methodological test of a theory involved its ability successfully to make surprising predictions.[7] Absent agreement on what "the scientific method" amounted to, demarcationists were scarcely in a position to argue persuasively that what individuated science was its method.

This approach was further embarrassed by a notorious set of ambiguities surrounding several of its key components. Specifically, many of the methodological rules proposed were much too ambiguous for one to tell when they were being followed and when breached. Thus, such common methodological rules as "avoid *ad hoc* hypotheses," "postulate simple theories," "feign no hypotheses," and "eschew theoretical entities" involved complex conceptions which neither scientists nor philosophers of the period were willing to explicate. To exacerbate matters still further, what most philosophers of science of the period offered up as an account of "the scientific method" bore little resemblance to the methods actually used by working scientists, a point made with devastating clarity by Pierre Duhem in 1908.[8]

As one can see, the situation by the late nineteenth century was more than a little ironic. At precisely that juncture when science was beginning to have a decisive impact on the lives and institutions of Western man, at precisely that time when "scientism" (i.e., the belief that science and science alone has the answers to all our answerable questions) was gaining ground, in exactly that quarter-century when scientists were doing battle in earnest with all manner of "pseudo-scientists" (e.g., homeopathic physicians, spiritualists, phrenologists, biblical geologists), scientists and philosophers found themselves empty-handed. Except at the rhetorical level, there was no longer any consensus about what separated science from anything else.

Surprisingly (or, if one is cynically inclined, quite expectedly), the absence of a plausible demarcation criterion did not stop *fin de siècle* scientists and philosophers from haranguing against what they regarded as pseudo-scientific nonsense (any more than their present-day counterparts are hampered by a similar lack of consensus); but it did make their protestations less compelling than their confident denunciations of "quackery" might otherwise suggest. It is true, of course, that there was still much talk about "the scientific method"; and doubtless many hoped that the methods of science could play the demarcationist role formerly assigned to certainty. But, leaving aside the fact that agreement was lacking about precisely what the scientific method was, there was no very good reason as yet to prefer any one of the proposed "scientific methods" to any purportedly "non-scientific" ones, since no one had managed to show either that any of the candidate "scientific methods" qualified them as "knowledge" (in the traditional sense of the term) or, even more minimally, that those methods were epistemically superior to their rivals.

3. A METAPHILOSOPHICAL INTERLUDE

Before we move to consider and to assess some familiar demarcationist proposals from our own epoch, we need to engage briefly in certain metaphilosophical preliminaries. Specifically, we should ask three central questions: (1) What conditions of adequacy should a proposed demarcation criterion satisfy? (2) Is the criterion under consideration offering necessary or sufficient conditions, or both, for scientific status? (3) What actions or judgments are implied by the claim that a certain belief or activity is "scientific" or "unscientific"?

(1) Early in the history of thought it was inevitable that characterizations of "science" and "knowledge" would be largely stipulative and *a priori*. After all, until as late as the seventeenth century, there were few developed examples of empirical sciences which one could point to or whose properties one could study; under such circumstances, where one is working largely *ab initio,* one can be uncompromisingly legislative about how a term like "science" or "knowledge" will be used. But as the sciences developed and prospered, philosophers began to see the task of formulating a demarcation criterion as no longer a purely stipulative undertaking. Any proposed dividing line between science

and non-science would have to be (at least in part) explicative and thus sensitive to existing patterns of usage. Accordingly, if one were today to offer a definition of "science" which classified (say) the major theories of physics and chemistry as non-scientific, one would thereby have failed to reconstruct some paradigmatic cases of the use of the term. Where Plato or Aristotle need not have worried if some or even most of the intellectual activities of their time failed to satisfy their respective definitions of "science," it is inconceivable that we would find a demarcation criterion satisfactory which relegated to unscientific status a large number of the activities we consider scientific or which admitted as sciences activities which seem to us decidedly unscientific. In other words, the quest for a latter-day demarcation criterion involves an attempt to render explicit those shared but largely implicit sorting mechanisms whereby most of us can agree about paradigmatic cases of the scientific and the non-scientific. (And it seems to me that there is a large measure of agreement at this paradigmatic level, even allowing for the existence of plenty of controversial problem cases.) A failure to do justice to these implicit sortings would be a grave drawback for any demarcation criterion.

But we expect more than this of a *philosophically* significant demarcation criterion between science and non-science. Minimally, we expect a demarcation criterion to identify the *epistemic* or *methodological* features which mark off scientific beliefs from unscientific ones. We want to know what, if anything, is special about the knowledge claims and the modes of inquiry of the sciences. Because there are doubtless many respects in which science differs from non-science (e.g., scientists may make larger salaries, or know more mathematics than non-scientists), we must insist that any philosophically interesting demarcative device must distinguish scientific and non-scientific matters in a way which exhibits a surer epistemic warrant or evidential ground for science than for non-science. If it should happen that there is no such warrant, then the demarcation between science and non-science would turn out to be of little or no philosophic significance.

Minimally, then, a philosophical demarcation criterion must be an adequate explication of our ordinary ways of partitioning science from non-science and it must exhibit epistemically significant differences between science and non-science. Additionally, as we have noted before, the criterion must have sufficient precision that we can tell whether various activities and beliefs whose status we are investigating do or do not satisfy it; otherwise it is no better than no criterion at all.

(2) What will the formal structure of a demarcation criterion have to look like if it is to accomplish the tasks for which it is designed? Ideally, it would specify a set of individually necessary and jointly sufficient conditions for deciding whether an activity or set of statements is scientific or unscientific. As is well known, it has not proved easy to produce a set of necessary and sufficient conditions for science. Would something less ambitious do the job? It seems unlikely. Suppose, for instance, that someone offers us a characterization which purports to be a necessary (but not sufficient) condition for scientific status. Such a condition, if acceptable, would allow us to identify

certain activities as decidedly unscientific, but it would not help "fix our beliefs," because it would not specify which systems actually were scientific. We would have to say things like: "Well, physics *might* be a science (assuming it fulfills the stated necessary conditions), but then again it *might* not, since necessary but not sufficient conditions for the application of a term do not warrant application of the term." If, like Popper, we want to be able to answer the question, "when should a theory be ranked as scientific?"[9] then merely necessary conditions will never permit us to answer it.

For different reasons, merely sufficient conditions are equally inadequate. If we are only told: "satisfy these conditions and you will be scientific," we are left with no machinery for determining that a certain activity or statement is *unscientific*. The fact that (say) astrology failed to satisfy a set of *merely sufficient* conditions for scientific status would leave it in a kind of epistemic twilight zone—possibly scientific, possibly not. Here again, we cannot construct the relevant partitioning. Hence, if (in the spirit of Popper) we "wish to distinguish between science and pseudo-science,"[10] sufficient conditions are inadequate. The importance of these seemingly abstract matters can be brought home by considering some real-life examples. Recent legislation in several American states mandates the teaching of "creation science" alongside evolutionary theory in high school science classes. Opponents of this legislation have argued that evolutionary theory is authentic science, while creation science is not science at all. Such a judgment, and we are apt to make parallel ones all the time, would *not* be warranted by any demarcation criterion which gave only necessary *or* only sufficient conditions for scientific status. Without conditions which are both necessary and sufficient, we are never in a position to say "*this* is scientific: but *that* is unscientific." A demarcation criterion which fails to provide both sorts of conditions simply will not perform the tasks expected of it.

(3) Closely related to this point is a broader question of the purposes behind the formulation of a demarcation criterion. No one can look at the history of debates between scientists and "pseudo-scientists" without realizing that demarcation criteria are typically used as *machines de guerre* in a polemical battle between rival camps. Indeed, many of those most closely associated with the demarcation issue have evidently had hidden (and sometimes not so hidden) agendas of various sorts. It is well known, for instance, that Aristotle was concerned to embarrass the practitioners of Hippocratic medicine; and it is notorious that the logical positivists wanted to repudiate metaphysics and that Popper was out to "get" Marx and Freud. In every case, they used a demarcation criterion of their own devising as the discrediting device.

Precisely because a demarcation criterion will typically assert the epistemic superiority of science over non-science, the formulation of such a criterion will result in the sorting of beliefs into such categories as "sound" and "unsound," "respectable" and "cranky," or "reasonable" and "unreasonable." Philosophers should not shirk from the formulation of a demarcation criterion merely because it has these judgmental implications associated with it. Quite the reverse, philosophy at its best should tell us what is reasonable to believe and what is not. But the value-loaded character of the term "science" (and its cognates) in our culture should make us realize that the labeling of a certain activity as "scientific" or "unscientific" has social and political ramifications which go well beyond the taxonomic task of sorting beliefs into two piles. Although the cleaver that makes the cut may be largely epistemic in character, it has consequences which are decidedly non-epistemic. Precisely because a demarcation criterion will serve as a rationale for taking a number of *practical* actions which may well have far-reaching moral, social and economic consequences, it would be wise to insist that the arguments in favor of any demarcation criterion we intend to take seriously should be especially compelling.

With these preliminaries out of the way, we can turn to an examination of the recent history of demarcation.

4. THE NEW DEMARCATIONIST TRADITION

As we have seen, there was ample reason by 1900 to conclude that neither certainty nor generation according to a privileged set of methodological rules was adequate to denominate science. It should thus come as no surprise that philosophers of the 1920s and 1930s added a couple of new wrinkles to the problem. As is well known, several prominent members of the *Wiener Kreis* took a syntactic or logical approach to the matter. If, the logical positivists apparently reasoned, epistemology and methodology are incapable of distinguishing the scientific from the non-scientific, then perhaps the theory of meaning will do the job. A statement, they suggested, was scientific just in case it had a determinate meaning; and meaningful statements were those which could be exhaustively verified. As Popper once observed, the positivists thought that "verifiability, meaningfulness, and scientific character all coincide."[11]

Despite its many reformulations during the late 1920s and 1930s verificationism enjoyed mixed fortunes as a theory of meaning.[12] But as a would-be demarcation between the scientific and the non-scientific, it was a disaster. Not only are many statements in the sciences not open to exhaustive verification (e.g., all universal laws), but the vast majority of non-scientific and pseudo-scientific systems of belief have verifiable constituents. Consider, for instance, the thesis that the Earth is flat. To subscribe to such a belief in the twentieth century would be the height of folly. Yet such a statement is verifiable in the sense that we can specify a class of possible observations which would verify it. Indeed, every belief which has ever been rejected as a part of science because it was "falsified" is (at least partially) verifiable. Because verifiable, it is thus (according to the "mature positivists'" criterion) both meaningful and scientific.

A second familiar approach from the same period is Karl Popper's "falsificationist" criterion, which fares no better. Apart from the fact that it leaves ambiguous the scientific status of virtually every singular existential statement, however well supported (e.g., the claim that there are atoms, that there is a planet closer to the sun than the Earth, that there is a missing link), it has the untoward consequence of countenancing as "scientific" every crank claim which makes ascertainably false assertions. Thus flat Earthers, biblical creationists, proponents of laetrile or orgone boxes, Uri Geller devotees, Bermuda Triangulators, circle squarers, Lysenkoists, charioteers of the gods, *perpetuum mobile* builders, Big Foot searchers, Loch Nessians, faith healers, polywater dabblers, Rosicrucians, the-world-is-about-to-enders, primal screamers, water diviners, magicians, and astrologers all turn out to be scientific on Popper's criterion—just so long as they are prepared to indicate some observation, however improbable, which (if it came to pass) would cause them to change their minds.

One might respond to such criticisms by saying that scientific status is a matter of degree rather than kind. Sciences such as physics and chemistry have a high degree of testability, it might be said, while the systems we regard as pseudo-scientific are far less open to empirical scrutiny. Acute technical difficulties confront this suggestion, for the only articulated theory of degrees of testability (Popper's) makes it impossible to compare the degrees of testability of two distinct theories *except when one entails the other.* Since (one hopes!) no "scientific" theory entails any "pseudo-scientific" one, the relevant comparisons cannot be made. But even if this problem could be overcome, and if it were possible for us to conclude (say) that the general theory of relativity was more testable (and thus by definition more scientific) than astrology, it would not follow that astrology was any less worthy of belief than relativity— for testability is a semantic rather than an epistemic notion, which entails nothing whatever about belief-worthiness.

It is worth pausing for a moment to ponder the importance of this difference. I said before that the shift from the older to the newer demarcationist orientation could be described as a move from epistemic to syntactic and semantic strategies. In fact, the shift is even more significant than that way of describing the transition suggests. The central concern of the older tradition had been to identify those ideas or theories which were worthy of belief. To judge a statement to be scientific was to make a *retrospective* judgment about how that statement had stood up to empirical scrutiny. With the positivists and Popper, however, this retrospective element drops out altogether. Scientific status, on their analysis, is not a matter of evidential support or belief-worthiness, for all sorts of ill-founded claims are testable and thus scientific on the new view.

The failure of the newer demarcationist tradition to insist on the necessity of retrospective evidential assessments for determining scientific status goes some considerable way to undermining the practical utility of the demarcationist enterprise, precisely because most of the "cranky" beliefs about which one might incline to be dismissive turn out to be "scientific" according to falsificationist or (partial) verificationist criteria. The older demarcationist tradition, concerned with actual epistemic warrant rather than potential epistemic scrutability, would never have countenanced such an undemanding sense of the "scientific." More to the point, the new tradition has had to pay a hefty price for its scaled-down expectations. Unwilling to link scientific status to any evidential warrant, twentieth-century demarcationists have been forced into characterizing the ideologies they oppose (whether Marxism, psychoanalysis or creationism) as unstable in principle. Very occasionally, that label is appropriate. But more often than not, the views in question can be tested, have been tested, and have failed those tests. But such failures cannot impugn their (new) scientific status: quite the reverse, *by virtue of failing the epistemic tests to which they are subjected, these views guarantee that they satisfy the relevant semantic criteria for scientific status!* The new demarcationism thus reveals itself as a largely toothless wonder, which serves neither to explicate the paradigmatic usages of "scientific" (and its cognates) nor to perform the critical stable-cleaning chores for which it was originally intended.

For these, and a host of other reasons familiar in the philosophical literature, neither verificationism nor falsificationism offers much promise of drawing a useful distinction between the scientific and the non-scientific.

Are there other plausible candidates for explicating the distinction? Several seem to be waiting in the wings. One might suggest, for instance, that scientific claims are well tested, whereas non-scientific ones are not. Alternatively (an approach taken by Thagard),[13] one might maintain that scientific knowledge is unique in exhibiting progress or growth. Some have suggested that scientific theories alone make surprising predictions which turn out to be true. One might even go in the pragmatic direction and maintain that science is the sole repository of useful and reliable knowledge. Or, finally, one might propose that science is the only form of intellectual system-building, which proceeds cumulatively, with later views embracing earlier ones, or at least retaining those earlier views as limiting cases.[14]

It can readily be shown that none of these suggestions can be a necessary and sufficient condition for something to count as "science," at least not as that term is customarily used. And in most cases, these are not even plausible as necessary conditions. Let me sketch out some of the reasons why these proposals are so unpromising. Take the requirement of well-testedness. Unfortunately, we have no viable over-arching account of the circumstances under which a claim may be regarded as well tested. But even if we did, is it plausible to suggest that all the assertions in science texts (let alone science journals) have been well tested and that none of the assertions in such conventionally non-scientific fields as literary theory, carpentry or football strategy are well tested? When a scientist presents a conjecture which has not yet been tested and is such that we are not yet sure what would count as a robust test of it, has that scientist ceased doing science when he discusses his conjecture? On the other side of the divide, is anyone prepared to say that we have no convincing evidence for such "non-scientific" claims as that "Bacon did not write the plays attributed to Shakespeare," that "a miter joint is stronger than a flush joint," or that "off-side kicks are not usually fumbled"? Indeed, are we not entitled to say that all these claims are much better supported by the evidence than many of the "scientific" assumptions of (say) cosmology or psychology?

The reason for this divergence is simple to see. Many, perhaps most, parts of science are highly speculative compared with many non-scientific disciplines. There seems good reason, given from the historical record, to suppose that most scientific theories are false; under the circumstances, how plausible can be the claim that science is the repository of all and only reliable or well-confirmed theories?

Similarly, cognitive progress is not unique to the "sciences." Many disciplines (e.g., literary criticism, military strategy, and perhaps even philosophy) can claim to know more about their respective domains than they did 50 or 100 years ago. By contrast, we can point to several "sciences" which, during certain periods of their history, exhibited little or no progress.[15] Continuous, or even sporadic, cognitive growth seems neither a necessary nor a sufficient condition for the activities we regard as scientific. Finally, consider the requirement of cumulative theory transitions as a demarcation criterion. As several authors[16] have shown, this will not do even as a necessary condition for marking off scientific knowledge, since many scientific theories—even those in the so-called "mature sciences"—do not contain their predecessors, not even as limiting cases.

I will not pretend to be able to prove that there is no conceivable philosophical reconstruction of our intuitive distinction between the scientific and the non-scientific. I do believe, though, that we are warranted in saying that none of the criteria which have been offered thus far promises to explicate the distinction.

But we can go further than this, for we have learned enough about what passes for science in our culture to be able to say quite confidently that it is not all cut from the same epistemic cloth. Some scientific theories are well tested; some are not. Some branches of science are presently showing high rates of growth; others are not. Some scientific theories have made a host of successful predictions of surprising phenomena; some have made few if any such predictions. Some scientific hypotheses are *ad hoc;* others are not. Some have achieved a "consilience of inductions"; others have not. (Similar remarks could be made about several non-scientific theories and disciplines.) *The evident epistemic heterogeneity of the activities and beliefs customarily regarded as scientific should alert us to the probable futility of seeking an epistemic version of a demarcation criterion.* Where, even after detailed analysis, there appear to be no epistemic invariants, one is well advised not to take their existence for granted. But to say as much is in effect to say that the problem of demarcation—the very problem which Popper labeled "the central problem of epistemology"—is spurious, for that problem *presupposes* the existence of just such invariants.

In asserting that the problem of demarcation between science and non-science is a pseudo-problem (at least as far as philosophy is concerned), I am manifestly not denying that there are crucial epistemic and methodological questions to be raised about knowledge claims, whether we classify them as scientific or not. Nor, to belabor the obvious, am I saying that we are never entitled to argue that a certain piece of science is epistemically warranted and that a certain piece of pseudo-science is not. It remains as important as it ever was to ask questions like: When is a claim well confirmed? When can we regard a theory as well tested? What characterizes cognitive progress? But once we have answers to such questions (and we are still a long way from that happy state!), there will be little left to inquire into which is epistemically significant.

One final point needs to be stressed. In arguing that it remains important to retain a distinction between reliable and unreliable knowledge, I am not trying to resurrect the science/non-science demarcation under a new guise.[17] However we eventually settle the question of reliable knowledge, the class of statements falling under that rubric will include much that is not commonly regarded as "scientific" and it will exclude much that is generally considered "scientific." This, too, follows from the epistemic heterogeneity of the sciences.

5. CONCLUSION

Through certain vagaries of history, some of which I have alluded to here, we have managed to conflate two quite distinct questions: What makes a belief well founded (or heuristically fertile)? And what makes a belief scientific? The first set of questons is philosophically interesting and possibly even tractable; the second question is both uninteresting and, judging by its checkered past, intractable. If we would stand up and be counted on the side of reason, we ought to drop terms like "pseudo-science" and "unscientific" from our vocabulary; they are just hollow phrases which do only emotive work for us. As such, they are more suited to the rhetoric of politicians and Scottish sociologists of knowledge than to that of empirical researchers.[18] Insofar as our concern is to protect ourselves and our fellows from the cardinal sin of believing what we wish were so rather than what there is substantial evidence for (and surely that is what most forms of "quackery" come down to), then our focus should be squarely on the empirical and conceptual credentials for claims about the world. The "scientific" status of those claims is altogether irrelevant.

NOTES

*I am grateful to NSF and NEH for support of this research. I have profited enormously from the comments of Adolf Grünbaum, Ken Alpern and Andrew Lugg on an earlier version of this paper.

1. See especially his *To Save The Phenomena* (Chicago: University of Chicago Press, 1969).

2. This shifting in orientation is often credited to the emerging emphasis on the continuity of the crafts and the sciences and to Baconian-like efforts to make science "useful." But such

an analysis surely confuses agnosticism about first causes—which is what really lay behind the instrumentalism of medieval and Renaissance astronomy—with a utilitarian desire to be practical.

3. For much of the supporting evidence for this claim, see the early chapters of Laudan, *Science and Hypothesis* (Dordrecht: D. Reidel, 1981).

4. See especially Chapter 8 of *Science and Hypothesis.*

5. E. V. Davis, writing in 1914.

6. See the discussions of this concept by Kavaloski, Hodge, and R. Laudan.

7. For an account of the history of the concept of surprising predictions, see Laudan, *Science and Hypothesis,* Chapters 8 and 10.

8. See Duhem's classic *Aim and Structure of Physical Theory* (New York: Atheneum, 1962).

9. Karl Popper, *Conjectures and Refutations* (London: Routledge and Kegan Paul, 1963), p. 33.

10. Ibid.

11. Ibid., p. 40.

12. For a very brief historical account, see C. G. Hempel's classic, "Problems and Changes in the Empiricist Criterion of Meaning," *Revue Internationale de Philosophie* 11 (1950), 41-63.

13. See, for instance, Paul Thagard, "Resemblance, Correlation and Pseudo-Science," in M. Hanen *et al., Science, Pseudo-Science and Society* (Waterloo, Ont.: W. Laurier University Press, 1980), pp. 17-28.

14. For proponents of this cumulative view, see Popper, *Conjectures and Refutations;* Hilary Putnam, *Meaning and the Moral Sciences* (London: Routledge and Kegan Paul, 1978); Wladyslaw Krajewski, *Correspondence Principle and Growth of Science* (Dordrecht, Boston: D. Reidel, 1977); Heinz Post, "Correspondence, Invariance and Heuristics," *Studies in History and Philosophy of Science* 2 (1971), 213-55; and L. Szumilewicz, "Incommensurability and the Rationality of Science," *British Journal for the Philosophy of Science* 28 (1977), 348ff.

15. Likely tentative candidates: acoustics from 1750 to 1780; human anatomy from 1900 to 1920; kinematic astronomy from 1200 to 1500; rational mechanics from 1910 to 1940.

16. See, among others: T. S. Kuhn, *Structure of Scientific Revolutions* (Chicago: University of Chicago Press, 1962); A. Grünbaum, "Can a Theory Answer More Questions than One of Its Rivals?", *British Journal for the Philosophy of Science* 27 (1976), 1-23; L. Laudan, "Two Dogmas of Methodology," *Philosophy of Science* 43 (1976), 467-72; L. Laudan, "A Confutation of Convergent Realism," *Philosophy of Science* 48 (1981), 19-49.

17. In an excellent study ["Theories of Demarcation Between Science and Metaphysics," in *Problems in the Philosophy of Science* (Amsterdam: North-Holland, 1968), 40ff)], William Bartley has similarly argued that the (Popperian) demarcation problem is not a central problem of the philosophy of science. Bartley's chief reason for devaluing the importance of a demarcation criterion is his conviction that it is less important whether a system is empirical or testable than whether a system is "criticizable." Since he thinks many non-empirical systems are nonetheless open to criticism, he argues that the demarcation between science and non-science is less important than the distinction between the revisable and the non-revisable. I applaud Bartley's insistence that the empirical/non-empirical (or, what is for a Popperian the same thing, the scientific/non-scientific) distinction is not central; but I am not convinced, as Bartley is, that we should assign pride of place to the revisable/non-revisable dichotomy. Being willing to change one's mind is a commendable trait, but it is not clear to me that such revisability addresses the central *epistemic* question of the well-foundedness of our beliefs.

18. I cannot resist this swipe at the efforts of the so-called Edinburgh school to recast the sociology of knowledge in what they imagine to be the "scientific image." For a typical example of the failure of that group to realize the fuzziness of the notion of the "scientific," see David Bloor's *Knowledge and Social Imagery* (London: Routledge and Kegan Paul, 1976), and my criticism of it, "The Pseudo-Science of Science?" *Philosophy of the Social Sciences* 11 (1981), 173-198.

22

Science at the Bar—Causes for Concern

Larry Laudan

In the wake of the decision in the Arkansas Creationism trial *(McLean v. Arkansas)*, the friends of science are apt to be relishing the outcome. The creationists quite clearly made a botch of their case and there can be little doubt that the Arkansas decision may, at least for a time, blunt legislative pressure to enact similar laws in other states. Once the dust has settled, however, the trial in general and Judge William R. Overton's ruling in particular may come back to haunt us; for, although the verdict itself is probably to be commended, it was reached for all the wrong reasons and by a chain of argument which is hopelessly suspect. Indeed, the ruling rests on a host of misrepresentations of what science is and how it works.

The heart of Judge Overton's Opinion is a formulation of "the essential characteristics of science." These characteristics serve as touchstones for contrasting evolutionary theory with Creationism; they lead Judge Overton ultimately to the claim, specious in its own right, that since Creationism is not "science," it must be religion. The Opinion offers five essential properties that demarcate scientific knowledge from other things: "(1) It is guided by natural law; (2) it has to be explanatory by reference to natural law; (3) it is testable against the empirical world; (4) its conclusions are tentative, i.e., are not necessarily the final word; and (5) it is falsifiable."

These fall naturally into two families: properties (1) and (2) have to do with lawlikeness and explanatory ability; the other three properties have to do with the fallibility and testability of scientific claims. I shall deal with the second set of issues first, because it is there that the most egregious errors of fact and judgment are to be found.

From *Science, Technology, & Human Values* 7, no. 41 (1982): 16-19. Copyright © 1982. Reprinted by permission of John Wiley & Sons, Inc.

At various key points in the Opinion, Creationism is charged with being untestable, dogmatic (and thus non-tentative), and unfalsifiable. All three charges are of dubious merit. For instance, to make the interlinked claims that Creationism is neither falsifiable nor testable is to assert that Creationism makes no empirical assertions whatever. That is surely false. Creationists make a wide range of testable assertions about empirical matters of fact. Thus, as Judge Overton himself grants (apparently without seeing its implications), the creationists say that the earth is of very recent origin (say 6,000 to 20,000 years old); they argue that most of the geological features of the earth's surface are diluvial in character (i.e., products of the postulated worldwide Noachian deluge); they are committed to a large number of factual historical claims with which the Old Testament is replete; they assert the limited variability of species. They are committed to the view that, since animals and man were created at the same time, the human fossil record must be paleontologically co-extensive with the record of lower animals. It is fair to say that no one has shown how to reconcile such claims with the available evidence —evidence which speaks persuasively to a long earth history, among other things.

In brief, these claims are testable, they have been tested, and they have failed those tests. Unfortunately, the logic of the Opinion's analysis precludes saying any of the above. By arguing that the tenets of Creationism are neither testable nor falsifiable, Judge Overton (like those scientists who similarly charge Creationism with being untestable) deprives science of its strongest argument against Creationism. Indeed, if any doctrine in the history of science has ever been falsified, it is the set of claims associated with "creation-science." Asserting that Creationism makes no empirical claims plays directly, if inadvertently, into the hands of the creationists by immunizing their ideology from empirical confrontation. The correct way to combat Creationism is to confute the empirical claims it does make, not to pretend that it makes no such claims at all.

It is true, of course, that some tenets of Creationism are not testable in isolation (e.g., the claim that man emerged by a direct supernatural act of creation). But that scarcely makes Creationism "unscientific." It is now widely acknowledged that many scientific claims are not testable in isolation, but only when embedded in a larger system of statements, some of whose consequences can be submitted to test.

Judge Overton's third worry about Creationism centers on the issue of revisability. Over and over again, he finds Creationism and its advocates "unscientific" because they have "refuse[d] to change it regardless of the evidence developed during the course of the[ir] investigation." In point of fact, the charge is mistaken. If the claims of modern-day creationists are compared with those of their nineteenth-century counterparts, significant shifts in orientation and assertion are evident. One of the most visible opponents of Creationism, Stephen Gould, concedes that creationists have modified their views about the amount of variability allowed at the level of species change. Creationists do, in short, change their minds from time to time. Doubtless they would create these shifts to their efforts to adjust their views to newly emerging evidence, in what they imagine to be a scientifically respectable way.

Perhaps what Judge Overton had in mind was the fact that some of Creationism's core assumptions (e.g., that there was a Noachian flood, that man did not evolve from lower animals, or that God created the world) seem closed off from any serious modification. But historical and sociological researches on science strongly suggest that the scientists of any epoch likewise regard some of their beliefs as so fundamental as not to be open to repudiation or negotiation. Would Newton, for instance, have been tentative about the claim that there were forces in the world? Are quantum mechanicians willing to contemplate giving up the uncertainty relation? Are physicists willing to specify circumstances under which they would give up energy conservation? Numerous historians and philosophers of science (e.g., Kuhn, Mitroff, Feyerabend, Lakatos) have documented the existence of a certain degree of dogmatism about core commitments in scientific research and have argued that such dogmatism plays a constructive role in promoting the aims of science. I am not denying that there may be subtle but important differences between the dogmatism of scientists and that exhibited by many creationists; but one does not even begin to get at those differences by pretending that science is characterized by an uncompromising open-mindedness.

Even worse, the *ad hominem* charge of dogmatism against Creationism egregiously confuses doctrines with the proponents of those doctrines. Since no law mandates that creationists should be invited into the classroom, it is quite irrelevant whether they themselves are close-minded. The Arkansas statute proposed that Creationism be taught, not that creationists should teach it. What counts is the epistemic status of Creationism, not the cognitive idiosyncrasies of the creationists. Because many of the theses of Creationism are testable, the mind set of creationists has no bearing in law or in the fact on the merits of Creationism.

What about the other pair of essential characteristics which the *McLean* Opinion cites, namely, that science is a matter of natural law and explainable by natural law? I find the formulation in the Opinion to be rather fuzzy; but the general idea appears to be that it is inappropriate and unscientific to postulate the existence of any process or fact which cannot be explained in terms of some known scientific laws—for instance, the creationists' assertion that there are outer limits to the change of species "cannot be explained by natural law." Earlier in the Opinion, Judge Overton also writes "there is no scientific explanation for these limits which is guided by natural law," and thus concludes that such limits are unscientific. Still later, remarking on the hypothesis of the Noachian flood, he says, "A worldwide flood as an explanation of the world's geology is not the product of natural law, nor can its occurrence be explained by natural law." Quite how Judge Overton knows that a worldwide flood "cannot" be explained by the laws of science is left opaque; and even if we did not know how to reduce a universal flood to the familiar laws of physics, this requirement is an altogether inappropriate standard for ascertaining whether a claim is scientific. For centuries scientists have recognized a difference between establishing the existence of a phenomenon and explaining that phenomenon in a lawlike way. Our ultimate goal, no doubt, is to do both. But to suggest, as the *McLean* Opinion does repeatedly, that an existence claim (e.g., there was a worldwide flood) is unscientific until we have found the laws on which the alleged phenomenon depends is simply outrageous. Galileo and Newton took themselves to have established the existence of gravitational phenomena, long before anyone was able to give a causal or explanatory account of gravitation. Darwin took himself to have established the existence of natural selection almost a half-century before geneticists were able to lay out the laws of heredity on which natural selection depended. If we took the *McLean* Opinion criterion seriously, we should have to say that Newton and Darwin were unscientific; and, to take an example from our own time, it would follow that plate tectonics is unscientific because we have not yet identified the laws of physics and chemistry which account for the dynamics of crustal motion.

The real objection to such creationist claims as that of the (relative) in-variability of species is not that such invariability has not been explained by scientific laws, but rather that the evidence for invariability is less robust than the evidence for its contrary, variability. But to say as much requires renunciation of the Opinion's order charge—to wit, that Creationism is not testable.

I could continue with this tale of woeful fallacies in the Arkansas ruling, but that is hardly necessary. What is worrisome is that the Opinion's line of reasoning—which neatly coincides with the predominant tactic among scientists who have entered the public fray on this issue—leaves many loopholes for the creationists to exploit. As numerous authors have shown, the require-ments of testability, revisability, and falsifiability are exceedingly *weak* require-ments. Leaving aside the fact that (as I pointed out above) it can be argued that Creationism already satisfies these requirements, it would be easy for a creationist to say the following: "I will abandon my views if we find a living specimen of a species intermediate between man and apes." It is, of course, extremely unlikely that such an individual will be discovered. But, in that statement the creationist would satisfy, in one fell swoop, all the formal requirements of testability, falsifiability, and revisability. If we set very weak standards for scientific status—and, let there be no mistake, I believe that all of the Opinion's last three criteria fall in this category—then it will be quite simple for Creationism to qualify as "scientific."

Rather than taking on the creationists obliquely and in wholesome fashion by suggesting that what they are doing is "unscientific" *tout court* (which is doubly silly because few authors can even agree on what makes an activity scientific), we should confront their claims directly and in piecemeal fashion by asking what evidence and arguments can be marshaled for and against each of them. The core issue is not whether Creationism satisfies some un-demanding and highly controversial definitions of what is scientific; the real question is whether the existing evidence provides stronger arguments for evolutionary theory than for Creationism. Once that question is settled, we will know what belongs in the classroom and what does not. Debating the scientific status of Creationism (especially when "science" is construed in such an unfortunate manner) is a red herring that diverts attention away from the issues that should concern us.

Some defenders of the scientific orthodoxy will probably say that my reservations are just nit-picking ones, and that—at least to a first order of approximation—Judge Overton has correctly identified what is fishy about Creationism. The apologists for science, such as the editor of *The Skeptical Inquirer,* have already objected to those who criticize this whitewash of science "on arcane, semantic grounds . . . [drawn] from the most remote reaches of the academic philosophy of science."[1] But let us be clear about what is at stake. In setting out in the *McLean* Opinion to characterize the "essential" nature of science, Judge Overton was explicitly venturing into philosophical terrain. His *obiter dicta* are about as remote from well-founded opinion in the philosophy of science as Creationism is from respectable geology. It simply will not do for the defenders of science to invoke philosophy of science when it suits them (e.g., their much-loved principle of falsifiability comes directly from the philosopher Karl Popper) and to dismiss it as "arcane" and "remote" when it does not. However noble the motivation, bad philosophy makes for bad law.

The victory in the Arkansas case was hollow, for it was achieved only at the expense of perpetuating and canonizing a false stereotype of what science is and how it works. If it goes unchallenged by the scientific community, it will raise grave doubts about that community's intellectual integrity. No one familiar with the issues can really believe that anything important was settled through anachronistic efforts to revive a variety of discredited criteria for distinguishing between the scientific and the non-scientific. Fifty years ago, Clarence Darrow asked, *à propos* the Scopes trial, "Isn't it difficult to realize that a trial of this kind is possible in the twentieth century in the United States of America?" We can raise that question anew, with the added irony that, this time, the pro-science forces are defending a philosophy of science which is, in its way, every bit as outmoded as the "science" of the creationists.

NOTE

1. "The Creationist Threat: Science Finally Awakens," *The Skeptical Inquirer* 6, no. 3 (Spring 1982): 2 5.

23

Pro Judice

Michael Ruse

As always, my friend Larry Laudan writes in an entertaining and provocative manner, but, in his complaint against Judge William Overton's ruling in *McLean v. Arkansas,*[1] Laudan is hopelessly wide of the mark. Laudan's outrage centers on the criteria for the demarcation of science which Judge Overton adopted, and the judge's conclusion that, evaluated by these criteria, creation-science fails as science. I shall respond directly to this concern—after making three preliminary remarks.

First, although Judge Overton does not need defense from me or anyone else, as one who participated in the Arkansas trial, I must go on record as saying that I was enormously impressed by his handling of the case. His written judgment is a first-class piece of reasoning. With cause, many have criticized the State of Arkansas for passing the "Creation-Science Act," but we should not ignore that, to the state's credit, Judge Overton was born, raised, and educated in Arkansas.

Second, Judge Overton, like everyone else, was fully aware that proof that something is not science is not the same as proof that it is religion. The issue of what constitutes science arose because the creationists claim that their ideas qualify as genuine science rather than as fundamentalist religion. The attorneys developing the American Civil Liberties Union (ACLU) case believed it important to show that creation-science is not genuine science. Of course, this demonstration does raise the question of what creation-science really is. The plaintiffs claimed that creation-science always was (and still is) religion. The plaintiffs' lawyers went beyond the negative argument (against

From *Science, Technology, & Human Values* 7, no. 41 (1982): 19–23. Copyright © 1982. Reprinted by permission of John Wiley & Sons, Inc.

science) to make the positive case (for religion). They provided considerable evidence for the religious nature of creation-science, including such things as the creationists' explicit reliance on the Bible in their various writings. Such arguments seem about as strong as one could wish, and they were duly noted by Judge Overton and used in support of his ruling. It seems a little unfair, in the context, therefore, to accuse him of "specious" argumentation. He did not adopt the naive dichotomy of "science or religion but nothing else."

Third, whatever the merits of the plaintiffs' case, the kinds of conclusions and strategies apparently favored by Laudan are simply not strong enough for legal purposes. His strategy would require arguing that creation-science is weak science and therefore ought not to be taught:

> The core issue is not whether Creationism satisfies some undemanding and highly controversial definitions of what is scientific; the real question is whether the existing evidence provides stronger arguments for evolutionary theory than for Creationism. Once that question is settled, we will know what belongs in the classroom and what does not.[2]

Unfortunately, the U.S. Constitution does not bar the teaching of weak science. What it bars (through the Establishment Clause of the First Amendment) is the teaching of religion. The plaintiffs' tactic was to show that creation-science is less than weak or bad science. It is not science at all.

Turning now to the main issue, I see three questions that must be addressed. Using the five criteria listed by Judge Overton, can one distinguish science from non-science? Assuming a positive answer to the first question, does creation-science fail as genuine science when it is judged by these criteria? And, assuming a positive answer to the second, does the Opinion in *McLean* make this case?

The first question has certainly tied philosophers of science in knots in recent years. Simple criteria that supposedly give a clear answer to every case— for example, Karl Popper's single stipulation of falsifiability[3]—will not do. Nevertheless, although there may be many gray areas, white does seem to be white and black does seem to be black. Less metaphorically, something like psychoanalytic theory may or may not be science, but there do appear to be clear-cut cases of real science and of real non-science. For instance, an explanation of the fact that my son has blue eyes, given that both parents have blue eyes, done in terms of dominant and recessive genes and with an appeal to Mendel's first law, is scientific. The Catholic doctrine of transubstantiation (i.e., that in the Mass the bread and wine turn into the body and blood of Christ) is not scientific.

Furthermore, the five cited criteria of demarcation do a good job of distinguishing the Mendelian example from the Catholic example. Law and explanation through law come into the first example. They do not enter the second. We can test the first example, rejecting it if necessary. In this case, it is tentative, in that something empirical might change our minds. The case of transubstantiation is different. God may have His own laws, but neither scientist nor priest can tell us about those which turn bread and wine into flesh and blood. There is no explanation through law. No empirical evidence is pertinent to the miracle. Nor would the believer be swayed by any empirical facts. Microscopic examination of the Host is considered irrelevant. In this sense, the doctrine is certainly not tentative.

One pair of examples certainly do not make for a definitive case, but at least they do suggest that Judge Overton's criteria are not quite as irrelevant as Laudan's critique implies. What about the types of objections (to the criteria) that Laudan does or could make? As far as the use of law is concerned, he might complain that scientists themselves have certainly not always been that particular about reference to law. For instance, consider the following claim by Charles Lyell in his *Principles of Geology* (1830/3): "We are not, however, contending that a real departure from the antecedent course of physical events cannot be traced in the introduction of man."[4] All scholars agree that in this statement Lyell was going beyond law. The coming of man required special divine intervention. Yet, surely the *Principles* as a whole qualify as a contribution to science.

Two replies are open: either one agrees that the case of Lyell shows that science has sometimes mingled law with non-law; or one argues that Lyell (and others) mingled science and non-science (specifically, religion at this point). My inclination is to argue the latter. Insofar as Lyell acted as scientist, he appealed only to law. A century and a half ago, people were not as conscientious as today about separating science and religion. However, even if one argues the former alternative—that some science has allowed place for non-lawbound events—this hardly makes Laudan's case. Science, like most human cultural phenomena, has evolved. What was allowable in the early nineteenth century is not necessarily allowable in the late twentieth century. Specifically, science today does not break with law. And this is what counts for us. We want criteria of science for today, not for yesterday. (Before I am accused of making my case by fiat, let me challenge Laudan to find one point within the modern geological theory of plate tectonics where appeal is made to miracles, that is, to breaks with law. Of course, saying that science appeals to law is not asserting that we know all of the laws. But, who said that we did? Not Judge Overton in his Opinion.)

What about the criterion of tentativeness, which involves a willingness to test and reject if necessary? Laudan objects that real science is hardly all that tentative: "[H]istorical and sociological researches on science strongly suggest that the scientists of any epoch likewise regard some of their beliefs as so fundamental as not to be open to repudiation or negotiation."[5]

It cannot be denied that scientists do sometimes—frequently—hang on to their views, even if not everything meshes precisely with the real world. Nevertheless, such tenacity can be exaggerated. Scientists, even Newtonians, have been known to change their minds. Although I would not want to say that the empirical evidence is all-decisive, it plays a major role in such mind changes. As an example, consider a major revolution of our own time, namely, that which occurred in geology. When I was an undergraduate in 1960, students were taught that continents do not move. Ten years later, they were told that they do move. Where is the dogmatism here? Furthermore, it was the new empirical evidence—e.g., about the nature of the sea-bed—which persuaded geologists. In short, although science may not be as open-minded as Karl Popper thinks it is, it is not as close-minded as, say, Thomas Kuhn[6] thinks it is.

Let me move on to the second and third questions, the status of creation-science and Judge Overton's treatment of the problem. The slightest acquaintance with the creation-science literature and Creationism movement shows that creation-science fails abysmally as science. Consider the following passage, written by one of the leading creationists, Duane T. Gish, in *Evolution: The Fossils Say No!:*

> CREATION. By creation we mean the bringing into being by a supernatural Creator of the basic kinds of plants and animals by the process of sudden, or fiat, creation.
>
> We do not know how the Creator created, what processes He used, *for He used processes which are not operating anywhere in the natural universe.* This is why we refer to creation as Special Creation. We cannot discover by scientific investigations anything about the creative processes used by the Creator.[7]

The following similar passage was written by Henry M. Morris, who is considered to be the founder of the creation-science movement:

> ... it is ... quite impossible to determine anything about Creation through a study of present processes, because present processes are not created in character. If man wishes to know anything about Creation (the time of Creation, the duration of Creation, the order of Creation, the methods of Creation, or anything else) his sole source of true information is that of divine revelation. God was there when it happened. We were not there ... therefore, we are completely limited to what God has seen fit to tell us, and this information is in His written Word. This is our textbook on the science of Creation![8]

By their own words, therefore, creation-scientists admit that they appeal to phenomena not covered or explicable by any laws that humans can grasp as laws. It is not simply that the pertinent laws are not yet known. Creative processes stand outside law as humans know it (or could know it) on Earth—at least there is no way that scientists can know Mendel's laws through observation and experiment. Even if God did use His own laws, they are necessarily veiled from us forever in this life, because Genesis says nothing of them.

Furthermore, there is nothing tentative or empirically checkable about the central claims of creation-science. Creationists admit as much when they join the Creation Research Society (the leading organization of the movement). As a condition of membership applicants must sign a document specifying that they now believe and will continue to believe:

(1) The Bible is the written Word of God, and because we believe it to be inspired throughout, all of its assertions are historically and scientifically true in all of the original autographs. To the student of nature, this means that the account of origins in Genesis is a factual presentation of simple historical truths. (2) All basic types of living things, including man, were made by direct creative acts of God during Creation Week as described in Genesis. Whatever biological changes have occurred since Creation have accomplished only changes within the original created kinds. (3) The great Flood described in Genesis, commonly referred to as the Noachian Deluge, was an historical event, worldwide in its extent and effect. (4) Finally, we are an organization of Christian men of science, who accept Jesus Christ as our Lord and Savior. The account of the special creation of Adam and Eve as one man and one woman, and their subsequent fall into sin, is the basis for our belief in the necessity of a Savior for all mankind. Therefore, salvation can come only thru accepting Jesus Christ as our Savior.[9]

It is difficult to imagine evolutionists signing a comparable statement, that they will never deviate from the literal text of Charles Darwin's *On the Origin of Species*. The non-scientific nature of creation-science is evident for all to see, as is also its religious nature. Moreover, the quotes I have used above were all used by Judge Overton, in the *McLean* Opinion, to make exactly the points I have just made. Creation-science is not genuine science, and Judge Overton showed this.

Finally, what about Laudan's claim that some parts of creation-science (e.g., claims about the Flood) are falsifiable and that other parts (e.g., about the originally created "kinds") are revisable? Such parts are not falsifiable or revisable in a way indicative of genuine science. Creation-science is not like physics, which exists as part of humanity's common cultural heritage and domain. It exists solely in the imaginations and writing of a relatively small group of people. Their publications (and stated intentions) show that, for example, there is no way they will relinquish belief in the Flood, whatever the evidence.[10] In this sense, their doctrines are truly unfalsifiable.

Furthermore, any revisions are not genuine revisions, but exploitations of the gross ambiguities in the creationists' own position. In the matter of origins, for example, some elasticity could be perceived in the creationist position, given the conflicting claims about the possibility of (degenerative) change within the originally created "kinds." Unfortunately, any open-mindedness soon proves illusory for creationists have no real idea about what God is supposed to have created in the beginning, except that man was a separate species. They rely solely on the Book of Genesis:

And God said, Let the waters bring forth abundantly the moving creature that hath life, and the fowl that may fly above the earth in the open firmament of heaven.

And God created great whales, and every living creature that moveth, which the waters brought forth abundantly, after their kind, and every winged fowl after his kind: and God saw that it was good.

And God blessed them, saying Be fruitful, and multiply, and fill the waters in the seas, and let fowl multiply in the earth.

And the evening and the morning were the fifth day.

And God said, Let the earth bring forth the living creature after his kind, cattle, and creeping thing, and beast of the earth after his kind: and it was so.

And God made the beast of the earth after his kind, and cattle after their kind, and everything that creepeth upon the earth after his kind: and God saw that it was good.[11]

But the *definition* of "kind," what it really is, leaves creationists as mystified as it does evolutionists. For example, creationist Duane Gish makes this statement on the subject:

[W]e have defined a basic kind as including all of those variants which have been derived from a single stock . . . We cannot always be sure, however, what constitutes a separate kind. The division into kinds is easier the more the divergence observed. It is obvious, for example, that among invertebrates the protozoa, sponges, jellyfish, worms, snails, trilobites, lobsters, and bees are all different kinds. Among the vertebrates, the fishes, amphibians, reptiles, birds, and mammals are obviously different basic kinds.

Among the reptiles, the turtles, crocodiles, dinosaurs, pterosaurs (flying reptiles), and ichthyosaurs (aquatic reptiles) would be placed in different kinds. Each one of these major groups of reptiles could be further subdivided into the basic kinds within each.

Within the mammalian class, duck-billed platypus, bats, hedgehogs, rats, rabbits, dogs, cats, lemurs, monkeys, apes, and men are easily assignable to different basic kinds. Among the apes, the gibbons, orangutans, chimpanzees, and gorillas would each be included in a different basic kind.[12]

Apparently, a "kind" can be anything from humans (one species) to trilobites (literally thousands of species). The term is flabby to the point of inconsistency. Because humans are mammals, if one claims (as creationists do) that evolution can occur within but not across kinds, then humans could have evolved from common mammalian stock—but because humans themselves are kinds such evolution is impossible.

In brief, there is no true resemblance between the creationists' treatment of their concept of "kind" and the openness expected of scientists. Nothing can be said in favor of creation-science or its inventors. Overton's judgment emerges unscathed by Laudan's complaints.

NOTES

1. For the text of Judge Overton's Opinion, see *Science, Technology, & Human Values* 40 (Summer 1982): 28–42; and Marcel LaFollette, *Creationism, Science, and the Law* (The MIT Press, 1983).

2. Larry Laudan, "Commentary: Science at the Bar—Causes for Concern," p. 351 of this volume.

3. Karl Popper, *The Logic of Scientific Discovery* (London: Hutchinson, 1959).

4. Charles Lyell, *Principles of Geology,* Volume I (London: John Murray, 1830), p. 162.

5. Laudan, *op. cit.,* p. 17.

6. Thomas Kuhn, *The Structure of Scientific Revolutions* (Chicago, Ill.: University of Chicago Press, 1962).

7. Duane Gish, *Evolution: The Fossils Say No!,* 3rd edition (San Diego, Calif.: Creation-Life Publishers, 1979), p. 40 (his italics).

8. Henry M. Morris, *Studies in the Bible and Science* (Philadelphia, Penn.: Presbyterian and Reformed Publishing Company, 1966), p. 114.

9. Application form for the Creation Research Society, reprinted in Plaintiffs' trial briefs, *McLean v. Arkansas* (1981).

10. See, for instance, Henry M. Morris, *Scientific Creationism* (San Diego, Calif.: Creation-Life Publishers, 1974); and my own detailed discussion in Michael Ruse, *Darwinism Defended: A Guide to the Evolution Controversies* (Reading, Mass.: Addison-Wesley, 1982).

11. Genesis, Book I, Verses 20-25.

12. Gish, *op. cit.,* pp. 34-35.

24

More on Creationism

Larry Laudan

Michael Ruse is distressed that I have taken exception to Judge William Overton's opinion in *McLean v. Arkansas.* Where I saw that ruling as full of sloppy arguments and non sequiturs, he hails it as "a first-class piece of reasoning." Since Ruse has claimed that my reservations are "hopelessly wide of the mark," I feel obliged to enter the fray once again, in the hope that reiteration will achieve what my initial argument has evidently failed to pull off, namely, to convince knee-jerk demarcationists like Ruse that things are more complicated than they have conceded.

In my short commentary, I sought to show: (1) that the criteria which Judge Overton offered as "essential conditions" of science are nothing of the sort, since many parts of what we all call "science" fail to satisfy those conditions; (2) that several of Overton's criteria constitute extremely *weak* demands from an epistemic point of view, so weak that if Creationism does not already satisfy them (which I believe it manifestly does), it would be child's play for creationists to modify their position slightly—thus making their enterprise (by Overton's lights) "scientific"; and (3) that Overton's preoccupation with the dogmatism and closemindedness of the advocates of creation-science has led him into a chronic confusion of doctrines and their advocates.

Ruse makes no reply to the second point. Quite why is unclear since, standing entirely alone, it is more than sufficient to give one pause about the worrying precedents set in *McLean v. Arkansas.* But Ruse does deal, after a fashion, with points (1) and (3). Since we do not see eye-to-eye about these matters, I shall try to redirect his gaze.

From *Science, Technology, & Human Values* 8 no. 42 (1983): 36–38. Copyright © 1983. Reprinted by permission of John Wiley & Sons, Inc.

THE LOGIC OF "ESSENTIAL CONDITIONS"

Consider the following parable: Suppose that some city dweller said that the "essential conditions" for something to be a sheep were that it be a medium-sized mammal and that it invariably butt into any human beings in its vicinity. A country fellow might try to suggest that his city cousin evidently did not understand what a sheep was. He might show, for instance, that there are plenty of things we call sheep which never butt into anything, let alone human beings. He might go further to say that what the city fellow is calling a sheep is what all the rest of us regard as a goat. Suppose, finally, that a second city fellow, on hearing his town friend abused by the bucolic bumpkin, entered the discussion saying "I once knew a sheep that butted into human beings without hesitation, and besides I once saw a goat which never bothered human beings. Accordingly, it is correct to say that the essential conditions of being a sheep are exactly what my friend said they were!"

Confronted by Michael Ruse's efforts to defend Overton's definition of science in the face of my counterexamples, I find myself as dumbfounded as the mythical farm boy. Overton offered five "essential characteristics of science." I have shown that there are respectable examples of science which violate *each* of Overton's desiderata, and moreover that there are many activities we do not regard as science which satisfy many of them. Stepping briskly to Overton's defense, Ruse points to *one* example of a scientific principle (Mendel's first law) which does fit Overton's definition and to one example of a non-science (the thesis of transubstantiation) which does not. "You see," Ruse seems to conclude, "Laudan is simply wrongheaded and Overton got it basically right."

At the risk of having to tell Ruse that he does not know how to separate the sheep from the goats, I beg to differ. To make his confusion quite explicit, I shall drop the parable and resort to some symbols. Whenever someone lists a set of conditions, $C_1, C_2, \ldots, C_n$, and characterizes them as "essential characteristics" of a concept (say, science), that is strictly equivalent to saying that each condition, C_1, taken individually is a *necessary* condition for scientific status. One criticizes a proposed set of necessary or essential conditions by showing that some things which clearly fall under the concept being explicated fail to satisfy the proposed necessary conditions. In my short essay, and elsewhere, I have offered plausible counterexamples to each of Overton's five characteristics.

Ruse mounts no challenge to those counterexamples. Instead, he replies by presenting instances (actually only one, but it would have been no better if he had given a hundred) of science which do satisfy Overton's demands. This is clearly to no avail because I was not saying that *all* scientific claims were (to take but one of Overton's criteria) untestable, only that *some* were. Indeed, so long as there is but one science that fails to exemplify Overton's features, then one would be ill-advised to use his demarcation criterion as a device for separating the "scientific" from the "non-scientific." Ruse fails to see the absolute irrelevance to my argument of his rehearsing examples that "fit" Overton's analysis. Similarly, I did not say that *all* non-scientific claims were testable, only that *some* were. So once again, Ruse is dangling a red herring before us when he reminds us that the thesis of transubstantiation is untestable. Finding untestable bits of non-science to buttress the claim that testability is the hallmark of science is rather like defending the claim that butting humans is essential to being a sheep by pointing out that there are many non-sheep (e.g., tomatoes) which fail to butt humans.

BELIEFS AND BELIEVERS

There is a more interesting—if equally significant—confusion running through much of Ruse's discussion, a confusion revealing a further failure to come to terms with the case I was propounding in "Science at the Bar." I refer to his (and Overton's) continual slide between assessing doctrines and assessing those who hold the doctrines. Ruse reminds us (and this loomed large in the *McLean* opinion as well) that many advocates of creation-science tend to be dogmatic, slow to learn from experience, and willing to resort to all manner of ad hoc strategies so as to hold onto their beliefs in the face of counterevidence. For the sake of argument, let all that be granted; let us assume that the creationists exhibit precisely those traits of intellectual dishonesty which the friends of science scrupulously and unerringly avoid. Ruse believes (and Judge Overton appears to concur) that, if we once establish these traits to be true of creationists, then we can conclude that Creationism is untestable and unfalsifiable (and "therefore unscientific").

This just will not do. Knowing something about the idiosyncratic mind-set of various creationists may have a bearing on certain practical issues (such as "Would you want your daughter to marry one?"). But we learned a long time ago that there is a difference between *ad hominem* and *ad argumentum*. Creationists make assertions about the world. Once made, those assertions take on a life of their own. Because they do, we can assess the merits or demerits of creationist theory without having to speculate about the unsavoriness of the mental habits of creationists. What we do, of course, is to examine the empirical evidence relevant to the creationist claims about earth history. If those claims are discredited by the available evidence (and by "discredited" I mean impugned by the use of rules of reasoning which legal and philosophical experts on the nature of evidence have articulated), then Creationism can safely be put on the scrap heap of unjustified theories.

But, intone Ruse and Overton, what if the creationists *still* do not change their minds, even when presented with what most people regard as thoroughly compelling refutations of their theories? Well, that tells us something interesting about the psychology of creationists, but it has no bearing whatever on an assessment of their doctrines. After all, when confronted by comparable problems in other walks of life, we proceed exactly as I am proposing, that is, by distinguishing beliefs from believers. When, for instance, several experiments turn out contrary to the predictions of a certain theory, we do not care whether the scientist who invented the theory is prepared to change his mind. We do not say that his theory cannot be tested, simply because he refuses to accept the results of the test. Similarly, a jury may reach the conclusion, in light of the appropriate rules of evidence, that a defendant who pleaded innocent is, in fact, guilty. Do we say that the defendant's assertion "I am innocent" can be tested only if the defendant himself is prepared to admit his guilt when finally confronted with the *coup de grâce?*

In just the same way, the soundness of creation-science can and must be separated from all questions about the dogmatism of creationists. Once we make that rudimentary separation, we discover both (a) that creation-science is testable and falsifiable, and (b) that creation-science has been tested and falsified—insofar as any theory can be said to be falsified. But, as I pointed out in the earlier essay, that damning indictment cannot be drawn so long as we confuse Creationism and creationists to such an extent that we take the creationists' mental intransigence to entail the immunity of creationist theory from empirical confrontation.

25

The Philosopher of Science as Expert Witness

Philip L. Quinn

When I was invited to contribute to this volume, the editors expressed the hope that the essays in it would paint "an overall picture of where philosophy of science is going in the near future." I hesitated. Since I am no prophet, could there be any profit in my speculating on the future of philosophy of science? In the course of the past generation or so, philosophy of science in the anglophone world has achieved professional autonomy. It has its own specialist journals and professional meetings; independent programs and departments in history and philosophy of science have taken root and flourished in major universities. Mastery of a formidable body of technical background knowledge has become a prerequisite for speaking with authority about a wide range of problems in philosophy of physics, philosophy of biology, philosophy of psychology, and philosophy of economics. Ties with other, more traditional areas of philosophy have been loosened if not dissolved; new jargon has sprung up like weeds. In short, philosophy of science has acquired a turf of its own to defend, become inaccessible to outsiders, and taken on the trappings of a typical academic specialization. So it would seem that the agenda for philosophy of science in the near future will be set by the insiders in each of its subspecializations. Whatever problems they find tractable will dominate discussion in the journals and at meetings. And since tractability is relative to the tools that happen to be at hand, unpredictable opportunism would appear to be the preferred tactic for philosophers of science, as it has often proved to be for scientists.

From *Science and Reality: Recent Work in the Philosophy of Science*, edited by James T. Cushing, C. F. Delaney, and Gary M. Gutting. Copyright © 1984 by University of Notre Dame Press. Reprinted by permission.

Yet this can hardly be the whole story. Philosophy of science is not so hermetically sealed off, not yet anyway, from the broader currents of contemporary intellectual life that it is completely insensitive to influences and pressures from the outside. Like other areas of philosophy, philosophy of science has not yet become nothing but a glass bead game. Indeed, external pressure may be producing movement in quite the opposite direction. Contemporary philosophy, so we are told, has taken an applied turn during the past twenty years. Starting with a renewal of interest in normative questions in ethics and political philosophy in the late sixties, the process has been gathering momentum ever since. Now there are philosophical societies, journals, programs or meetings devoted to medical ethics, legal ethics, business ethics, engineering ethics, environmental ethics, ethics and animals, and agricultural ethics; and the end of the process is nowhere in sight. It is, of course, no accident that the process began at a time when the United States was bogged · down in a despicable war in Southeast Asia. Nor is it entirely coincidental that it has continued during a period of hard times in the academic marketplace. Philosophers are no less prone than others to looking for a justification of their activities in terms of social utility and for a match between their intellectual interests and economic demand. What was good enough for the scientific community in the forties, when Vannevar Bush wrote his ideological masterpiece *Science: The Endless Frontier,* was not to be despised by the philosophical community of the seventies and eighties. And so the word went out that philosophers too had their bit of expertise to contribute to the formulation and implementation of public policy. Could philosophers of science join the ethicists in taking the applied turn? Should they?

At first glance, it would appear that philosophers of science were ideally positioned to jump on the bandwagon. We live, we are told, in an age of science. The present division of intellectual labor confers unprecedented cognitive authority on what is taken to be reputable scientific thinking. It is a mark of enlightened lay opinion to defer to what the scientific experts say on an immense variety of topics. If we are told that it is scientifically respectable to believe that the universe originated in a big bang billions of years ago, we nod assent without further ado, even if we have a very imperfect grasp of what is actually being asserted and no idea at all about what evidence could be adduced to support such an assertion. If we are told that it is not scientifically respectable to believe that ancient astronauts had an airfield constructed on the plain of Nazca, we hasten to agree, even if the hypothesis of ancient astronauts seems intuitively plausible. And if we are told that respectable scientific opinion is divided about whether there is a safe level of radiation exposure, we think it rationally incumbent upon ourselves to try not to panic immediately if we learn we have been exposed to low-level radiation. But whence comes the authority of scientific opinion? And how is the boundary around respectable scientific opinion to be drawn? Most scientists would be hard pressed to give decent answers to such questions. Like the practitioners of other crafts, scientists tend to be too busy doing science to spend much time reflecting upon or articulating the epistemological presuppositions of their characteristic activities. Textbook clichés about the scientific method, imperfectly remembered, serve as proxies for serious thought about such questions in the minds of many scientists. By contrast, philosophers of science are supposed to be experts on such second-order questions. From historical case studies, they are supposed to have learned what has differentiated scientific thinking from superstition and other modes of thought. From rational reconstructions of the achievements of contemporary science, they are supposed to have derived insight into the grounds of science's epistemic authority. So philosophers of science ought to be able to provide politicians and granting agencies with good advice about which proposals have the earmarks of respectable science and deserve funding, and they ought to be able to explain to the educated lay public even better than most scientists why confidence in respectable scientific opinion is justified and what counts as respectable scientific opinion. Thus it would seem that philosophers of science are well placed to take the applied turn and to grab a piece of the policy-making action.

As a matter of fact, philosophers of science have been rather slow to inject themselves into the process of creating public policy. By way of illustration, consider the proceedings of a conference held in 1977, sponsored by the Philosophy of Science Association and funded by the National Science Foundation, on critical research problems in philosophy of science.[1] These proceedings contain papers on methodological approaches to philosophy of science, philosophy of science and other disciplines, meta-science topics, and philosophical foundations of various scientific disciplines. Apart from some sensible but rather mundane suggestions in one paper about incorporating some results from philosophy of science into science education, they contain no discussion of whether philosophy of science has any bearing on what goes on outside the groves of academe. Only within the past two years have we seen a case in which philosophy of science has played a major and well-documented role in a significant policy decision. This case has all the makings of an interesting precedent. It has already generated polite but fierce controversy among philosophers of science both in print and in discussion and is likely to provoke further debate. It illustrates vividly some of the pitfalls which await philosophers of science who apply their expertise to issues in the political arena. So if we subject this case to close scrutiny, we may be able to arrive at some tentative conclusions about whether an applied turn in philosophy of science is likely to lead into the corridors of power or only into a dead-end street. And these conclusions may help us make up our minds about whether we think it a good idea to respond to whatever pressures there may be toward an applied turn in philosophy of science by giving in to them or by resisting them. We shall then be in a better position to address the question of whether this is a direction in which philosophy of science ought to go in the near future.

The case I have in mind is, of course, the notorious legal case *McLean v. Arkansas.* On March 19, 1981, the Governor of Arkansas signed into law Act 590 of 1981, entitled the "Balanced Treatment of Creation-Science and

Evolution-Science Act," which required balanced treatment of creation science and evolutionary biology in the public schools of Arkansas. On May 27, 1981, a group of plaintiffs mounted a challenge to the Act's constitutional validity by bringing suit in U.S. District Court against the Arkansas Board of Education. The plaintiffs alleged that Act 590 violated the Establishment Clause of the First Amendment. Trial began on December 7, 1981. The Court's Opinion, handed down by Judge William R. Overton on January 5, 1982, held the Act to be unconstitutional. Michael Ruse, a prominent philosopher of biology, provided a number of position papers for the plaintiffs, and testified as an expert witness for the plaintiffs, on the defining characteristics of science. Judge Overton's Opinion acknowledges that its statement of the essential characteristics of science is based in large part on Ruse's testimony. On the basis of this statement of the essential characteristics of science, the Opinion argues that creation science is not real science. Subsequent to the Court's decision, Larry Laudan, another prominent philosopher of science, severely criticized the argumentation of the Opinion. Ruse responded with a spirited defense of Judge Overton's reasoning and, incidentally, of his own views on the nature of science. Later Laudan argued that Ruse had not successfully replied to his original criticisms.

In this paper, I propose to do a detailed analysis of this controversy. In the course of the discussion it will become clear that I agree with many but not all of Laudan's critical points. But I think more needs to be said in criticism both of Judge Overton's Opinion and of Ruse's philosophical views. First, I shall lay out the main lines of the relevant portion of the argument in Judge Overton's Opinion and try to show that a crucial part of the argument is unsound. Then I shall consider the question of whether the argument of the Opinion accurately reflects the views Ruse has expressed in print. I shall argue that, although judging by what Ruse says, there is some ambiguity about what views he actually holds, the Opinion is in essential respects a good representation of his published views. From these arguments, I shall extract some lessons about the dangers philosophers of science let themselves in for when they assume the role of expert witness. Because I know Ernan McMullin has a long-standing interest in both the methodological question I shall be discussing and the problem of the proper relation between theology and science raised by *McLean v. Arkansas,* I hope he will find the story I have to tell engaging and, perhaps, somewhat amusing.

JUDGE OVERTON'S OPINION

According to the Opinion, a statute violates the Establishment Clause if it fails any part of the following three-pronged test:

> First, the statute must have a secular legislative purpose; second, its principal or primary effect must be one that neither advances nor inhibits religion . . . ; finally, the statute must not foster "an excessive government entanglement with religion."[2]

Though a proof that Act 590 failed on any one part of this three-pronged test would have been sufficient to show that it was unconstitutional, the Opinion argues that the Act fails on all three parts. Since the issues of importance from the point of view of philosophy of science all arise in the course of the argument that the Act fails the second part of the test, I shall confine my analysis to the Opinion's reasoning to the conclusion that the Act's primary effect is the advancement of religion. Because I am going to be very critical of that reasoning, I should say at the outset that I find the argument for the conclusion that Act 590 fails the first part of the test entirely convincing. The Opinion cites precedents to show that courts are not bound to consider only legislative statements of purpose or legislative disclaimers in determining the legislative purpose of a statue but may also consider such factors as the historical context of a statute, the specific sequence of events leading up to passage of the statute, departures from normal procedural sequences, substantive departures from the normal, and contemporaneous statements of the legislative sponsor. Appealing to evidence which is relevant to such contextual factors, the Opinion argues that Act 590 was passed by the Arkansas General Assembly with the specific purpose of advancing religion and so fails to have a secular legislative purpose. I agree. And since failure on this part of the three-pronged test is, as noted above, by itself sufficient to show the Act to be in violation of the Establishment Clause, I also concur with Judge Overton's ultimate conclusion that Act 590 is unconstitutional. If the Opinion had argued no more than this, which would have been enough to dispose of Act 590, I would have no quarrel with it. Unfortunately, the Opinion showed no such admirable restraint.

Judge Overton's argument to the conclusion that Act 590 has the advancement of religion as its primary effect may be schematically reconstructed as follows:

(1) Act 590 does have the advancement of religion as a major effect.
(2) Act 590 does not have the advancement of science as an effect.
(3) Act 590 does not have the advancement of any other thing as an effect.
(4) Hence, Act 590 has the advancement of religion as its only effect.
(5) Whence, Act 590 has the advancement of religion as its primary effect.

The argument is intuitively valid: (4) follows from (1)-(3), and (5) follows from (4). But is it sound? It would appear that each of (1)-(3) needs further justification.

It is curious that Judge Overton says nothing at all to justify (3). Indeed, the Opinion shows no indication that he is aware of the need to assume (3), or something very much like it, to insure the validity of its argument. Speaking to this point, Laudan takes Judge Overton to task for making "the claim, specious in its own right, that since Creationism is not 'science' it must be religion."[3] In response, Ruse tries to defend Judge Overton against the charge of having adopted "the naive dichotomy of 'science or religion but nothing else.'"⁴ But Judge Overton does not make the specious claim Laudan attributes to him, and Ruse's reply does nothing to show that Judge Overton is not committed to a somewhat different but equally dubious thesis. Both Laudan and Ruse have missed the point; neither has seen where the real problem with this part of Judge Overton's argument lies.

Judge Overton first argues that creation science is, or at least is inspired by, religion. This argument supports (1), and I will have more to say about it later on. He then gives an independent argument to show that creation science is not science. This argument supports (2), and I will later analyze it in great detail. Finally, Judge Overton couples the conclusions of the previous arguments in this passage:

> The conclusion that creation science has no scientific merit or educational value as science has legal significance in the light of the Court's previous conclusion that creation science has, as one major effect, the advancement of religon. The second part of the three-pronged test for establishment reaches only those statutes having as their *primary* effect the advancement of religion. Secondary effects which advance religion are not constitutionally fatal. Since creation science is not science, the conclusion is inescapable that the *only* real effect of Act 590 is the advancement of religion.⁵

But the conclusion is very far indeed from being inescapable. Though it does not rest on the specious claim that since creation science is not science it must be religion, it does rest on the claim that since creation science advances religion but not science it advances only religion. This claim is certainly not obvious and needs to be supported by argument. To argue for it would be to argue in support of something like (3) which would rule out other possibilities. This the Opinion does not do, and the failure to do it is a serious flaw in the Opinion. As long as the possibility that creation science is, above and beyond being religion and not science, some other thing as well (perhaps, for example, speculative philosophy) has not been ruled out, it remains an open question whether Act 590 also has the advancement of some such other thing as a major effect, and hence it remains an open question whether Act 590 has the advancement of this other thing as its primary effect and the advancement of religion only as its secondary effect. Since, as Judge Overton says, secondary effects which advance religion are not constitutionally fatal, such questions cannot be left open. The Opinion, however, contains no arguments for a principle like (3) which would close them. Though, for all I know, (3) may be true and so, for all I have said so far, the argument in which it is a premise may be sound, that argument is unconvincing because (3) is not obviously true and the Opinion furnishes no evidence to support it. Perhaps the best that can be done by way of a defense of Judge Overton's reasoning in the passage quoted above is to suppose that, since the defendants had not claimed that Act 590 has the advancement of something other than science as an effect,

Judge Overton thought there was no need for him to consider possibilities that had not come up in the course of the trial. If Judge Overton did think something like this, I would say he made a mistake, for his failure to rule out such possibilities makes his argument less than rationally compelling.

As I mentioned before, the Opinion does contain an argument for (1). It begins with a citation from Section 4 of Act 590:

> *Definitions:* As used in this Act:
> (a) "Creation-science" means the scientific evidences for creation and inferences from those scientific evidences. Creation-science includes the scientific evidences and related inferences that indicate: (1) Sudden creation of the universe, energy, and life from nothing; (2) The insufficiency of mutation and natural selection in bringing about development of all living kinds from a single organism; (3) Changes only within fixed limits of originally created kinds of plants and animals; (4) Separate ancestry for man and apes; (5) Explanation of the earth's geology by catastrophism, including the occurrence of a world-wide flood; and (6) A relatively recent inception of the earth and living kinds.[6]

According to the Opinion, "Section 4(a) is unquestionably a statement of religion, with the exception of 4(a)(2) which is a negative thrust aimed at what the creationists understand to be the theory of evolution."[7] As I understand Judge Overton's reasoning, he deploys two distinct arguments in support of this claim. The more general argument aims to establish that the reference to creation from nothing in 4(a)(1) is religious. To show that the concept of creation from nothing is the concept of creation by God, Judge Overton appeals to uniformity of testimony. He notes, first, that all the theologians who testified, including defense witnesses, were of the opinion that the concept of creation from nothing is the concept of creation by God and, second, that leading creationist writers are of the same opinion. To show that the concept of creation by God is religious, Judge Overton appeals to legal precedent. He cites from *Malnak v. Yogi* the opinion that concepts concerning a supreme being are religious and do not cease to be so in virtue of being presented as philosophy or science. Since (i) the concept of creation from nothing is the concept of creation by God, and (ii) the concept of creation by God is the concept of creation by a supreme being, and (iii) any concept concerning a supreme being is a religious concept, the reference to creation from nothing in 4(a)(1) is therefore religious. The more specific argument aims to show both that Section 4(a) as a whole is congruent with the literal interpretation of Genesis favored by fundamentalist Christians and that the ideas in 4(a)(1) in particular are identical to the ideas in the Genesis creation story as literally interpreted. To support the congruence claim Judge Overton enumerates extensive and specific parallels between Section 4(a) and Genesis. They include: (i) the parallel between 4(a)(1) and Genesis 1; (ii) the parallel between 4(a)(5) and Genesis 7 and 8; and (iii) the parallel between the use of the term "kinds" in 4(a)(2), 4(a)(3) and 4(a)(6) and its use in Genesis. To support the identity claim Judge Overton says only that the ideas of 4(a)(1) are parallel to the creation story in Genesis 1 and are not parallel to any other creation story. From the combination of the two arguments, Judge Overton draws the conclusion that there is "no doubt that a major effect of the Act is the advancement of particular religious beliefs."[8]

Because my purpose in this paper is not to deal in depth with the tangled knot of issues in theology and scriptural exegesis raised by this part of the Opinion, I have taken the liberty of stating Judge Overton's arguments briefly and informally. Even so, some of his claims should not go unchallenged. Three points deserve special emphasis. First, though there may be a technical, legal sense in which any use of the concept of a supreme being counts as religious, this is not a sense in which any such use advances or promotes religion. After all, that concept is used in purely theoretical discussions in works of metaphysics that have no tendency to promote or advance religious belief or practice by any individual or group. So there is plenty of room to doubt that Judge Overton's more general argument does anything at all to support the conclusion quoted above. Second, despite the assertions of some fundamentalist Christians, it is unlikely that the best literal interpretation of the creation story in Genesis 1 makes reference to creation from nothing. The authors of that creation story seem to think of God's creative activity as consisting of imposing form and structure on a pre-existent formless stuff, which scholars identify with the primordial ocean of ancient Semitic cosmogonies.[9] And so there is also plenty of room to doubt Judge Overton's claim that the ideas of 4(a)(1) are identical, and not merely similar, to the ideas in the Genesis creation story as literally interpreted. Third, though there are extensive parallels between Section 4(a) and Genesis, it is doubtful that this by itself is sufficient to show that a major effect of Act 590 is to advance particular religious beliefs. After all, there are also extensive parallels between parts of the criminal law and Deuteronomy. Both the Decalogue in Deuteronomy 5 and the criminal law prohibit such things as murder, theft and perjury. But surely this does not suffice to show that a major effect of the criminal law is to advance particular religious beliefs. Hence, I think there are solid rational grounds for doubting that Judge Overton's argument for (1) is successful.

As I have said before, the Opinion also contains an argument for (2). Because this is the argument to which philosophy of science is supposed to make some contribution, I am going to analyze it in detail. Judge Overton begins with a statement of what he takes to be the essential characteristics of science:

(1) It is guided by natural law;
(2) It has to be explanatory by reference to natural law;
(3) It is testable against the empirical world;
(4) Its conclusions are tentative, i.e., are not necessarily the final word; and
(5) It is falsifiable (Testimony of Ruse and other science witnesses).[10]

The Opinion does not make it clear whether Judge Overton takes these characteristics to be conditions both necessary and sufficient for scientific status. However, the argument Judge Overton bases on them requires, if it is to be valid, that they be necessary conditions, but does not require that they also be sufficient conditions, for scientific status. So I shall proceed to reconstruct the argument by attributing to Judge Overton *only* the claim that the characteristics in question are necessary conditions for scientific status. Since Judge Overton is trying to show that creation science is not science but not that evolution science is science, he does not need to take a stand on the issue of which conditions are sufficient to scientific status.

The strategy of the Opinion is to argue that each of the positive theses of Section 4(a) of Act 590 lacks one or more of the relevant characteristics and is for that reason unscientific. With respect to 4(a)(1), Judge Overton claims that the hypothesis of creation from nothing "is not science because it depends upon a supernatural intervention which is not guided by natural law. It is not explanatory by reference to natural law, is not testable, and is not falsifiable."[11] Section 4(a)(2) is dismissed as "an incomplete negative generalization directed at the theory of evolution."[12] With reference to 4(a)(3), Judge Overton complains that it "fails to conform to the essential characteristics of science for several reasons. First, there is no scientific definition of 'kinds' and none of the witnesses was able to point to any scientific authority which recognized the term or knew how many 'kinds' existed. . . . Second, the assertion appears to be an effort to establish outer limits of changes within species. There is no scientific explanation for these limits which is guided by natural law and the limitations, whatever they are, cannot be explained by natural law."[13] About 4(a)(4) Judge Overton says only that "it explains nothing and refers to no scientific fact or theory."[14] According to Judge Overton, 4(a)(5) "completely fails as science." Because the flood referred to in 4(a)(5) is the Noachian flood and even creationist writers concede that it depends upon supernatural intervention, such a "worldwide flood as an explanation of the world's geology is not the product of natural law, nor can its occurrence be explained by natural law."[15] And, as for 4(a)(6), Judge Overton thinks it too "fails to meet the standards of science" because the phrase "relatively recent inception" has "no scientific meaning"; it can only be given meaning in the context of Act 590 by reference to creationist estimates that the earth is between 6,000 and 20,000 years old, and such a procedure is "not the product of natural law; not explainable by natural law; nor is it tentative."[16] But if each of the positive theses of Section 4(a) is unscientific, then creation science as defined by Act 590 is not science, and hence Act 590 does not have the advancement of science as an effect.

Judge Overton's argument can therefore be adequately represented in the following schematic manner:

(6) If any statement *S* is scientific, then *S* either is a natural law or is explainable by a natural law and is testable, tentative and falsifiable.

(7) The statement in 4(a)(1) is neither a natural law nor explainable by a natural law and is neither testable nor falsifiable.

(8) Hence, the statement in 4(a))(1) is not scientific.

(9) The statement in 4(a)(3) is neither a natural law nor explainable by a natural law.

(10) Hence, the statement in 4(a)(3) is not scientific.

(11) The statement in 4(a)(4) is neither a natural law nor explainable by a natural law.

(12) Hence, the statement in 4(a)(4) is not scientific.

(13) The statement in 4(a)(5) is neither a natural law nor explainable by a natural law.

(14) Hence, the statement in 4(a)(5) is not scientific.

(15) The statement in 4(a)(6) is neither a natural law nor explainable by a natural law and is not tentative.

(16) Hence, the statement in 4(a)(6) is not scientific.

(17) Thus, none of the positive statements definitive of creation science in Section 4(a) of Act 590 is scientific.

(18) Whence, creation science as defined by Act 590 is not scientific.

(19) If creation science as defined by Act 590 is not scientific, then Act 590 does not have the advancement of science as an effect.

(20) Therefore, Act 590 does not have the advancement of science as an effect.

I have deliberately constructed this argument in such a way that it is intuitively valid. Unfortunately, it is all too clear that it is unsound. The problem is that (6) is demonstrably false. None of the characteristics it alleges to be necessary conditions for an individual statement to have scientific status is, in fact, a necessary condition of scientific status of an individual statement.

Consider first the condition of either being a natural law or being explainable by a natural law. As Laudan has pointed out, scientists have for a long time understood the difference between establishing the existence of phenomena and explaining them by natural laws.[17] Darwin, for instance, established the existence of natural selection nearly half a century before the discovery of the laws of heredity which help to explain it. If, contrary to fact, there had turned out to be no laws to explain natural selection, Darwin's achievement would still have been scientific. To be sure, as Ruse notes, science looks for explanatory laws.[18] But if there are no laws to be found, scientists are prepared to settle for less and can do so without forfeiting the scientific status of their achievements. Certain statements about individual events in the quantum domain are not laws and have no known explanations in terms of laws; moreover, they can have no explanation in terms of laws if contemporary quantum theory is correct, as it seems to be. But they will remain scientific statements even if contemporary quantum theory is correct. Hence, either being a natural law or being explainable by a natural law is not a necessary condition for scientific status. Thus, the arguments for (10), (12), and (14) fail.

Consider next the conditions of testability and falsifiability. As a result of the work of Pierre Duhem, it has been known to philosophers of science for three-quarters of a century that many scientific statements are neither test-able nor falsifiable individually and in isolation but only conjunctively and in corporate bodies. Hence, being testable and being falsifiable are not necessary for individual statements to have scientific status, and the argument for (8) fails too. Moreover, it would not strengthen Judge Overton's argument to retreat to the more plausible claim that only in the case of whole theories, and not on the level of each individual statement, do testability and falsifiability count as necessary conditions for scientific status. Creation science as defined in Section 4(a) of Act 590 and as further interpreted by Judge Overton himself clearly satisfies these conditions. For example, the statements in 4(a)(1) and 4(a)(6), as Judge Overton interprets them, together imply that there is no matter on earth more than 20,000 years old. The trouble with this claim is not that it is untestable or unfalsifiable. Its problem is rather that it has been repeatedly tested and is so highly disconfirmed that, for all practical purposes, it has been falsified.

Unfortunately, the patently false claim that creation science is neither testable nor falsifiable seems well on its way to becoming, for some evolutionary biologists, a rhetorical stick with which to belabor their creationist opponents. In a recent collection of essays, Stephen Jay Gould claims that "'scientific creationism' is a self-contradictory nonsense phrase precisely because it cannot be falsified."[19] And in another essay in the same collection Gould has this to say about creationists:

> They present no testable alternative but fire a volley of rhetorical criticism in the form of unconnected, shaky factual claims—a potpourri (literally, a rotten pot, in this case) of nonsense that beguiles many people because it masquerades in the guise of fact and trades upon the false prestige of supposedly pure observation.[20]

Ironically, in the next sentence Gould goes on to contradict himself by asserting that "the individual claims are easy enough to refute with a bit of research."[21] Indeed, some of them are! But since they are easily refuted by research, they are after all falsifiable and, hence, testable. This glaring inconsistency is the tip-off to the fact that talk about testability and falsifiability functions as verbal abuse and not as a serious argument in Gould's anticreationist polemics.

Finally, consider the condition of tentativeness. Unlike the other conditions, tentativeness is not a structural or methodological condition on the content of a body of beliefs but is a psychological condition on the attitudes of believers. But whether a belief is held tentatively or dogmatically is completely irrelevant to whether or not it is scientifically or in any other way epistemically respectable. Laudan puts the point very well:

> Since no law mandates that creationists should be invited into the classroom, it is quite irrelevant whether they themselves are close-minded. The Arkansas statute proposed that Creationism be taught, not that creationists should teach

it. What counts is the epistemic status of Creationism, not the cognitive idiosyncrasies of the creationists. Because many of the theses of Creationism are testable, the mind set of creationists has no bearing in law or fact on the merits of Creationism.[22]

Being held tentatively is not a necessary condition for scientific status which the beliefs of creationists fail to satisfy. No matter how dogmatically some contemporary physicists believe the principle of the conservation of mass-energy nothing is done thereby to impugn its scientific status. So the argument for (16) also fails. And since the arguments for (8), (10, (12), (14) and (16) are all unsuccessful, the argument for (17), (18) and (20) also collapses.

Ruse makes two last-ditch attempts to defend Judge Overton's argumentative strategy. Neither of them succeeds.

The first involves citing one scientific case in which, so we are to suppose, (6) is true and one non-scientific case in which, so we are told, (6) is also true. Ruse says:

> For instance, explanation of the fact that my son has blue eyes, given that both parents have blue eyes, done in terms of dominant and recessive genes and with an appeal to Mendel's first law, is scientific. The Catholic doctrine of transubstantiation (i.e., that in the Mass the bread and wine turn into the body and blood of Christ) is not scientific.[23]

In the Mendelian example, we are to suppose that each statement is either a natural law or is explainable by a natural law and is testable, falsifiable and tentatively held by the proponent of the explanation. In the other example, we are to suppose that the doctrine of transubstantiation is neither a natural law nor explainable by a natural law and is neither testable, nor falsifiable nor tentatively held by its proponents. But supposing all this gets us exactly nowhere. Two positive instances cannot prove a universal generalization which covers other cases, as Ruse himself admits. But a single counterexample can falsify a universal generalization, and between us Laudan and I have provided more than one counterexample to (6).[24] So this line of defense does nothing at all to help Judge Overton's argument.

Ruse's second ploy is to suggest that for legal purposes Judge Overton had to argue that creation science is not science at all because he could not have held Act 590 in violation of the Establishment Clause if he had merely shown that creation science, though testable, has been tested and massively disconfirmed, and is therefore bad or weak science.[25] But this suggestion is mistaken on two counts. First, as I noted above, Judge Overton could have held Act 590 in violation of the Establishment Clause without even addressing the question of the scientific status of creationism merely by arguing, as he in fact did, that Act 590 fails part of the three-pronged test. Second, if Judge Overton had been able to show that Act 590 has as a major effect the advancement of religion, then he could at least have tried to argue from the premise that creation science is bad science to the conclusion that Act 590 has the advancement of science only as a minor effect at best. And if he had successfully done this and also shown that Act 590 has no other major effects, then he would have been entitled to conclude that Act 590 has the advancement of religion as its primary effect, which is all he needed to establish in order to show that Act 590 fails the second part of the three-pronged test.

There are two conclusions I wish to draw from this portion of my discussion. First, because Judge Overton's Opinion does not provide adequate justification for any of (1)-(3), it is not a rationally persuasive argument for the claim that Act 590 fails the second part of the three-pronged test. Second, because the argument (6)-(20) is demonstrably unsound, philosophy of science contributed nothing of positive value to the quality of Judge Overton's Opinion. One can only hope that if other judges cite *McLean v. Arkansas* as a precedent, they will have the good sense to ignore the material in Section IV(C).

RUSE'S OPINIONS

It is interesting to ask how we should assign responsibility for the manifold errors contained in Judge Overton's argument for (2). Did Judge Overton simply fail to grasp what Ruse and other witnesses told him about how to demarcate science from non-science? Fortunately we are in a position to say something about this question if we examine Ruse's published views on creation science. I consider it fair to proceed in this way because Ruse himself says of his published discussion that it "is the same as what I provided for the plaintiffs in a number of position papers. It also formed the basis of my testimony in court, and, as can be seen from Judge Overton's ruling, was accepted by the court virtually verbatim."[26] As we shall soon see, there appear to be some differences between Ruse's published views and Judge Overton's Opinion. But I shall argue that the real differences are relatively minor and insignificant.

As I pointed out above, Judge Overton must construe the five essential characteristics of science on his list as necessary conditions for scientific status if his argument that any thesis of creation science which lacks one or more of them is for that reason unscientific is to be valid. But Ruse begins by saying: "It is simply not possible to give a neat definition—specifying necessary and sufficient characteristics—which separates all and only those things that have ever been called 'science.'"[27] So it might seem at the outset that Judge Overton misinterpreted Ruse when he treated the characteristics on his list as necessary conditions for scientific status. However, things are not quite so simple. Ruse quickly proceeds to muddy the waters. He also says, "Creation-science is not science because there is absolutely no way in which creationists will budge from their position."[28] When push comes to shove and an argument against creationism has to be made, Ruse does treat holding one's views tentatively as a necessary condition for the views held to be scientific. As

I argued above, and as Laudan had argued before me, this is a mistake. But when Judge Overton makes it, he displays no misunderstanding of Ruse's real position on creationism. Indeed, it seems he understands it perfectly well.

Ruse's search for "defining features"[29] of science appears, at first glance, to yield a list somewhat different from Judge Overton's. Ruse tells us that "science involves a search for order. More specifically, science looks for unbroken, blind, natural regularities (*laws.*)"[30] He goes on to say that "a major part of the scientific enterprise involves the use of law to effect explanation,"[31] including prediction and retrodiction. These are modest enough and sweetly reasonable statements. One tells us only that science looks for laws; it offers no guarantee that science can always succeed in finding them. The other tells us only that a major part of the scientific enterprise as a whole is explanation in terms of law. Neither says or implies that every scientific statement either is a natural law or is explainable by a natural law. But, again, when the crunch comes, Ruse abandons modesty and sweet reason. In his response to Laudan, when he tries to deal with the fact that Charles Lyell in his *Principles of Geology* explicitly left open the possibility that divine intervention is required to account for human origins, Ruse pushes a harder line:

> Science, like most human cultural phenomena, has evolved. What was allowable in the early nineteenth century is not necessarily allowable in the late twentieth century. Specifically, science today does not break with law. And this is what counts for us. We want criteria of science for today, not for yesterday.[32]

So Ruse's concession that there are no neat necessary and sufficient conditions for all and only the things that have ever been called "science" cuts no ice in practice. All that really matters when Ruse wants a stick with which to beat the creationists is that in the late twentieth century science allows, so Ruse alleges, no break with law. Hence, once again, Judge Overton displays no real misunderstanding of Ruse's position when he lays it down for purposes of a decision in 1982 that every scientific statement either is a natural law or is explainable by a natural law. And again, as I argued above, the view in question is seriously in error.

To these characteristics, Ruse adds testability, which encompasses both confirmability and falsifiability. Concerning testability in general, he says that "a genuine scientific theory lays itself open to check against the real world."[33] And with respect to falsifiability, he says that "a body of science must be *falsifiable.*"[34] Hence, although Ruse is not committed, as Judge Overton is, to the false claim that each individual scientific statement is testable and falsifiable in isolation, he is committed to the view that testability and falsifiability are necessary conditions of scientific status for theories as wholes. But, as I pointed out before, creation science as defined in Section 4(a) of Act 590 satisfies these conditions. Ruse disagrees. He asserts that "creation scientists do little or nothing by way of genuine test."[35] Even if true, this remark is completely irrelevant. The requirement is that a scientific theory be testable, not that its proponents actually test it. However, Ruse goes on to say, even if creation scientists do expose their theories to tests, "when new counter-empirical evidence is discovered, creation scientists appear to pull back, refusing to allow their position to be falsified."[36] This remark too, even if accurate, is utterly beside the point. The requirement is that a theory *be* falsifiable by empirical evidence, not that its adherents admit that it has been falsified if and when it has been. Once creation scientists make testable assertions, as they have, it is up to the evidence and not to them whether those assertions are disconfirmed to the point of being falsified. Hence, Ruse's main reasons for considering creation science untestable and unfalsifiable turn out to be, upon inspection, nothing more than two irrelevant *ad hominem* arguments.

My discussion of Ruse's position thus far yields two conclusions. First, though some of Ruse's statements are more cautious and nuanced than parallel statements by Judge Overton, when it comes time to mount an argument against creation science both of them are committed to the untenable view that being either a natural law or explainable by a natural law and being held tentatively by its adherents are conditions necessary for a statement to be scientific. Second, although Ruse does not make the mistake Judge Overton does in assuming that the criteria of testability and falsifiability apply to all individual statements of science, this is of no help in constructing a case against the scientific status of creation science because creation science as defined in Act 590 satisfies both these criteria. Both Ruse's arguments against this conclusion are irrelevant because fallaciously *ad hominem*.

There is one characteristic on Ruse's list of the defining features of science that is not precisely reflected anywhere on Judge Overton's list. This is what Ruse thinks of as professional integrity. Describing this characteristic, Ruse says:

> A scientist should not cheat or falsify data or quote out of context or do any other thing that is intellectually dishonest. Of course, as always, some individuals fail; but science as a whole disapproves of such actions. Indeed, when transgressors are detected, they are usually expelled from the community.[37]

In the light of what has already been said, it should be tolerably clear that integrity, like tentativeness, is a characteristic of persons either individually or as groups. As such, it has no bearing on the scientific status of their ideas or assertions. I have no wish to defend, mitigate or excuse the intellectual dishonesty one finds in the writings of some creationists. But I do detect a certain unctuous tone in Ruse's praise of the intellectual honesty of scientists in general at a time when scientific fraud has become something of a national scandal in the United States. All over the country institutions are busy putting in place formal and explicit policies for dealing with dishonesty in research, and clearly the impetus to do so comes mainly from the numerous and well-publicized cases of scientific fraud that have occurred in recent years. The fact that institutions find it necessary to make formal policy on such matters is a good indication that the informal self-policing mechanisms of the scientific community are not doing their job.[38]

We are now in a position to answer the question asked at the beginning of this section. Judge Overton did not seriously misunderstand Ruse's position. Although some of Ruse's views are carefully qualified when abstractly stated, the qualifications are omitted when it comes time to argue that creation science is not real science, and the results are philosophical mistakes which match in most important respects the errors in Judge Overton's Opinion. And even where there are differences between their views, as in the case of the unit to which the criteria of testability and falsifiability are properly to be applied, the only result is that they reach similar false conclusions by means of somewhat different bad arguments.

In a way, this is a pity. By allowing the issue to turn entirely on whether creation science is or is not real science, Judge Overton missed other opportunities to argue that Act 590 fails the second part of the three-pronged test. A remark by Laudan will help us focus our attention on one such possibility:

> Rather than taking on the creationists obliquely and in wholesale fashion by suggesting that what they are doing is "unscientific" *tout court* (which is doubly silly because few authors can even agree on what makes an activity scientific), we should confront their claims directly and in piecemeal fashion by asking what evidence and arguments can be marshaled for and against each of them. The core issue is not whether Creationism satisfies some undemanding and highly controversial definitions of what is scientific; the real question is whether the existing evidence provides stronger arguments for evolutionary theory than for Creationism.[39]

The question is not whether creation science fails to accord with some dubious and probably ephemeral theories about what is necessary for counting as science. The real issue is whether creation science, whatever it may be, now has high epistemic status as compared to its rivals for credibility in the empirical domain. Since it does not, the following argument seems promising:

(21) Act 590 does have the advancement of religion as a major effect.

(22) Act 590 does not have the advancement of empirical knowledge as a major effect.

(23) Act 590 does not have the advancement of any other aim as a major effect.

(24) Hence, Act 590 has the advancement of religion as its only major effect.

(25) Whence, Act 590 has the advancement of religion as its primary effect.

Assuming that a better argument for (21) than the one Judge Overton gave could be found, the arguments for (22) and (23) would be relatively straightforward. In outline, here is the argument for (22). Because creation science does make empirically testable claims, it is appropriate to judge its epistemic status by how well those claims have stood up under testing. But those of the empirical claims made by creation science which have been subjected to testing have not fared well at all; they have failed to be confirmed and many have been highly disconfirmed. Hence, creation science contributes at best indirectly and in a minor way to our empirical knowledge. And, thus, Act 590 does not have the advancement of empirical knowledge as a major effect. Here in brief is the argument for (23). Because creation science makes only religious and empirical claims, it makes no major contribution to other fields of study. Whence, Act 590 does not have the advancement of any other aim as a major effect.

Of course, even this line of argument is not without its problems. Perhaps it could be argued that exposing students to creation science would be useful in teaching them to criticize unwarranted empirical claims, and so maybe creation science could be made to serve some legitimate pedagogical purpose, though hardly the purpose envisaged by those who favored Act 590. Still, it seems more promising than the line of argument actually pursued in Judge Overton's Opinion. It is too bad that myopic focus on whether creation science is science led to the neglect of this and similar argumentative strategies in *McLean v. Arkansas,* and it is a shame that none of the expert witnesses brought such possibilities to Judge Overton's attention.

CONCLUSIONS

Scientists and their friends should derive little comfort from the outcome of *McLean v. Arkansas.* Victory was indeed achieved at the wholly unnecessary "expense of perpetuating and canonizing a false stereotype of what science is and how it works."[40] Philosophers of science may hope to derive from this a cautionary tale.

Like other academic specialists, philosophers of science run multiple risks when they bring their expertise into the policy-making arena. Three substantial ones spring to mind. One is the risk of failure to communicate. If the expert is utterly unable to communicate with the other participants in the policy-making process or to show how his or her expert opinion bears on the policy question under consideration, then he or she will fail to affect the outcome of the process. Clearly this was not a problem in *McLean v. Arkansas.* Judge Overton understood pretty thoroughly where Ruse's arguments were tending and how Ruse's conclusions could be applied to the question he had to decide. Another is the risk of being misunderstood. If the other participants in the policy-making process do not grasp all the nuances of the expert's opinion, they are likely to interpret it in a way that leads to mistaken inferences and to policies contrary to those the expert's opinion is intended to support or can reasonably be construed as supporting. Somewhat by accident, this turned out not to be the major problem in *McLean v. Arkansas* either. To be sure, Judge Overton did not appreciate the distinction between applying the criteria of testability and falsifiability at the level of individual statements and applying those criteria at the level of whole theories, and on that account he made some mistakes in arguing for the conclusion that creation science is not science. But Ruse arrived at the same conclusion by a different, though equally fallacious, route. And obviously Ruse's opinion that creation science is not science was intended to support and can reasonably be construed as supporting the decision to declare Act 590 unconstitutional. The third risk is that of misrepresentation. If the expert's views are not representative of a settled consensus of opinion in the relevant community of scholars, then policy based on those views will lack credibility within that community, and the members of that community are likely to regard such lack of credibility as discrediting the policy in question. This was the major problem in *McLean v. Arkansas*. Ruse's views do not represent a settled consensus of opinion among philosophers of science. Worse still, some of them are clearly false and some are based on obviously fallacious arguments. If they do not suffice to discredit the decision in *McLean v. Arkansas* among philosophers of science, it is only because it is at least arguable that Judge Overton made no mistake in reasoning to the conclusion that Act 590 fails to pass the first, and perhaps also the third, part of the three-pronged test. But, as I suggested at the end of the previous section, even this problem might have been avoided by a philosopher of science who focused his expert testimony on the real philosophical and scientific defects in creation science.

So I see no objection in principle to philosophers of science taking the applied turn by serving as expert witnesses. There are substantial risks, to be sure, and many philosophers of science would be reluctant to run them. The moral I draw from *McLean v. Arkansas* is that philosophers of science should not underestimate the risks but should proceed with care and caution to minimize them. One bad precedent, particularly one so extensively publicized and so apt to arouse passionate feelings, is already one too many.

NOTES

1. Peter D. Asquith and Henry E. Kyburg, Jr., eds., *Current Research in Philosophy of Science* (East Lansing, Mich.: Philosophy of Science Association, 1979). In the present context, what is significant about this volume is not what it says, which is of high quality, but what is left unsaid.

2. William R. Overton, "Opinion in *McLean v. Arkansas*," *Science, Technology & Human Values* 7, no. 40 (Summer 1982):29.

3. Larry Laudan, "Commentary: Science at the Bar—Causes for Concern," *Science, Technology & Human Values* 7, no. 41 (Fall 1982):16.

4. Michael Ruse, "Response to the Commentary: *Pro Judice*," *Science, Technology & Human Values* 7, no. 41 (Fall 1982):20.

5. Overton, "Opinion in *McLean v. Arkansas*," p. 40.

6. Ibid., pp., 33-34.

7. Ibid., p. 34.

8. Ibid., p. 35.

9. See the translation of Genesis in *The New American Bible* (New York: P. J. Kenedy & Sons, 1970) and especially the explanatory footnote to Genesis 1:2.

10. Overton, "Opinion in *McLean v. Arkansas,*" p. 36.

11. Ibid.

12. Ibid.

13. Ibid.

14. Ibid.

15. Ibid.

16. Ibid., pp. 36-37.

17. Laudan, "Commentary," p. 18.

18. Michael Ruse, "Creation Science is Not Science," *Science, Technology & Human Values* 7, no. 40 (Summer 1982):73.

19. Stephen Jay Gould, *Hen's Teeth and Horse's Toes* (New York & London: W. W. Norton & Company, 1983) p. 256. The essay from which this sentence is quoted originally appeared in *Discover*.

20. Ibid., pp. 384-385. The essay from which this passage is quoted originally appeared in *Natural History*.

21. Ibid., p. 385.

22. Laudan, "Commentary," p. 17.

23. Ruse, "Response," p. 20.

24. See also Larry Laudan, "More on Creationism," *Science, Technology & Human Values* 8, no. 1 (Winter 1983):37.

25. Ruse, "Response," p. 20.

26. Ruse, "Creation Science," p. 77.

27. Ibid., p. 72.

28. Ibid., p. 76.

29. Ibid., p. 72.

30. Ibid., p. 73.

31. Ibid.

32. Ruse, "Response," p. 21.

33. Ruse, "Creation Science," p. 73.

34. Ibid.

35. Ibid., p. 75.

36. Ibid.

37. Ibid., p. 74.

38. Nicholas Wade, "Madness in Their Method," *The New Republic* (June 27, 1983): 13-17.

39. Laudan, "Commentary," p. 18. For further discussion along the same lines, see Larry Laudan, "The Demise of the Demarcation Problem," *Physics, Philosophy and Psychoanalysis*, ed. by R. S. Cohen and L. Laudan (Dordrecht, Boston and Lancaster: D. Reidel, 1983), pp. 111-128.

40. Laudan, "Commentary," p. 19.

26

The Academic as Expert Witness

Michael Ruse

Among the expert witnesses called by the American Civil Liberties Union (ACLU) in 1981, in its successful attack on the Arkansas Creationism law, was a historian and philosopher of science, namely, me. As I have explained elsewhere, my task was to bridge the gap between the theological witnesses and the scientists.[1] In particular, I had to show, first, that the history of evolutionary theory is not a simplistic story of dogmatic atheistic science crushing sensitive religion and, second, that judged by proper criteria, whereas modern evolutionary theory qualifies as science, Creationism does not.

It was the second, philosophical part of my testimony that was the more crucial, and as it happens, the judge in the case, William R. Overton, made it part of his reasoning in denying the constitutionality of the Arkansas law. Listing some five criteria (provided by me) for the scientific status, Overton concluded that the law violated the First Amendment separation of Church and State.[2]

Shortly after Overton's judgment appeared, an eminent philosopher of science, Larry Laudan, harshly criticized my contribution, arguing that the criteria of demarcation I provided failed to separate science from non-science, and that Creationism (what its present-day supporters like to call "Creation-science") should be excluded from the classroom because it is bad science, rather than because it is religion (of whatever kind or value).[3] Now, another equally eminent philosopher of science, Philip Quinn, has joined the attack. On the one hand, he argues that the U.S. Constitution has been misrepresented—that it is not necessary to show that Creation-science is religion, no

From *Science, Technology, & Human Values* 11, no. 2 (1986): 68-73. Copyright © 1086. Reprinted by permission of John Wiley & Sons, Inc.

more and no less, in order to exclude it from the classroom. On the other hand, with Laudan, he argues that the Ruse/Overton line fails badly to make its case.

> Scientists and their friends should derive little comfort from the outcome of *McLean v. Arkansas*. Victory was indeed achieved at the wholly unnecessary "expense of perpetuating and canonizing a false stereotype of what science is and how it works." Philosophers of science may hope to derive from this a cautionary tale (p. 50).[4]

It would be simple enough for me, or for anyone sympathetic to my cause, to jump again into the fray, arguing that it is Quinn who fails to make *his* case. But, at best, this would lead to a technical in-house dispute between professional philosophers. At worst, we would have self-indulgence. Either way, the time of both author and reader would be better spent elsewhere.

Nevertheless, at a broader level, Quinn (and other critics) does merit a response, for his attack raises important general issues. Somewhat cautiously, Quinn allows that he sees "no objection in principle to philosophers of science taking the applied turn by serving as expert witnesses" (p. 51). (The real caution comes in the next qualifying sentence: "There are substantial risks, to be sure, and many philosophers of science would be reluctant to run them.)" However, Quinn hedges his conclusion with various criteria that he feels any academic must satisfy, if he/she is to play a proper role as an expert witness. For instance, Quinn feels particularly strongly about the problem of "misrepresentation," which he thinks was much at issue in the Creation trial.

> If the expert's views are not representative of a settled consensus of opinion in the relevant community of scholars, then policy based on those views will lack credibility within that community, and the members of that community are likely to regard such lack of credibility as discrediting the policy in question (p. 51).

(He adds: "Worse still, some of them are clearly false and some are based on obviously fallacious arguments.")

Responding, therefore, to the general spirit of Quinn's critique, in this discussion note I want to raise and answer three questions pertaining to an academic's participation in a trial as an expert witness. I am not trying to excuse myself. I was proud to be in the Arkansas trial, and I still feel that way. I am interested rather in the fundamental points at issue.

SHOULD ACADEMICS EVER BE EXPERT WITNESSES?

Assuming that the question of participation in a trial ultimately revolves on moral issues, one can certainly think of cases where academics should not be expert witnesses—especially those cases where probably no one at all should be a witness. Suppose one has an instance of a person deliberately using technicalities in the law, to drag out a case and thus to avoid a negative judgment. One can imagine a point beyond which virtually everyone would agree that morality is best served by ending a suit, and anyone standing in the way of such an end serves only the cause of wrong.

Equally, one can imagine cases where there is no question but that academics should be expert witnesses. A professor of pathology in a murder trial, for instance. But, this point even applies to those academics (like philosophers) not usually at the cutting edge of life's immediate problems. For instance, in a case of animal cruelty, if a philosopher like Peter Singer has something of professional value to offer about consciousness or animal sentience, then he/she should be prepared to offer it.[5] The situation is no different from that of an expert metallurgist testifying about the collapse of a bridge.

Moreover, in some cases, one might feel obliged to testify, despite qualms about a defendant, or a particular law, or some such like thing. A philosopher might be called upon to testify for a defendant in a case involving freedom of speech, even though he/she loathes that defendant, who has (truly) been accused of disseminating hate literature.

Of course, most cases—particularly those involving points of law—do not raise burning moral issues, either for or against. And, in such instances, one might argue that, in the absence of a pressing moral crusade, the academic has no right to allow him/herself to be used as a "hired gun," especially if substantial fees are involved. (My experience is that the reaction to compensation is especially apt to raise emotions in philosophers, who—from Socrates on—have had very ambivalent feelings about being paid for their expertise. Payment for the activity of philosophy is felt by many to be corrupting.)

Hastening to note that none of the expert witnesses in the Arkansas trial was paid a penny in fees, let me say that I see no reason in principle why an academic should not sell his/her expertise, and if this includes being an expert witness, then so be it. Apart from anything else, in America (and Canada and Britain) we have an adversarial court system, where two sides fight out a dispute, according to certain rules, before judges or juries. This is our system, and it has much to commend it. You have the right to try to change it, but until it is changed, we are stuck with it.

With such a court system, you need experts, and if academics (including philosophers) are needed, then a case can be made for saying that there is a general presumption that one ought to make oneself available. Or, at least, if one feels a disinclination to participate, one ought to support others in their involvement as witnesses. (Apparently, this need for a general sympathy is a more important point than you might think. A large number of philosophers of science refused to help the ACLU, simply because they did not want to appear in court.)

WHO IS RESPONSIBLE FOR LEGAL STRATEGY?

In order to tackle this second question (and to show why I felt the need to pose it), let me use the Arkansas case as a starting point. A large legal team was involved in the attack on the Creation law. (I would estimate that there were at least twenty and probably more lawyers, together with various associates.) Some of the lawyers were from the ACLU head office (in New York), some were from local Arkansas firms, and some were being lent by the large New York corporate firm, Skadden, Arps, Slate, Meager and Flom (these included one partner and several associates). In charge of my testimony (and also the advocate who led me through direct examination) was Jack Novik of the ACLU head office.

The case, as put to the expert witnesses for the ACLU, was that the Arkansas law had to be shown to be promoting the teaching of religion in state-supported schools. To this end, it had to be shown that Creation-science is not truly science, but religion. Again and again, Novik insisted that it would not be enough to show that Creation-science is bad science. The Constitution does not bar the teaching of bad science. It addresses the question of freedom of religion. Hence, any criteria of demarcation had to separate science from religion. You could have gray borderline cases, but there had to be real black, and there had to be real white.

Quinn disagrees with this interpretation of the Constitution. He argues that I was wrong in my understanding of the law, that (at least implicitly) the ACLU was wrong, and that Overton's judgment contains "manifold errors" (p. 45). Quinn denies that my suggested criteria of demarcation will separate science from non-science, and he seems unhappy with the prospect of any such criteria. Thus, he seems to think that something (specifically Creation-science) could be both scientific and religious, at the same time. This is not to deny that one aspect would predominate, and it is here that we find the wedge to hold the classroom door, with Creation-science on the outside. Quinn thinks that Creation-science is undoubtedly false. Hence, were one to teach it, one would obviously have other ends than the teaching of science in view, and these other ends would clearly be religious.

Thus, a law like that of Arkansas which mandates the teaching of Creation-science has the advancement of religion as a major effect and the advancement of science as only a minor effect. This is unconstitutional. Hence, we see that my criteria of demarcation were not needed, and the fact that (in Quinn's opinion) Creation-science qualifies as science under the criteria does not legitimate the teaching of Creation-science in the classroom.

I suspect that, if Quinn's case be well-taken, the fight against Creationism is now far steeper than he suggests: If Creation-science is indeed genuine science, then providing an argument showing that its primary purpose is nevertheless to advance religion, strikes me as a fairly onerous task. But here I shall pass by these matters. I am neither an American nor a lawyer, so were I to enter into detailed debate about the interpretation of the U.S. Constitution, you would have before you something midway between the presumptuous and the foolhardy.

Quinn's critique, however, raises an issue that does concern me. Who is to be responsible for the strategy in a law case, and what obligations fall on the expert witness—in our case, the academic as expert witness? In particular, does the academic (perhaps *qua* academic) have particular responsibility for the strategy, and must he/she therefore share in the praise or blame for strategy (or outcome)?

My immediate response is to quote that favorite subject for symbolic translation in logic textbooks: "The lawyer who defends himself has a fool for a client." In other words, let the lawyers get on with their job of finding and arguing the law, and let the rest of us—plaintiffs, defendants, witnesses, and others—sit back, out of their way. This attitude applies *particularly* if you have an interest, like a moral interest, in the outcome. In short, if you are an expert witness (and I include here any academic), then you should follow the law as it is taught to you. This is your job. This is your obligation. And, it could well be self-defeating if you do otherwise.

Of course, we (especially philosophers, who are good at this sort of thing) can think of possible counterexamples. Although you are a witness, perhaps you have some genuine legal knowledge, and were you to keep silent, there would be a miscarriage of justice. Or you might have reason to doubt the sincerity of your lawyer, even though he/she is arguing your case. Nevertheless, I would suggest that the immediate response is close to the true mark. All other things being equal, you should accept the law as it is taught to you. Nor should you worry that in so doing, you are simply acting like a modern-day Pontius Pilate, abrogating your responsibility by letting the lawyers do the thinking for you. Remember again that our courts are based on an adversary system. Your lawyer's strategy will not be taken as gospel, beyond argument. The other side can challenge your interpretation, and the judges (at the various levels) will undoubtedly have their say.

In the case of the Arkansas law, it might be added that, for all of Overton's "errors," the state did not challenge his judgment. Moreover, since that time Overton's judgment has been accepted as definitive for a similar Creation-science law in Louisiana.[6]

WHAT IS THE WITNESS'S RESPONSIBILITY?

The expert witness is not responsible for the law. However he/she is not an intellectual eunuch, with no responsibilities at all. You are called as a witness, because (supposedly) you have specialized knowledge and (especially in the case of an academic) you have official seals of approval testifying to this knowledge. By this latter, I refer to your degrees, your rank, your honors, and to your *curriculum vitae* generally. Because you are an *expert,* you are allowed to say certain things which other witnesses are not. Most particularly, "on the basis of your professional judgment," you can give an opinion. *I* cannot call you crazy, but a psychiatrist can.

It is here that we arrive at the nub of Quinn's criticism. On the one hand, he thinks I did a dreadful job. As a result, Overton's judgment is badly flawed. In the future, at best the fight against Creationism will be insincere, and at worst it will be unsuccessful. On the other hand, Quinn feels the philosophy of science community has been betrayed. "One bad precedent, particularly one so extensively publicized and so apt to arouse passionate feelings, is already one too many" (p. 51).

In line with the rest of this response, I shall say little about Quinn's attack on me. This is hardly necessary, for much of Quinn's attack parallels Laudan's, and I have responded already to the latter. Frankly, where Quinn goes beyond Laudan, the critique is at its least convincing. For instance, in criticizing my claim that scientific understanding requires reference to law, Quinn writes:

> Darwin, for instance, established the existence of natural selection nearly half a century before the discovery of the laws of heredity which help to explain it. If, contrary to fact, there had turned out to be no laws to explain natural selection, Darwin's achievement would still have been scientific (p. 42).

Such a comment shows appalling ignorance of recent Darwin studies. No one searched harder for laws than did Charles Darwin, and he and his supporters (and critics) always felt that, inasmuch as he had made any contribution to science, it was precisely because he had found laws, including those which lead to natural selection. To mix up Darwin's achievement with those of geneticists is historically and conceptually absurd.[7]

Let us turn, rather, to the underlying question that Quinn raises. What of the academic as expert witness's connection with his/her discipline at large? How should this guide the witness? How does his/her performance reflect on the discipline? Quinn suggests that the witness had better not open his/her mouth unless his/her views are "representative of a settled consensus of opinion in the relevant community of scholars" (p. 51). I want to challenge this claim strongly. Especially, I want to challenge the claim in the context of a subject like philosophy, where disagreement is almost our defining characteristic.

Begin with the obvious, the witness must not lie. What he/she says in court (and elsewhere, for instance when depositions are being taken) must be what he/she sincerely believes. Of course, there is always the problem of how much additional information should be offered, but one certainly has the moral obligation not so to truncate one's answers that they mislead. (In the case of the Arkansas trial, preferring to be judged sincere albeit stupid rather than subtle albeit insincere, I can state unequivocally that I always believed what I said.)

The emphasis must surely be on the fact that the *expert* gives his/her *professional opinion*. This statement does not mean that you can say anything you feel like saying—or even everything that you personally believe. What it does mean is that you, as a philosopher (or whatever), can say what you think reasonable, in the light of your specialized knowledge, as a philosopher (or whatever). It is not necessary that everyone in your discipline agree with you. Genuine, reasoned disagreements exist between professionals. One suspects that usually you will have some people in your field who agree with you, for the reasons that you would forward. But, it could be that your position is a very lonely one. (Presumably, were a philosophical maverick like Paul Feyerabend ever to testify, his would be an opinion of one.) The point is that one does not have to speak from unanimity, or even from a majority position. What is required is that one's views be reasoned and based on one's knowledge of one's field.

Remember, again, that we are in an adversarial position. The other side can call its experts, and can cross-examine you. If yours really is an idiosyncratic position, then the lawyers will soon point this out. In the Arkansas trial, for instance, I spent some three hours being cross-examined (as compared to one hour on direct examination). Much of the time was devoted to a discussion of why my assessment of evolutionary biology differs from that of (what all acknowledge to be) a far more important philosopher than I, Sir Karl Popper.[8] (In direct and in cross-examination I was able to stress that mine is a fairly conventional philosophical position, in fact, a condensed version of so-called logical empiricism, which undoubtedly still has many supporters.[9])

Finally, what of the expert witness and of his/her connection with his/her discipline? It should be clear by now that the witness is certainly not a spokesperson for the discipline. However, this is not to say that there is no connection, nor that the discipline does not in some way authenticate the witness. (What else would one mean by "expert"?) Clearly, the witness's authority derives from the discipline's seal of approval—the degrees, the honors, the refereed publications, and so forth. Inasmuch as the witness goes forward with these qualifications, the discipline stands behind him/her. Conversely, the witness's performance reflects on the discipline. A well-qualified philosopher who brazenly lied in court would discredit philosophy. Someone who brilliantly clarified a subject would credit his/her discipline.

What is not expected, by the profession, is that the witness present a consensus opinion simply because it is a consensus. When Thomas Kuhn wrote *The Structure of Scientific Revolutions,*[10] no one thought the worse of him because most people (i.e., most philosophers) found his conclusions unacceptable. He did not thereby discredit philosophy, or anything else, including himself. The same is true of a witness and his/her testimony. There is no discredit in offering a professionally qualified opinion, however unpopular it may be within one's discipline. Hence, Quinn is wide of the mark when he argues that the major problem in the Arkansas trial was that "Ruse's views do not represent a settled consensus of opinion among philosophers of science" (p. 51).

CONCLUDING REFLECTIONS

It is the issues, not the details or personalities, which are important. Let me end therefore by drawing a general conclusion from the Arkansas trial, and from the philosophical dispute which it has sparked. Those of us concerned to defend evolution in particular, and science and education in general, see ourselves in a desperate battle with the Creationists. There are good reasons for our feelings. The Arkansas trial was a major victory against the Creationists, but only the naive would think that it (or the recent Louisiana case) is the end of the war.

Because of these feelings, you might think (as several have suggested to me) that the very last thing we evolutionists should be doing is airing our linen in public. If Ruse and Laudan and Quinn want to disagree about the worth of Overton's judgment, and about the merits of Ruse's contribution, let them do so in private over a few drinks. Do not air the various views, publicly, in print. You may not find the differences compelling, but you can rest assured that the Creationists do.

I cannot sufficiently express the strength of my disagreement with this line of argument. To follow it puts you in league with the very forces we are fighting. The major reason why Creation-science is not genuine science is that its supporters have to believe, without question and without dispute, in the literal truth of Genesis. The reason why we are justified in fighting the Creationists is because we do subject our ideas to criticism. To cease doing this for tactical reasons is to sacrifice our moral cause on the altar of expediency.

However, the determination not to be bowed—the resolve forthrightly to go on doing good science or philosophy or whatever—does not preclude the taking of sensible precautions. It is just plain stupid to forget that the Creationists are waiting to pounce. It is, therefore, incumbent on us all to prepare our arguments with great care. Criticize each other, certainly. But make sure your criticisms are as strong and well-thought-through as possible.

At the risk of sounding pompous (Quinn accuses me of being "unctuous," so this is just one more failing), I fear the dispute between me and my critics has not always been of the highest caliber. I have referred already to Quinn's astounding views on Darwinism. He has every right to such a view, after he has consulted the relevant (and readily available) literature. Until then, better he were silent.

More generally, neither Laudan nor Quinn shows any evidence of having read one scrap of Creationist literature. I would certainly not ask them (or anyone) to read a great deal—apart from anything else, a monotonous sense of *deja vu* soon sets in. Nevertheless, despite the venerable philosophical tradition of pouncing *a priori* on empirical matters without ever having left one's study, this really is a case where a little background research is vital. For instance, I have myself at times argued that falsifiability as a criterion of demarcation between science and non-science is not all that has sometimes been suggested. But, reading the Creationist literature has convinced me that here is one point where the question of falsifiability is absolutely central. You simply cannot lay down a number of points before you begin, as do the Creationists, expect your supporters to sign a statement affirming the absolute truth of these points, and then claim to have science.

You may disagree with me. I ask that you inform yourself of the subject of disagreement. However, we are now starting to descend again to particulars. So, re-emphasizing that although we are free to argue what we will, we had better back our conclusions with good arguments, I will bring this discussion to an end.

NOTES

1. A full treatment of the Arkansas Creationism dispute can be found in *Science, Technology, & Human Values,* Volume 7, Number 40 (Summer 1982) and in *Creationism, Science, and the Law: The Arkansas Case,* Marcel C. La Follette, ed. (Cambridge, Mass.: The MIT Press, 1983). I give an account of my own participation in the trial in "A Philosopher at the Monkey Trial," *New Scientist* 932 (1982): 317–319.

2. The proposed law, Act 590 of 1981, *Acts of Arkansas,* and Overton's judgment, "McLean v. Arkansas: Opinion of William R. Overton," are given in full in the issue of *Science, Technology, & Human Values* and book cited in Note 1.

3. Larry Laudan, "Science at the Bar—Causes for Concern," *Science, Technology, & Human Values* 7, 41 (Fall 1982): 16–19. I respond to Laudan in *"Pro Judice,"* *Science, Technology, & Human Values,* 7, no. 41 (Fall 1982): 19–23.

4. Philip L. Quinn, "The Philosopher of Science as Expert Witness," in J. T. Cushing, C. F. Delaney, and G. M. Gutting, eds., *Science and Reality* (Notre Dame, Ind.: University of Notre Dame Press, 1984), pp. 32–53, quoting Laudan, *op. cit.,* p. 19.

5. Peter Singer is a leader in the animal rights movement. His *Animal Liberation* (New York: New York Review, 1975) has been most influential.

6. Colin Norman, "Creationists Lose Again," *Science* 229 (1985): 368–369.

7. For more details, see David Hull, *Darwin and His Critics* (Cambridge, Mass.: Harvard University Press, 1973): D. Oldroyd, *Darwinian Impacts* (Kensington, Australia: New South Wales University Press, 1980); and P. Bowler, *Evolution* (Berkeley, Calif.: University of California Press, 1984).

8. In *Objective Knowledge* (Oxford, England: Oxford University Press, 1972) and elsewhere, Karl Popper doubts the genuineness of Darwinism judged as science. I have tackled this claim in *Darwinism Defended: A Guide To The Evolution Controversies* (Reading, Mass.: Addison-Wesley, 1982). I understand that Popper is now more satisfied as to Darwinism's scientific status.

9. Explicitly, I relied on such works as E. Nagel, *The Structure of Science* (New York: Harcourt, Brace, and World, 1961), and C. G. Hempel, *The Philosophy of Natural Science* (Englewood Cliffs, N.J.: Prentice-Hall, 1966).

10. Thomas S. Kuhn, *The Structure of Scientific Revolutions* (Chicago, Ill.: Chicago University Press, 1962).

27

Creationism, Methodology, and Politics

Philip L. Quinn

In my published work on creationism,[1] I have treated it primarily as a source of political and, hence, legal problems. The methodological problems that are dear to the hearts of philosophers of science have, in my opinion, only secondary importance in the debate about so-called "creation-science." Methodological positions bear on the debate only to the extent that they can be made to serve as weapons in a political struggle, and methodologists should be interested in taking part in the debate only insofar as they wish to play a role in the policy-making arena.

I shall not on this occasion repeat the arguments found in my published work; they are complex and intricate, perhaps to a point verging on pedantry. What I shall do instead is simply restate two of the main conclusions I had previously arrived at; they will serve as starting points for the further thoughts I now wish to articulate. The first conclusion is this: Judge William R. Overton was right to hold Arkansas Act 590 of 1981, entitled the "Balanced Treatment of Creation-Science and Evolution-Science Act," in violation of the Establishment Clause of the First Amendment. Judge Overton cogently argued that Act 590 did not have a secular legislative purpose, and having a secular legislative purpose is a necessary condition for constitutionality according to the three-part test formulated in *Lemon v. Kurtzman,* which is often cited as a precedent in Establishment Clause cases these days. The second conclusion is this: Judge Overton argued unsoundly that Act 590 has the advancement of religion as its primary effect and thereby failed to show that it does not satisfy the second part of the *Lemon v. Kurtzman* test, according to which the principal or

Unpublished paper read at St. Mary's College, Indiana, June 1985.

primary effect of a statute must be one that neither advances nor inhibits religion. Judge Overton tried to establish that Act 590 does not have the advancement of science as an effect by showing that "creation-science" as defined by Act 590 is not scientific. In order to show that "creation-science" is not scientific, Judge Overton tried to prove that "creation-science" lacks each of five characteristics which are individually necessary for scientific status and which therefore serve to demarcate, at least in part, science from non-science. But this attempt at proof was a resounding failure. Each of the five characteristics on Judge Overton's list is such that either it is not a necessary condition for scientific status, as can be shown by counterexample, or, if it is when properly construed, it is possessed by "creation-science," as can be shown from admissions Judge Overton himself makes in his written opinion. Hence, an essential premise of the argument is false, and the argument as a whole is unsound.

In arguing for my second conclusion, I had to be severely critical of some views expressed by my fellow panelist Michael Ruse, because he was one of the expert witnesses upon whose testimony Judge Overton relied in formulating his list of essential characteristics of science. So I expect I shall have to catch some flak from Michael in the course of this evening's discussion. But rather than trying to anticipate that in these opening remarks, I shall focus on two other points. One is minor, but it is methodological. The other is, I think, rather deep, but it is political.

I hope it will be obvious that nothing I have said should be interpreted in such a way as to imply that I consider "creation-science" to be good science. If I may borrow part of the title of a recent paper by Philip Kitcher,[2] "creation-science" is, at best, not just bad science; it is dreadful science. As Kitcher observes, "creation-science" has been in full retreat for nearly two centuries, achieving no new explanations of its own. It is about as clear an example of a research program that is degenerating in every way as one could hope to find. During most of the same period its evolutionary rival has been splendidly progressive. And though, as Imré Lakatos often pointed out, progress and degeneration are reversible, at least in principle, I see no reason whatsoever to suggest that "creation-science" has the resources to stage a comeback. Indeed, recent evidence suggests that creationists, for the most part, are no longer even engaging in the competition.

Eugenie Scott and Henry Cole used SCI-SEARCH to do a three-year scan of a thousand scientific and technical journals, focusing on the names of editorial board members of the Creation Research Society, research associates and technical advisors of the Institute for Creation Research, and keywords such as "creationism," "special creation," and "scientific creationism." The yield was unimpressive! Of eighteen items discovered, four criticized scientific creationism as pseudoscience, five were editorials discussing the controversy over creationism, and nine were letters to the editor expressing a mixture of opinions on the merits of creationism versus evolution. Scott and Cole concluded that nothing resembling empirical or experimental evidence for "creation-science" had been discovered. In a more recent study, Scott and Cole looked at submissions to sixty-eight scientific, technical, and educational journals, which had received a hundred and thirty-five thousand submitted manuscripts over a three-year period. Only eighteen of them addressed scientific creationism, and all but three of these had been rejected by journal editors, the three exceptions being still under review at a science education journal. After examining the reviewers' reasons for recommending rejection, Scott and Cole concluded that it appears as if laymen rather than professional scientists are submitting the few articles that have surfaced during the last three years.[3] Certainly this evidence gives no indication that a creationist research program is bearing scientific fruit or even, for that matter, attracting appreciable scientific interest.

But creationism as a political issue is an altogether different kettle of fish. And this gives rise to a rather more interesting problem for philosophers. I shall call it the philosopher's dilemma because it is a special case of a problem rooted in the history of philosophy at least as far back as Plato. It is also a variation in a minor key on the issue political philosophers call the problem of "dirty hands." So, of course, it is not a problem uniquely for philosophers; it can confront other scholars who get involved with political activity.

Recently philosophers have been getting increasingly involved in activities related to policy making. To cite a couple of examples, professional philosophers have served on the staff of a major presidential commission and, as congressional fellows, on the staffs of several congresspersons. In the creationism controversy, one philosopher, Michael Ruse, served, as I have already mentioned, as an expert witness in the case of *McLean v. Arkansas,* and another, Philip Kitcher, has publicly debated a leading creationist spokesman. When they get involved in such activities, philosophers are usually not the political or legal decisionmakers, at least not *qua* professional philosophers. Their role is to furnish arguments to those who do decide—to advise, testify, or persuade. But doing such things in the policy-making arena is rather different from doing them in an academic setting; different norms and constraints apply.

In an academic setting, philosophers must follow the argument wherever it leads. This is what we were taught to do, and this is what we try to teach our students to do. Criticism must be taken seriously; counterarguments and counterevidence, if they have any merit, may never be ignored. If one's thesis has been refuted by counterexample, one must abandon or revise it. If one's argument has been shown to be unsound, one must look for a better argument. If the evidence is shown to be too weak to support one's conclusion, one must acknowledge this and reconsider the issue. And so forth! Of course, these are ideals, and we do not always succeed in living up to them. Nevertheless, they are normative for academic philosophy, as well as for other academic pursuits. Our having internalized them is part of what constitutes our intellectual integrity.

Philosophers who step outside the academic setting to get involved with policy making presumably do so with the noble *intent* of doing good in and for the political realm. They wish the right views, or at least rationally support-able views, to prevail and to help shape public policy. They want their advice, testimony, or arguments to be effective. Indeed, there would be something frivolous, self-defeating, and, perhaps, even morally sub par about getting involved in policy making without having such intentions and wishes. But there is the rub! It sometimes happens that the best arguments one can give in support of a view are not going to be effective and the most effective arguments one can give are not going to be good. After all, decision-makers are sometimes too busy to master complex arguments. Then, too, they can be prejudiced or even stupid. When one is aware that this is the situation—and I suspect this is rather common—then one confronts the philosopher's dilemma.

One horn looks roughly like this. Convinced of the overall rightness of one's position, one opts to present the effective bad argument. Each time one does this, one's hands get a little bit dirtier. At first one is painfully sensitive to even small compromises that one knows to be violations of one's intellectual integrity, but gradually numbness of conscience sets in. At last, when presenting the effective bad argument has become easy and habitual—second nature, as it were—one's hands have become dirty beyond all cleansing and one suffers from a thoroughgoing corruption of mind.

The other horn looks roughly like this. Concerned to preserve one's intellectual integrity at all costs, one resolves never to present the effective bad argument. One always presents the best argument one can for the position one thinks most nearly right, and one's hands remain clean. But frequently those good arguments fail to persuade or carry the day, and gradually one's credibility and effectiveness wane. At last, when one has an established track record of failure, the decisionmakers conclude that one is of no use to them, and one is unceremoniously cast aside.

Though it should be obvious that I have been exaggerating a bit for rhetorical effect, I think the hard choice between corruption and ineffectuality is sometimes real enough. That is the dilemma! Is there a way between its horns? Perhaps. My colleague, Dan Brock,[4] suggests that academic philosophers should only get involved in the policy-making arena on a temporary, short-term basis. Maybe this is a way in which we could manage to have our cake and eat it too. For a short period one might engage in giving bad effective arguments without being thoroughly corrupted. Then one could retreat back to the academy to wash one's moderately soiled hands. After having one's intellectual integrity restored and reinforced, one might then be ready to repeat the cycle.

The application of what I have been saying to the creationist controversy is straightforward. It seems to me that the attempts by creationists to foist their particular brand of dreadful science on public school curricula are pernicious. We should resist such attempts and resist them effectively in the political realm. But some of the creationists who are making such attempts are, to put it not too harshly, shysters. So there may well be circumstances in which only the bad effective argument will work against them in the political or legal arenas. If there are, then I think, though I come to this conclusion reluctantly, it is morally permissible for us to use the bad effective argument, provided we continue to have qualms of conscience about getting our hands soiled. But I also believe we must be very careful not to allow ourselves to slide all the way down the slippery slope to intellectual corruption. Perhaps, if we divide up the labor so that no one among us has to resort to the bad effective argument too frequently, we can succeed in resisting effectively without paying too high a price in terms of moral corruption.

NOTES

1. Philip L. Quinn, "The Philosopher of Science as Expert Witness," *Science and Reality*, ed. J. T. Cushing, C. F. Delaney, and G. M. Gutting (Notre Dame: University of Notre Dame Press, 1984).

2. Philip Kitcher, "Good Science, Bad Science, Dreadful Science, and Pseudoscience," *JCST* (December 1984/January 1985): 168-73.

3. This work is summarized in Roger Lewin, "Evidence for Scientific Creationism?" *Science* (17 May 1985): 837.

4. In a paper read at the 1985 APA Western Division Meeting in Chicago.

Glossary

ADAPTATION: A feature like the eye that helps an organism to survive and reproduce.

ALLELE: One of a number of variant genes that can occupy the same place on the chromosome.

ALLOMETRY: Comparative growth rates.

ALLOPATRIC: Groups that occur apart geographically. (*See also* sympatric.)

AMINO ACID: One of twenty organic compounds strung together by the RNA in particular order to make proteins.

ANAGENESIS: Speciation by transformation from one form to another.

BASE: One of a number of different molecules, the order of which determines the information in the DNA.

BAUPLAN: The basic blueprint of a set of organisms.

CATASTROPHISM: The belief that the earth has been molded by occasional upheavals (usually linked to a directional earth history).

CHROMOSOME: A stringlike body in a cell that carries genes. Sexual organisms have two sets of paired chromosomes.

CLADOGENESIS: Speciation by splitting.

CONSILIENCE OF INDUCTIONS: The bringing together of several disparate areas of science under one unifying hypothesis.

CREATIONISM OR CREATION SCIENCE: The belief that the world began as described in the early chapters of Genesis, and that this can be justified by science.

CRITERION OF DEMARCATION: A way of distinguishing science from nonscience. (Example: falsifiability.)

DARWINISM: The theory of evolution that makes natural selection central.

DEME: A separate, although potentially interbreeding, population.

DNA (DEOXYRIBONUCLEIC ACID): The macromolecule that carries the information of heredity; the molecular equivalent of the gene.

DOMINANT GENE OR ALLELE: An allele that masks the effects of its corresponding (paired) allele.

DROSOPHILA: Fruitflies. Popular subjects of evolutionary geneticists.

ELECTROPHORESIS: A technique for detecting proteins.

ENZYME: A protein that acts as a catalyst, speeding up reactions.

EPISTEMOLOGY: The study of theories of knowledge.

EVOLUTION: The claim that the origins and development (especially of organisms) followed a continuous natural, that is, law-bound, process, from their beginning forms to those extant today.

EXOBIOLOGY: The study of, or speculation about, life elsewhere in the universe.

FALLIBILISM: The belief that there can be no absolutely certain knowledge.

FALSIFIABILITY: The claim made by Karl Popper that the mark of genuine science is that it be open to check and, if necessary, rejection.

FINAL CAUSES: Teleology.

GENE: A unit of heredity.

GENETIC DRIFT: The alteration of gene ratios in (small) populations because of purely random factors, especially vagaries of mating and reproduction.

GENOME: The set of different chromosomes in a cell.

GENOTYPE: Genetic makeup.

GERM CELL OR GAMETE: A reproductive cell.

HETEROCHRONY: The effects on adults of changes in rates of growth.

HETEROZYGOTE: An organism having nonmatching alleles at some point on the paired chromosomes.

HOMOLOGY: A similarity of form not connected to function (as distinguished from "analogy," where similarity of form relates to function).

HYMENOPTERA: Ants, bees, and wasps.

HYPOTHETICO-DEDUCTIVE SYSTEM: A view of theories as axiom systems that start with a few hypotheses, with all else following deductively.

KARYOTYPIC: Having to do with the chromosomes.

LAMARCKISM: A view of evolution from monad to man, especially one connected with belief in the inheritance of acquired characteristics.

LEPIDOPTERA: Butterflies and moths.

LOCUS: The position that a gene occupies on a chromosome. (Alleles can be substituted at the same locus.)

MACROEVOLUTION: Evolution of major phenomena over time.

MICROEVOLUTION: Short-term (and time) evolution.

MUTATION: Change of gene from one form to another.

NATURAL SELECTION: The evolutionary mechanism that requires an ongoing, systematic differential between successful (fit) and unsuccessful (unfit) that occurs when more organisms are born than can survive and reproduce.

NATURAL THEOLOGY: Religious beliefs that come through reason.

NUCLEOTIDE: A compound formed from a base, a sugar, and phosphoric acid. (These sugars and phosphoric acids tie the bases together in the DNA molecule.)

ONTOGENY: The growth of the individual.

OVERDOMINANCE OR HETEROSIS: The result of selection that favors the heterozygote over both homozygotes.

PALEOANTHROPOLOGY: The study of human origins.

PANMICTIC: Uniformly and randomly interbreeding throughout a population or group.

PARAPATRIC: Groups touching each other, but not overlapping.

PHENOTYPE: Physical makeup.

PHYLETIC EVOLUTION: Gradual change within an evolving line.

PHYLOGENY: The path or history of a group.

PLEIOTROPISM: A condition produced by genes that affect more than one feature at once.

PROTEINS: Complex organic compounds that are the building blocks of the cell and, thus, of living things.

PUNCTUATED EQUILBRIA THEORY: The claim that evolution is not absolutely gradual, but jerky.

RECAPITULATION: The belief that the life of an organism (ontogeny) retraces the history of the group (phylogeny).

RECESSIVE GENE OR ALLELE: An allele that is masked by the effects of its corresponding (paired) allele.

REVEALED THEOLOGY: Religious beliefs that come through faith.

RNA (RIBONUCLEIC ACID): A macromolecule that takes the genetic information from the DNA and uses it to make proteins.

SALTATION: A jump across a generation from one organic form to another.

SPECIATION: The formation of one or more new species by anagenesis or cladogenesis.

SPONTANEOUS GENERATION: The one-step appearance of life from nonlife.

STASIS: A period of descent in which there is no change.

SYMPATRIC: Groups occurring together geographically. (Sympatric speciation involves splitting without geographical separation.)

SYNTHETIC THEORY OF EVOLUTION: Modern selection theory backed by today's understanding of heredity; also known as "neo-Darwinism."

TAUTOLOGY: A statement that is necessarily true simply by virtue of its form. (Example: "All men are men.")

TELEOLOGY: An understanding in terms of ends, closely connected with the notion of *design* (where things are produced for a particular *function*).

UNIFORMITARIANISM: The geological claim that the earth is in a steady state where the same forces always operate.

VERA CAUSA (TRUE CAUSE): The best kind of cause, according to the nineteenth-century philosophers. Interpreted by Herschel in terms of analogy and by Whewell in terms of consilience.

ZYGOTE: Fertilized ovum or female germ cell.

Further Reading

I would suggest you begin with a recently published collection, edited by the philosopher of science and Catholic priest, Ernan McMullin: *Evolution and Creation* (Notre Dame: University of Notre Dame Press, 1985). Many aspects of science, religion, and philosophy are considered carefully and sympathetically. The editor's introduction, tracing the history of the science/religion relationship, is an absolute gem. Also worth looking at for a comprehensive overview is *Zygon* 22 (1987): 131–248, on the topic "Creationism and Evolution." Contributors to both of these collections successfully attempt to look at all sides to the problems.

The best general history of evolutionary thought is Peter Bowler, *Evolution: The History of an Idea* (Berkeley: University of California Press, 1984). You really should read Darwin's *Origin of Species*. The Penguin paperback is the first edition (written before Darwin started to revise, often mistakenly, in the face of criticisms) and has an excellent introduction by John Burrow. A good recent collection of pertinent articles is *Darwinism and Divinity,* edited by John Durant (Oxford: Blackwell, 1985). And, always entertaining and stimulating in his views on evolution, past and present, is Stephen Jay Gould, especially in his *Ever Since Darwin* (New York: Norton, 1977).

As an introduction to modern evolutionary theory, my favorite is still John Maynard Smith's *The Theory of Evolution,* 3rd ed. (Harmondsworth, Middlesex: Penguin, 1978). An excellent collection (from which Ayala's article was taken) can be found in the *Scientific American* (September 1978). More of Popper's views on evolutionary theory can be found in his *Objective Knowledge* (Oxford: Oxford University Press, 1972). A clear introduction to the "punctuationist challenge," giving not only theory but also some of its history, is Niles Eldredge, *Time Frames: The Rethinking of Darwinian*

Evolution and the Theory of Punctuated Equilibria (New York: Simon and Schuster, 1985). Above all else, do not miss Richard Dawkins, *The Blind Watchmaker* (New York: Norton, 1986). If you do not accept Darwinism after reading this book, you never will!

The best critical treatments of creationism are by philosopher Philip Kitcher, *Abusing Science: The Case Against Creationism* (Cambridge, Mass.: MIT Press, 1982), and by biologist Douglas Futuyma, *Science on Trial: The Case for Evolution* (New York: Pantheon, 1983). If you are to make any kind of informed judgment, you must read some creationist material. I recommend either the book abstracted in this collection, *Scientific Creationism,* edited by Henry Morris (San Diego: Creation-Life Publishers, 1974) or Duane T. Gish's *Evolution: The Fossils Say No!* (San Diego: Creation-Life Publishers, 1973). The best, and in respects a deeply moving, account of the Arkansas trial is by the theologian Langdon Gilkey, *Creationism on Trial* (Minneapolis: Winston Press, 1985).

The *locus classicus* of recent worth on the attempt to demarcate science from nonscience is, of course, Popper's *Logic of Scientific Discovery* (London: Hutchinson, 1959). In respects, however, I prefer his *Conjectures and Refutations* (London: Routledge and Kegan Paul, 1963). Almost any text in the philosophy of science takes up the issue somewhere along the line, so I will conclude simply by recommending a little collection you might find stimulating: *Science, Pseudo-Science and Society,* ed. Marsha Hanen and others (Waterloo, Ont.: Wilfred Laurier University Press, 1980).